Making Sense of Microsoft Excel

Exercises may be downloaded at:
http://www.herzogsystems.com/resources

Making Sense of Microsoft Excel

First Edition

John P. Herzog

cognella® | ACADEMIC PUBLISHING

Bassim Hamadeh, CEO and Publisher
Jennifer Codner, Senior Field Acquisitions Editor
Michelle Piehl, Senior Project Editor
Jess Estrella, Senior Graphic Designer
Stephanie Kohl, Licensing Coordinator
Natalie Piccotti, Director of Marketing
Kassie Graves, Vice President of Editorial
Jamie Giganti, Director of Academic Publishing

Printed in the United States of America.

ISBN: 978-1-5165-3688-7 (pbk) / 978-1-5165-3689-4 (br)

cognella | ACADEMIC PUBLISHING

Contents

INTRODUCTION

Excel Spreadsheet Basics

BY THE END of this introduction, it is our objective that the user can have a better understanding of:

1. What a spreadsheet is.
2. Why we use spreadsheets.
3. Various terms regarding spreadsheets.
4. The capacity of a spreadsheet in **Microsoft Excel**.
5. What is meant by a formula versus a function and examples of them.
6. The different types of items that can be entered into a cell.
7. How dates are treated.

WHAT IS A SPREADSHEET?

A **spreadsheet** (also called a worksheet) is a computer program that allows the users to perform routine calculations easier, faster, and with greater accuracy than with a calculator. If you use a spreadsheet, you will likely not need to use a calculator much anymore. Just about anything you can do with a calculator can also be accomplished in a spreadsheet, but in a way that is much, much faster and more accurate.

One major reason that spreadsheets are so great is that, after you set them up, all you need to do is to enter data, and the math is performed for you automatically! If you make a mistake with your data entry, you can correct the one data entry error and the spreadsheet will automatically recalculate all of your numbers. On many calculators, if you enter a large number of figures, one mistake may require you to start over and reenter those numbers. With spreadsheets, those days are gone.

Microsoft Excel is a computer program within **Microsoft Office**. The means by which you open **Microsoft Office** and then **Microsoft Excel**

depends upon the version of the **Windows Operating System** that is installed on your computer and the way it has been installed. At this point, it is best that you open **Microsoft Excel** by whatever means your computer requires. When you do, you will see a screen that will look *similar to* what you see in Figure i-1.

There are things called **templates** that will likely be visible on the screen. A **template** is something that can be used to get you started on building a spreadsheet if it is one that is commonly used. For example, if you want to create a budget by using a spreadsheet, there is likely a budget spreadsheet already in existence in the template area for you to use that was created by **Microsoft Excel**. You can then modify the template to fit your needs. The ones shown in the first screen you see will depend upon several things, including what templates Microsoft has recently posted online for you to see and, if you are on a network, what templates your network administrator wants you to see. In our example in Figure i-1, we have shown only one for simplicity. At this point we will not be using the templates, but we will be discussing them later in this book.

At this time, in this screen, click **Blank Workbook** (Figure i-1, Arrow).

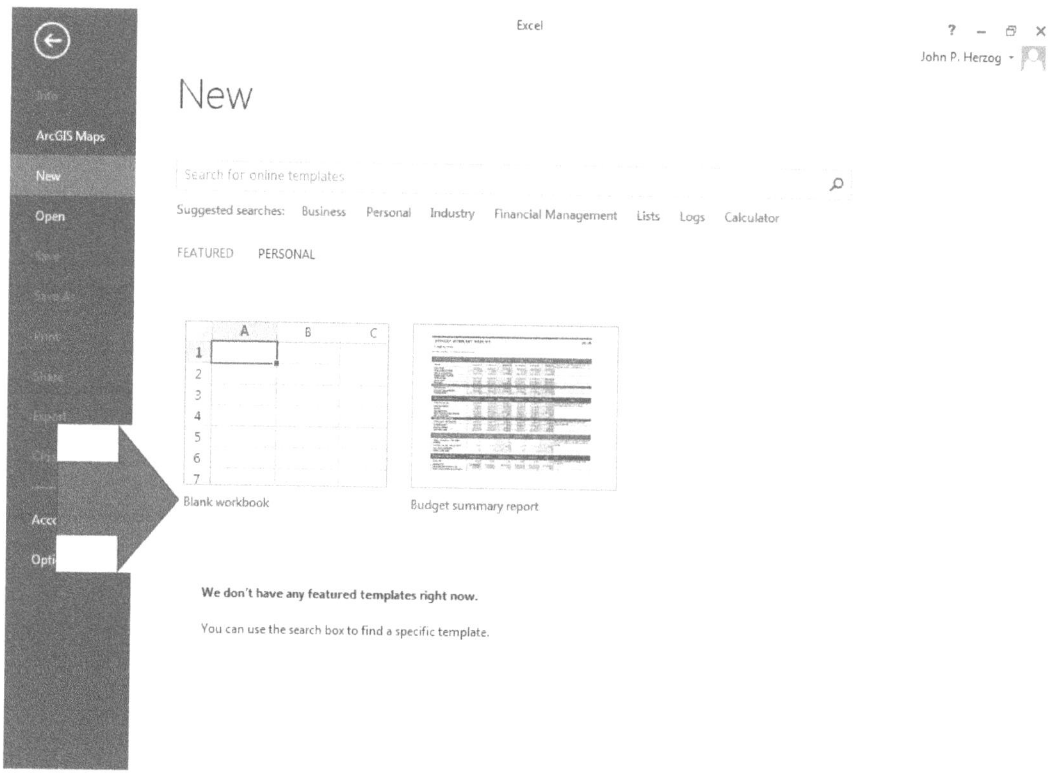

FIGURE i-1

That will take you to the screen that is seen in Figure i-2. Data is entered into what is called a **cell.** A cell is an intersection of a **column** and a **row** in a spreadsheet. In Figure i-2, you can see that columns and rows intersect in a place where computer data can be entered. For example, column A intersects at row 1; thus the cell it forms is known as cell A1 (Figure i-2, Arrow). Column B also intersects with row 1 at cell B1, and so on.

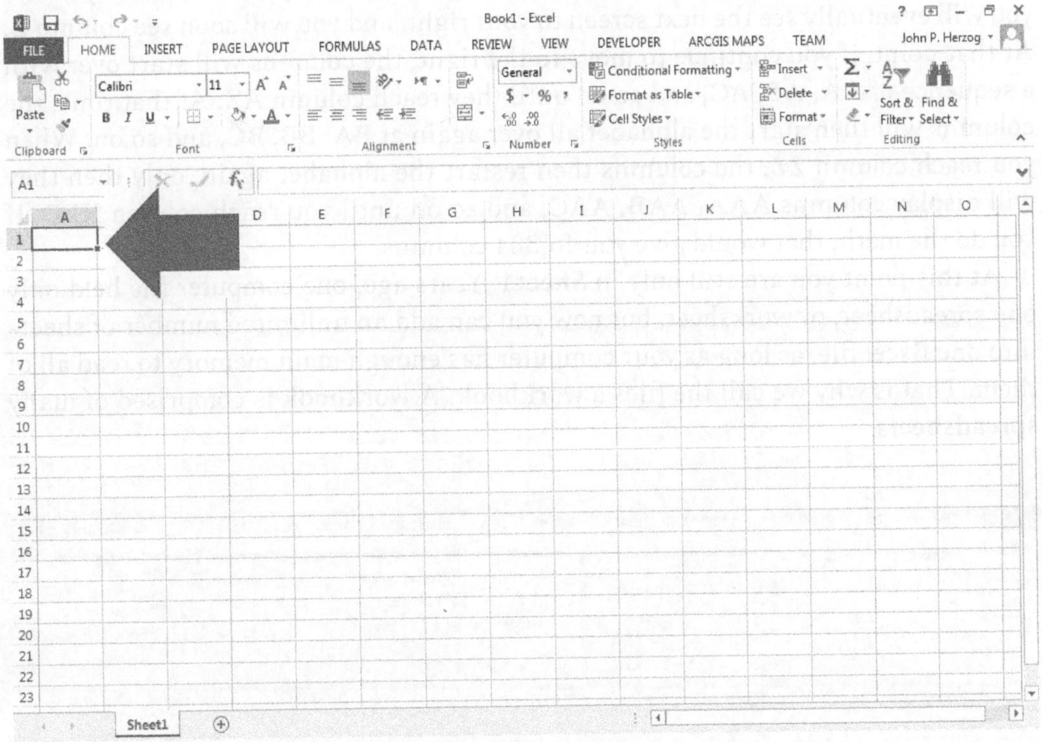

FIGURE i-2

HOW MUCH DATA CAN BE ENTERED INTO A SPREADSHEET?

There are 17,179,869,184 cells in each spreadsheet in **Microsoft Excel**. That is because there are 1,048,576 rows (as seen in Figure i-3, which shows the bottom row of a spreadsheet) and 16,384 columns (as seen in Figure i-4, which shows the last column of a spreadsheet). If you press and hold the **Ctrl key** on your keyboard and then press the **down arrow key** (↓), Excel will take your cursor to the bottom of the spreadsheet you are in, and you will see that it has as many rows as we have stated. If you press and hold the **Ctrl** key on your keyboard and then press the **right arrow key** (→), Excel will take your cursor to the far right of the spreadsheet you are in, which is column

XFD (which will be explained shortly). To return to cell A1, press and hold the **Ctrl** key and then press the **Home key** (**Ctrl Home**). The only columns seen in your particular view, seen earlier in Figure i-2, are columns A through O. The resolution on your computer and/or the level to which you have zoomed in or out of your spreadsheet will make this vary. It could make you see more or fewer columns. It could also make you see more or fewer rows.

If you press and hold the **Tab key** or the **right arrow key** (→) on your keyboard, you will eventually see the next screen to your right, and you will soon see column Z. At that point, if you continue to move to the right, the columns will start over with a sequence of AA, AB, AC, and so on until they reach column AZ. At that time, the columns will then start the alphabet all over again at BA, BB, BC, and so on. When you reach column ZZ, the columns then restart the alphabet again, only then they will display columns AAA, AAB, AAC, and so on until you reach column XFD. If you do the math, that would give you 16,384 columns.

At this point you are still only in **Sheet1**. Years ago, one computer file held only one spreadsheet, or worksheet, but now you can add an unlimited number of sheets into one Excel file, as long as your computer has enough main memory to read all of them. That is why we call the files a **workbook**. A **workbook** is comprised of many **spreadsheets**.

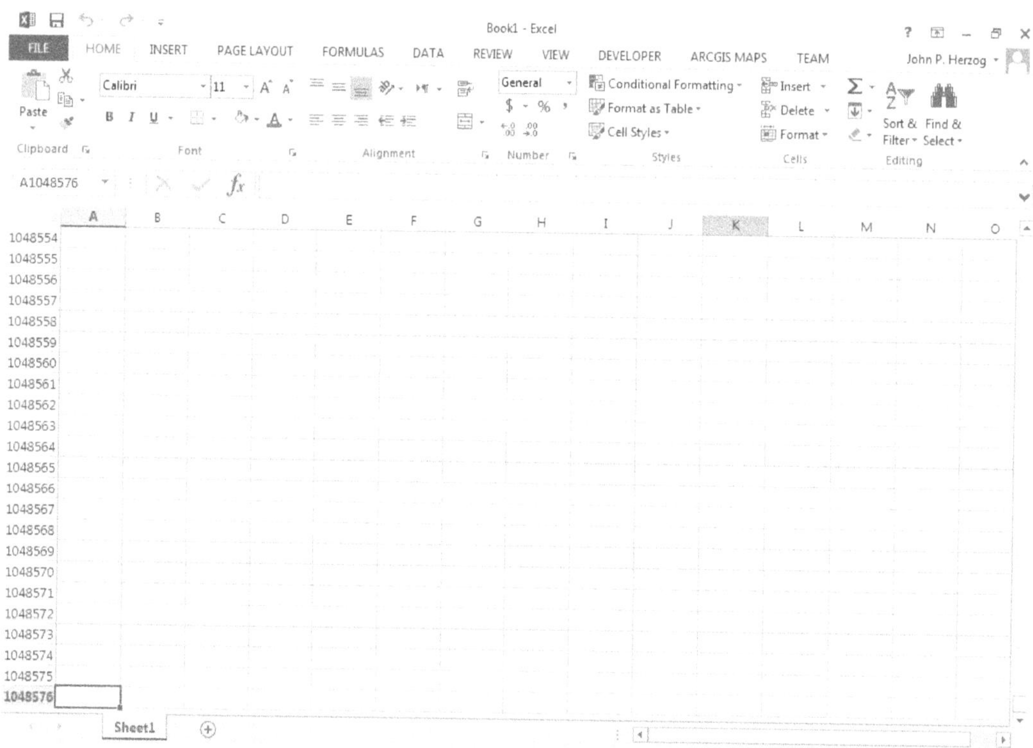

FIGURE i-3

FIGURE i-4

WHAT IS ENTERED INTO A SPREADSHEET?

The data that can be entered into the cells within each spreadsheet is one of four things:

1. **Text**: Words that describe the numbers and items you are entering.
2. **Text with numbers (alphanumeric)**: Words mixed with letters such as a street address.
3. **Numbers that are treated as text**: Numbers that cannot be calculated, such as a social security number or telephone number.
4. **Values**: Things that can be used in a calculation, such as
   a. Numbers
   b. **Formulas**: Cell entries that add, subtract, divide, or multiply a number or a cell with another number or cell.
   c. **Functions**: Cell entries that perform more sophisticated calculations than many formulas, and that display data in more useful ways.

CHAPTER ONE

How to Create the Spreadsheet and Make a Graph of the Data from that Spreadsheet

BY THE END of this chapter, it is our objective that the user will be able to:

1. Enter data into a spreadsheet.
2. Merge and center cells.
3. Enter formulas.
4. Enter the **SUM** and **IF** functions.
5. Copy cells and fill their content to other cells.
6. Use relative address and absolute address.
7. Set data formats.
8. Create table formats.

CHAPTER 1 TASK: TRACKING SALES

Suppose you want to track some simple data of sales in an organization. Suppose you also want to do an analysis of the data by finding things, such as totals and whether or not employees will be getting bonuses. In the next few steps you will create the *Herzog Appliances* **spreadsheet** that will make it easier to do such things. It will look like what you see in Figure 1-1 when you finish. You will also create a **chart** that graphs its data as you see in Figure 1-2.

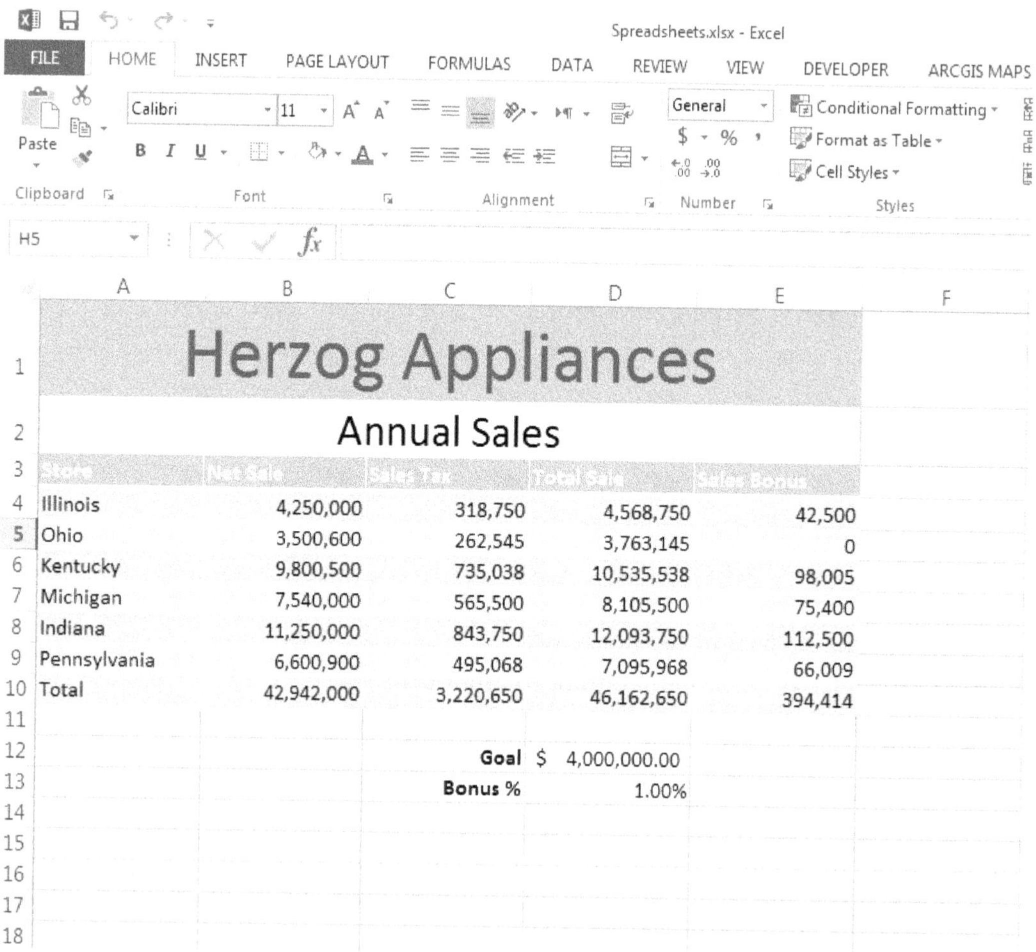

FIGURE 1-1

TO START A SPREADSHEET

Step 1: Open Microsoft Excel.

Step 2: Click **Blank Workbook** as shown in (Figure 1-3, Arrow).

That will give you a blank spreadsheet (Figure 1-4) where you can enter any data that you want.

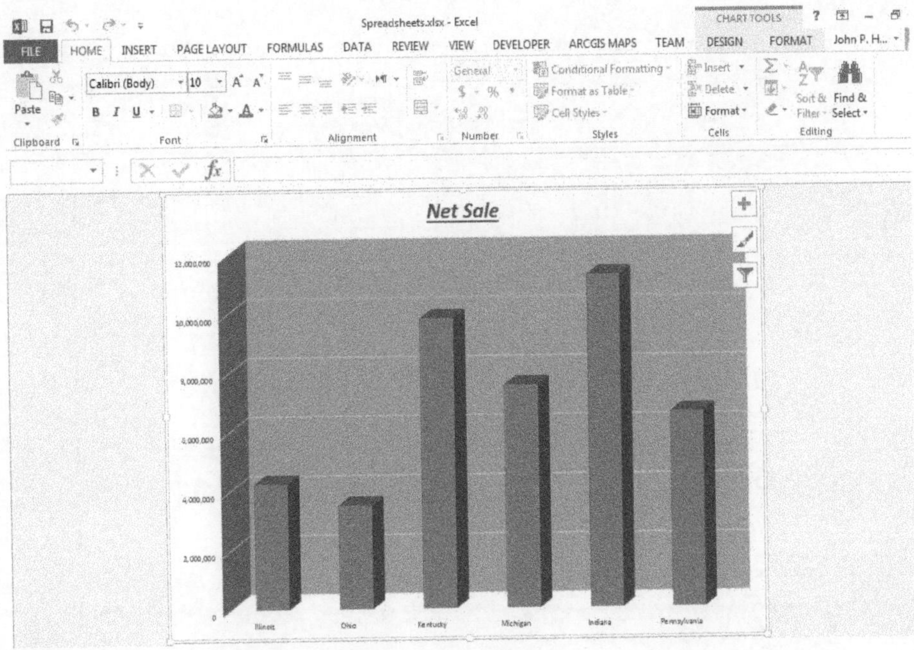

FIGURE 1-2

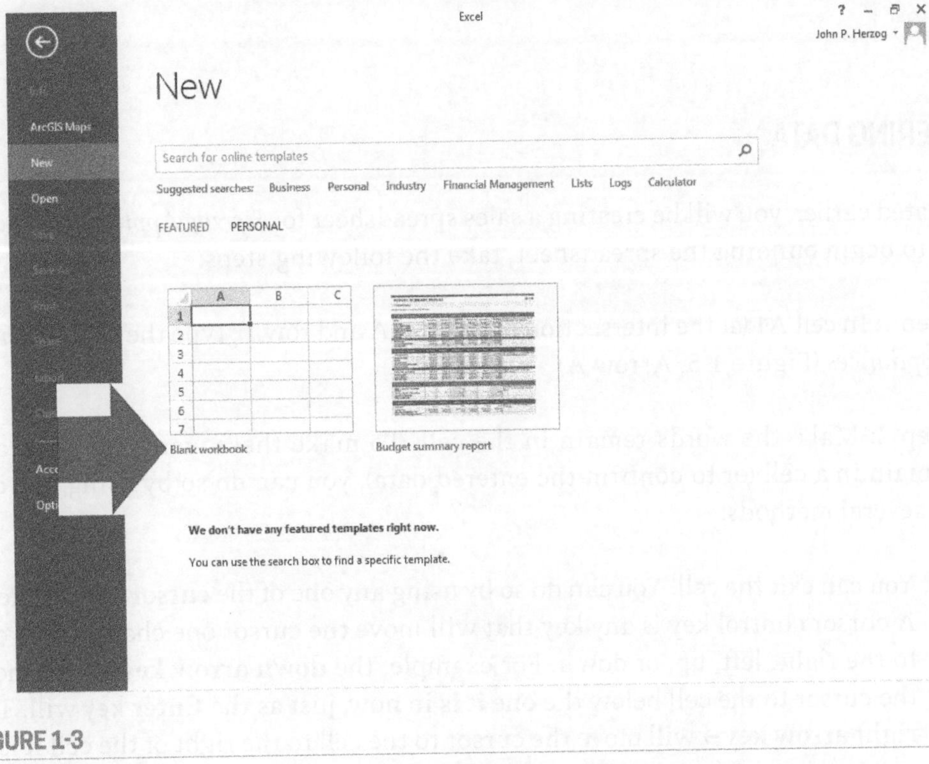

FIGURE 1-3

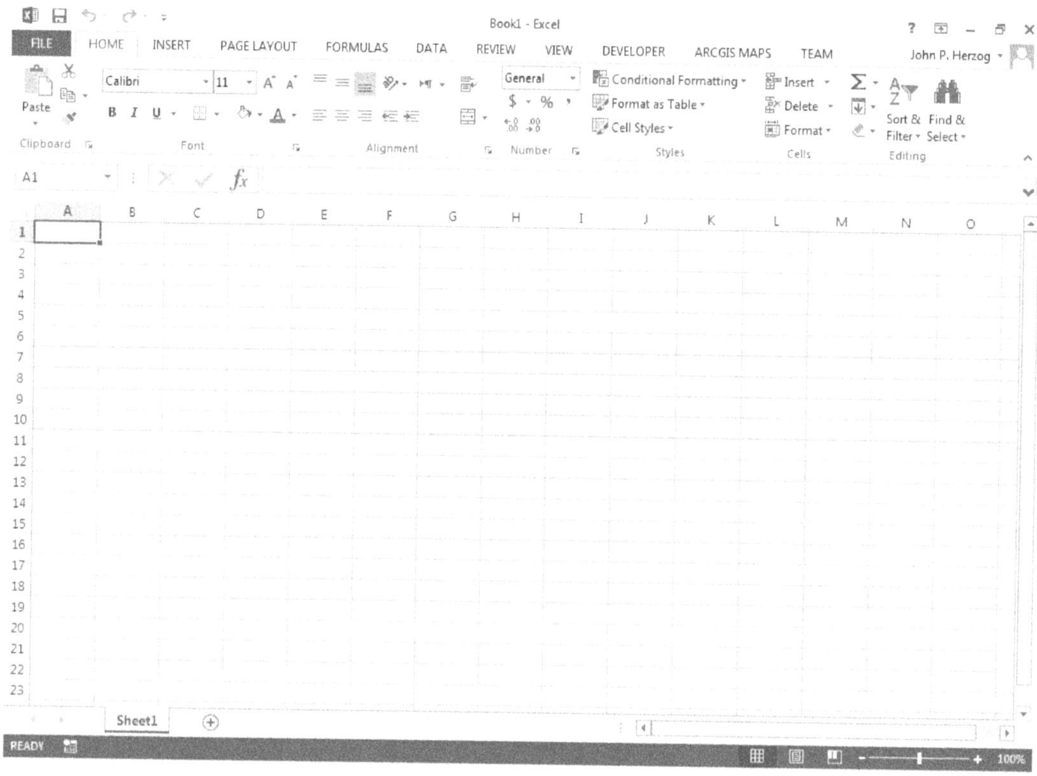

FIGURE 1-4

ENTERING DATA

As stated earlier, you will be creating a sales spreadsheet for *Herzog Appliances*. Therefore, to begin building the spreadsheet, take the following steps:

Step 1: In cell A1 (at the intersection of column A and row 1), type the words *Herzog Appliances* (Figure 1-5, Arrow A).

Step 2: Make the words remain in the cell. To make these words, or any data, remain in a cell (or to confirm the entered data), you can do so by using any one of several methods:

1. You can exit the cell. You can do so by using any one of the **cursor control keys**. A cursor control key is any key that will move the cursor one character or cell to the right, left, up, or down. For example, the **down arrow key** ↓ will move the cursor to the cell below the one it is in now, just as the **Enter** key will. The **right arrow key** → will move the cursor to the cell to the right of the cell it is in

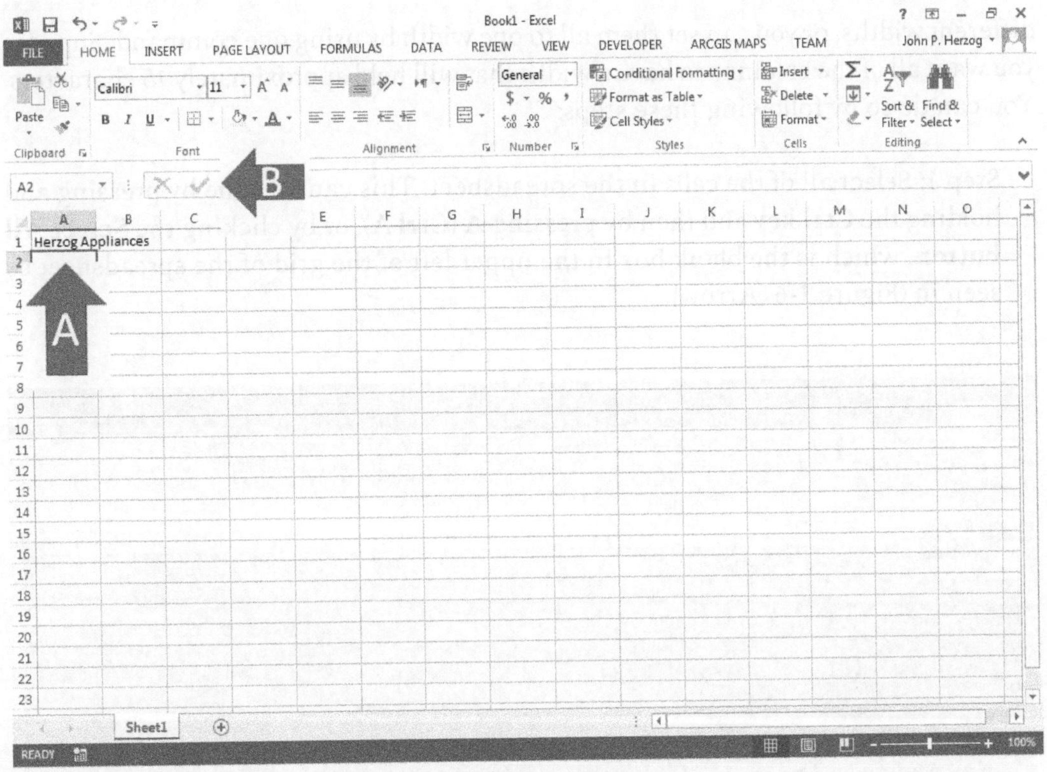

FIGURE 1-5

now, just as the **Tab key** will. All of the other arrow keys will also take you to the next cell in the direction that they are pointing. You can also exit the cell by pressing another cursor control key, such as the **PgUp** or **PgDn keys**. They will move you up and down an entire screen.

2. You can click the checkmark √ in what is called the **formula bar** (Figure 1-5, Arrow B). The formula bar has several functions. Later, when we discuss formulas, it will tell you the formula that generated the numbers that are now displayed in the cell. It also displays the means by which data can be canceled or entered, or which functions can be inserted. Please note that if you click the **X**, it will erase or cancel the data you just entered if you have not yet exited the cell or confirmed the data. It has the same effect as using the **Esc key**.

MAKING THE COLUMN WIDTHS THE DESIRED SIZE

Sometimes you may want your columns to be a different width than what the default widths are when you begin a spreadsheet. You can set them each individually at different widths, or you can set them all to one width by using one command. Suppose you want all of the columns to have a width that will hold approximately *15* characters. You can do so by following these steps:

Step 1: Select all of the cells in the spreadsheet. This can be done by pressing and holding the **Ctrl key** and then by pressing **A** (**Ctrl A**), or by clicking the **Select All button**, which is the blank box in the upper left of the grid of the spreadsheet as seen in (Figure 1-6, Arrow).

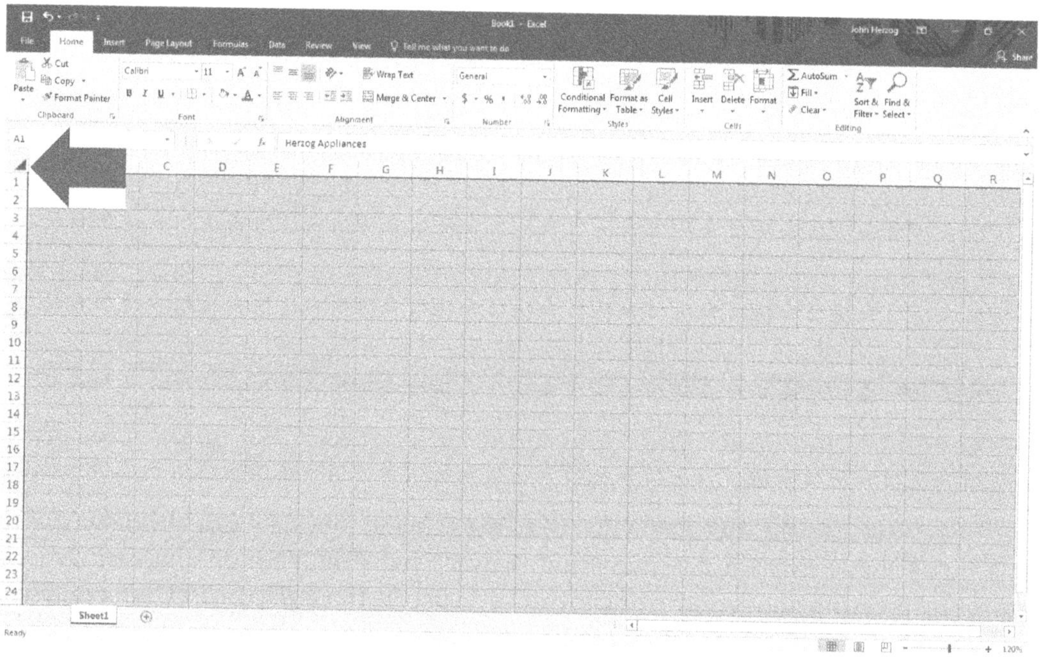

FIGURE 1-6

Step 2: When the spreadsheet goes dark, place your mouse on any one of the small lines that separate the **column headings**. It doesn't matter which one you choose. When you do, your mouse will appear as a tiny black cross (Figure 1-7, Arrow).

Step 3: Click and drag that line to the right until the width of the column is a size *15* as seen in Figure 1-8.

NOTE: Another way to change the **column width** to a size of *15* is to click **Format** (Figure 1-9, Arrow A) in the **Home Ribbon**, and then click **Column Width** as seen in Figure 1-9, Arrow B.

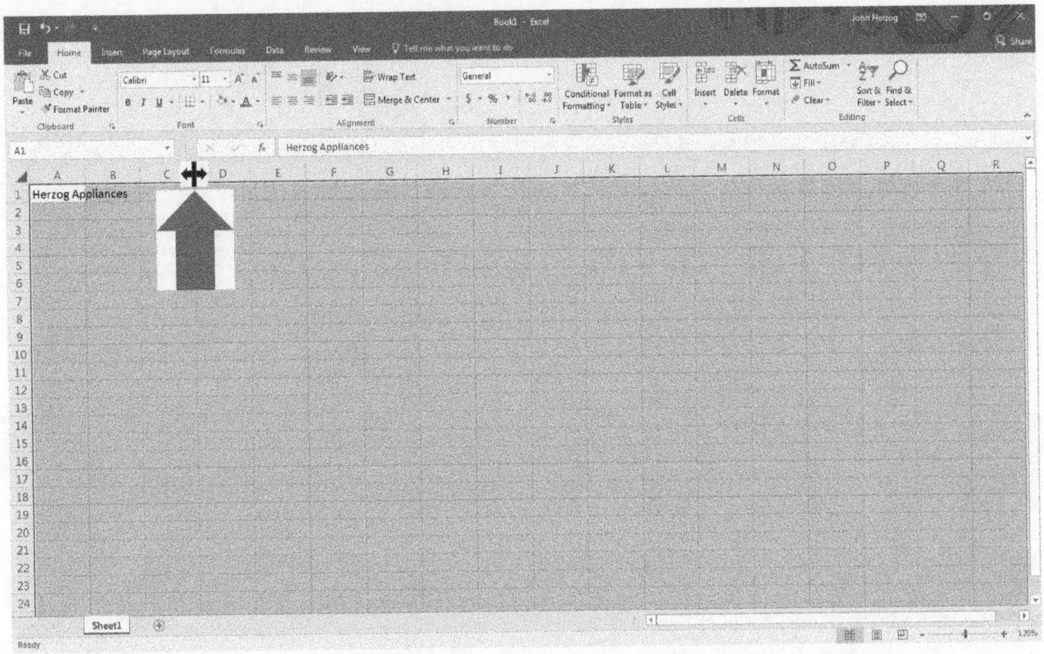

FIGURE 1-7

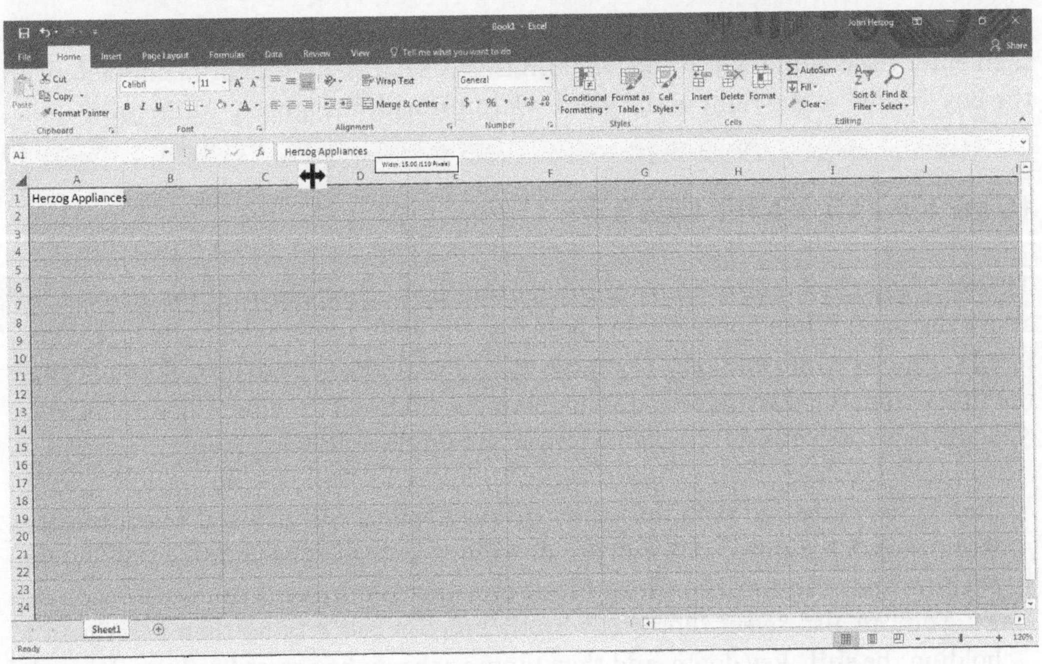

FIGURE 1-8

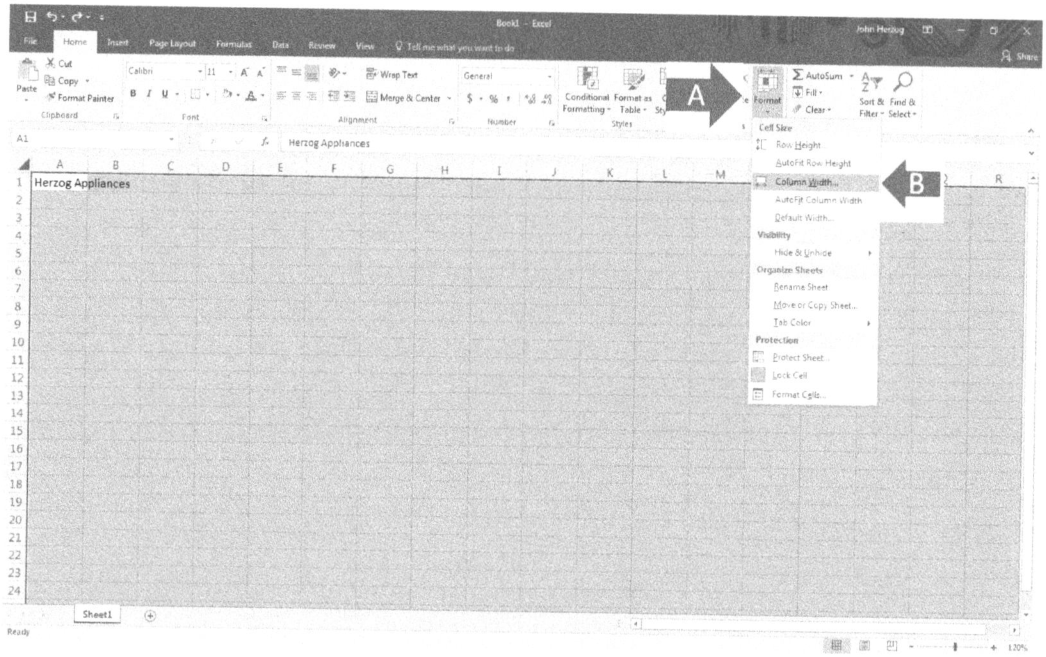

FIGURE 1-9

When you do, you will see the **pop-up menu**, shown below in Figure 1-10. You can then enter the column width you want (in this case, it is a width of *15* characters) in the box located to the right of the words **Column width** (Figure 1-10, Arrow).

TO MERGE AND CENTER CELLS

There are many who want to put a title at the top of a spreadsheet that is centered above the data. When doing so, it is best that the cells across which the title is centered are merged into one cell. To **merge** cell A1 with the cells to its right and **center** the title within the merged cells (in this example, we want to center the words *Herzog Appliances* across cells A1 through E1), you can do the following steps:

Step 1: Select the cell with the words *Herzog Appliances* in it AND select the cells B1 through E1 (Figure 1-11). You can do so by clicking in the middle of cell A1 and clicking and dragging the **white-block cross** across the cells until you reach cell E1. You can also select those cells by clicking on cell A1 and then pressing and holding the **shift key** down, and then tapping the **right arrow key** (→) to leave the trail of highlighting behind.

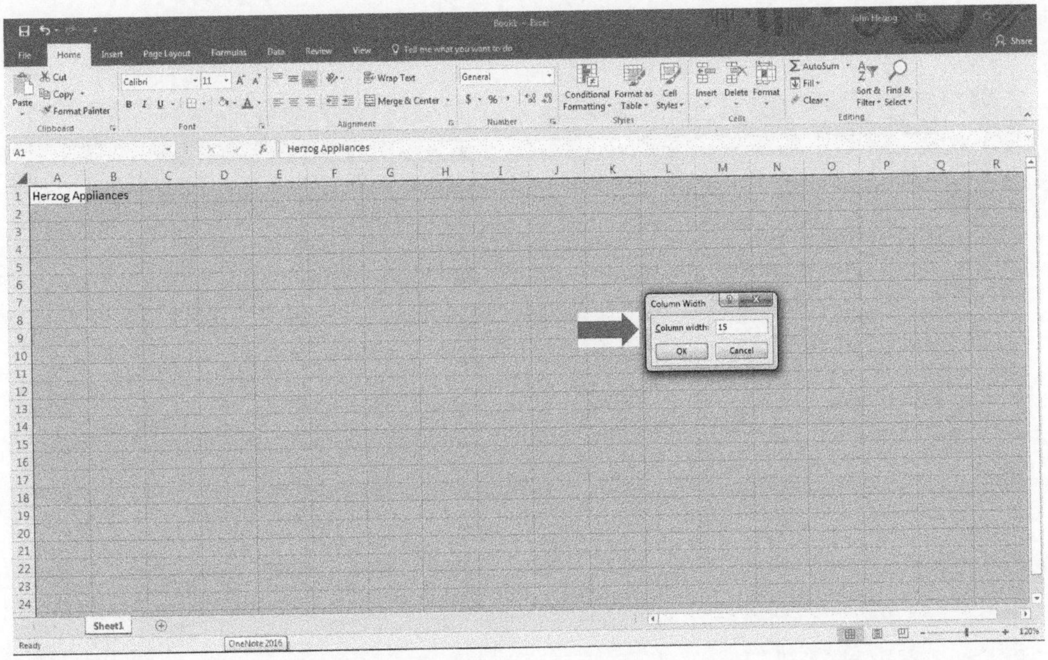

FIGURE 1-10

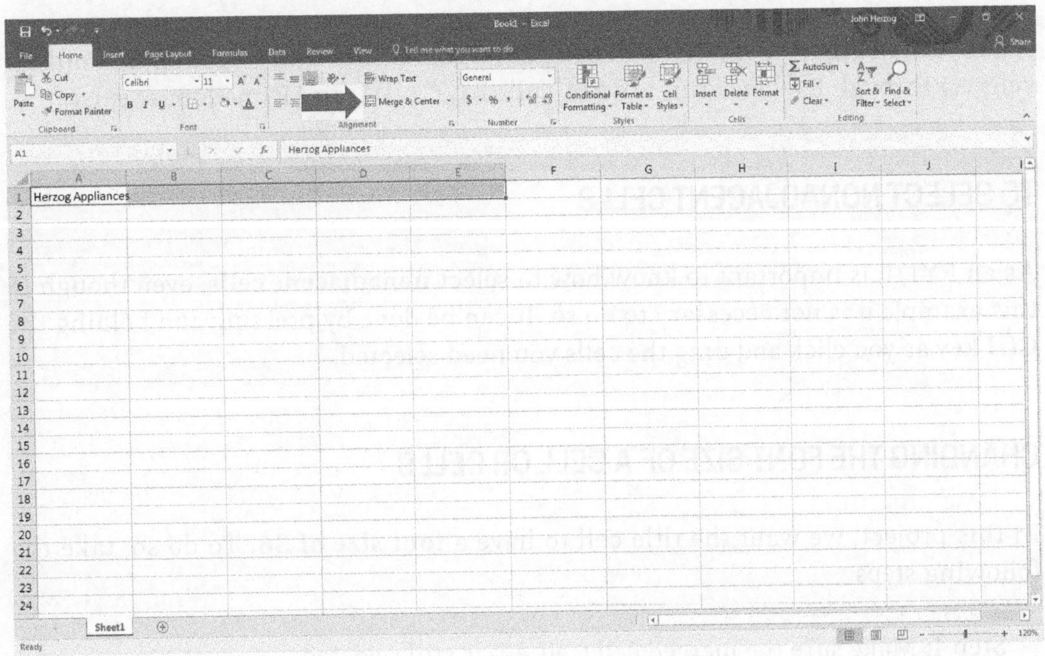

FIGURE 1-11

Step 2: Click **Merge & Center** in the **Home Ribbon** (Figure 1-11, Arrow).

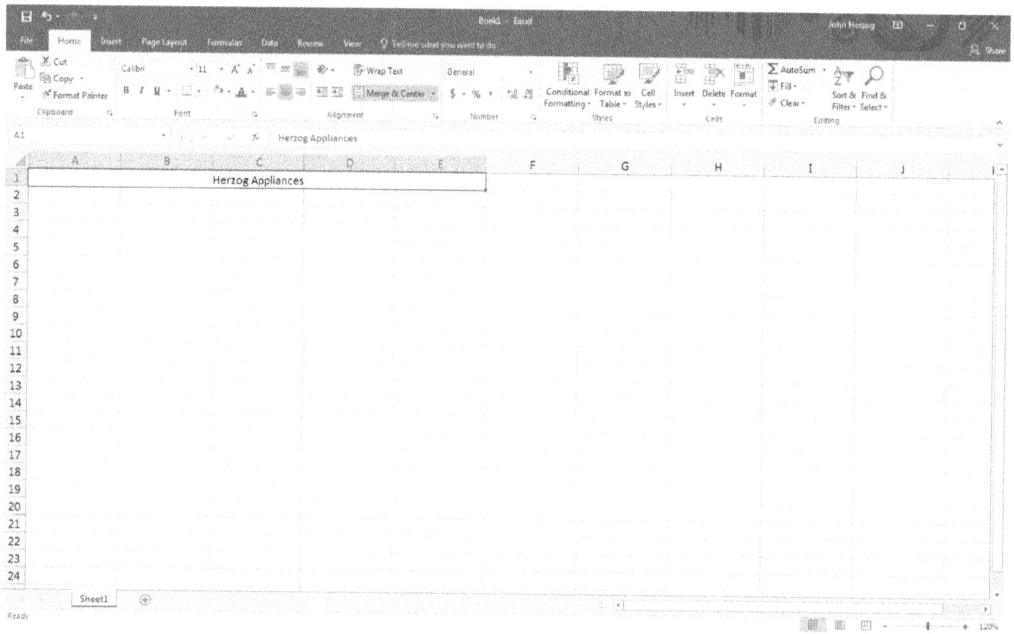

FIGURE 1-12

As seen in Figure 1-12, that will merge and center the *Herzog Appliances* words in cells A1 through E1. Once this is done, cells B1, C1, D1, and E1 no longer exist.

TO SELECT NONADJACENT CELLS

As an FYI, it is important to know how to select **nonadjacent cells**, even though in this example it is not necessary to do so. It can be done by pressing and holding the **Ctrl key** as you click and drag the cells you need selected.

CHANGING THE FONT SIZE OF A CELL OR CELLS

In this project, we want the title cell to have a **font size of 36**. To do so, take the following steps:

Step 1: Make sure the merged cell (cell A1) is still selected.

Step 2: Select the **font size of 36** in the font **drop-down list box** that is located in the **Home Ribbon** (Figure 1-13, Arrow).

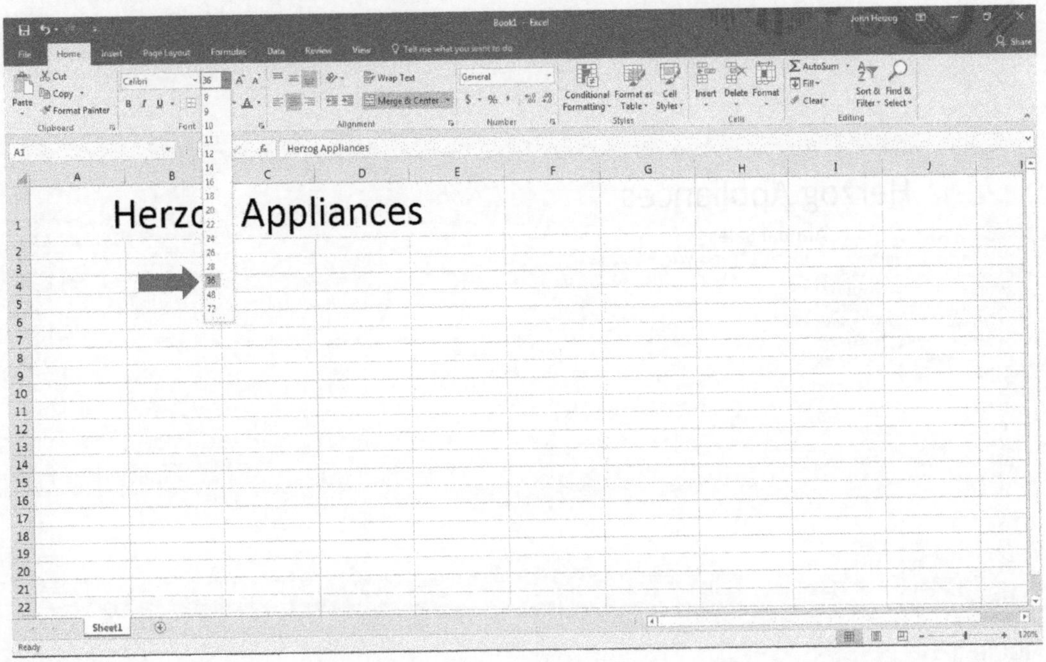

FIGURE 1-13

Now enter the words *Annual Sales* into cell A2. Repeat the process of merging and centering those words and set its **font size** to *22* points.

It is now necessary to type in the data that is needed as seen in Figure 1-14. At this point, all that is needed would **ONLY** be:

a. The stores' names.
b. The labels at the top of each column that describe the data.
c. The stores' sale figures.

All of the other numbers in this spreadsheet will be calculated by the use of formulas and functions, which will be discussed shortly.

NOTE: When entering the numbers or the signs of operation in a spreadsheet, it is always best to use the **numeric keypad**, which is way over on the right of your keyboard. It is designed just like the keyboard of a calculator. Using it will be easier that using the keys across the top of your keyboard. It will eliminate your need for having to occasionally press the shift key when you enter a plus sign (+). The numeric keypad will only work for you if the **Num Lock** is on.

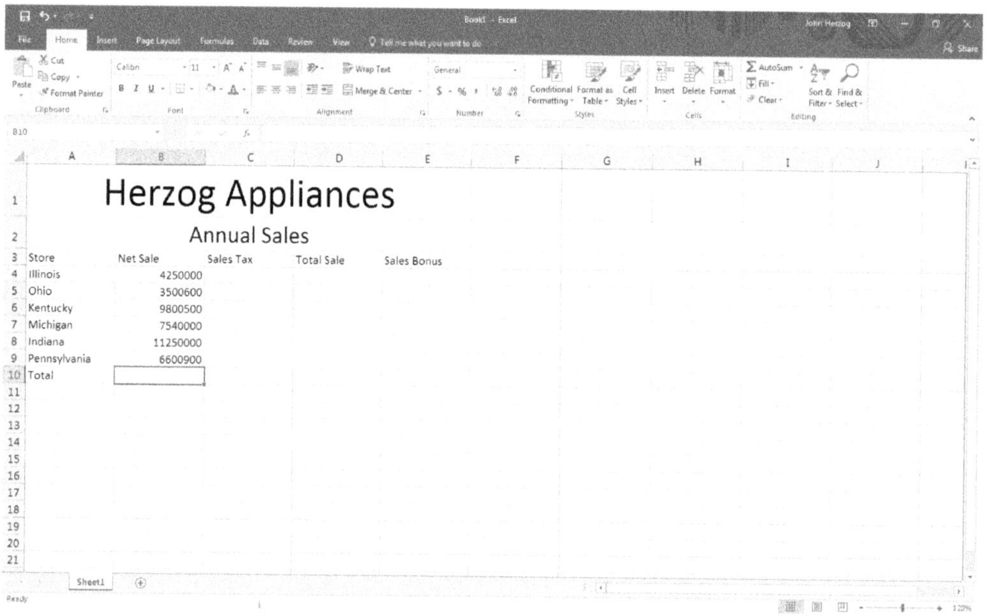

FIGURE 1-14

RENAMING A SHEET TAB

Normally, people want the tab at the bottom of a spreadsheet to describe its content. The default name for the tabs are **Sheet1** or **Sheet2**, depending upon how many sheets your Excel version is set to create initially when you begin to build the spreadsheet. To rename a sheet tab:

Step 1: Double-click the **sheet tab** (Figure 1-15, Arrow). It is currently named **Sheet1**, but after double-clicking it, Excel will put the sheet tab in a state that will allow you to edit it by typing over it. When it does, type the name you would like for the tab. In this example, it will be named *Sales* (Figure 1-15, Arrow).

ENTERING A FORMULA

Suppose the *sales tax* rate for the area in which *Herzog Appliances* does business is *7.5%*. To make the spreadsheet tally the *sales tax* for each sale, you can enter a formula by doing the following:

Step 1: Click on cell C4.

Step 2: Press the equal sign key (=) on your keyboard. Every formula or function in Excel begins with the equal sign key (=).

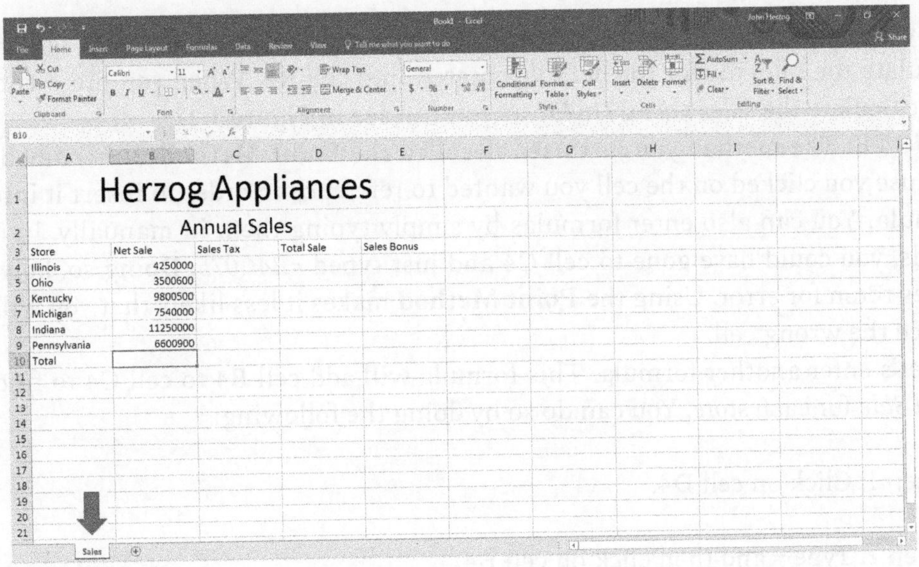

FIGURE 1-15

Step 3: Click on cell B4. Excel will automatically enter what is called a **cell reference** that reads =B4.

Step 4: Type **.075* into the cell (Figure 1-16, Arrow).

Step 5: Press **Enter.**

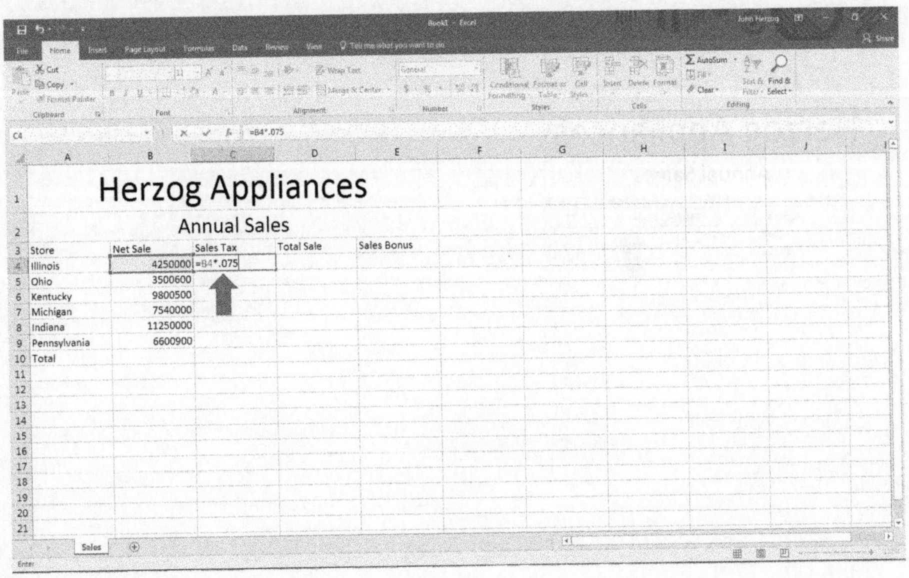

FIGURE 1-16

By doing so, you will have entered the formula that will read *=B4*.075*, which will calculate the *sales tax* for the *net sale* entered into cell B4. When you press **enter**, it will calculate the *sales tax* as *318750* as you can see in Figure 1-17.

NOTE: The method you just used is called the **Point Method**. It is named that because you clicked on the cell you wanted to reference in order to insert it into the formula. You can also enter formulas by simply typing them in manually. In other words, you could have gone to cell C4 and just typed *=B4*.075*. Doing so, however, leaves room for error. Using the **Point Method** makes it less likely that you will reference the wrong cell.

Let's enter another formula. This formula will add cell B4 to cell C4 to find the *Total Sale* for each *store*. You can do so by doing the following:

Step 1: Click on cell D4.

Step 2: Type = and then click on cell B4.

Step 3: Press + (found on the numeric keypad on the right of your keyboard), and then click on cell C4.

Your screen should look like Figure 1-17.

Step 4: Press **Enter**.

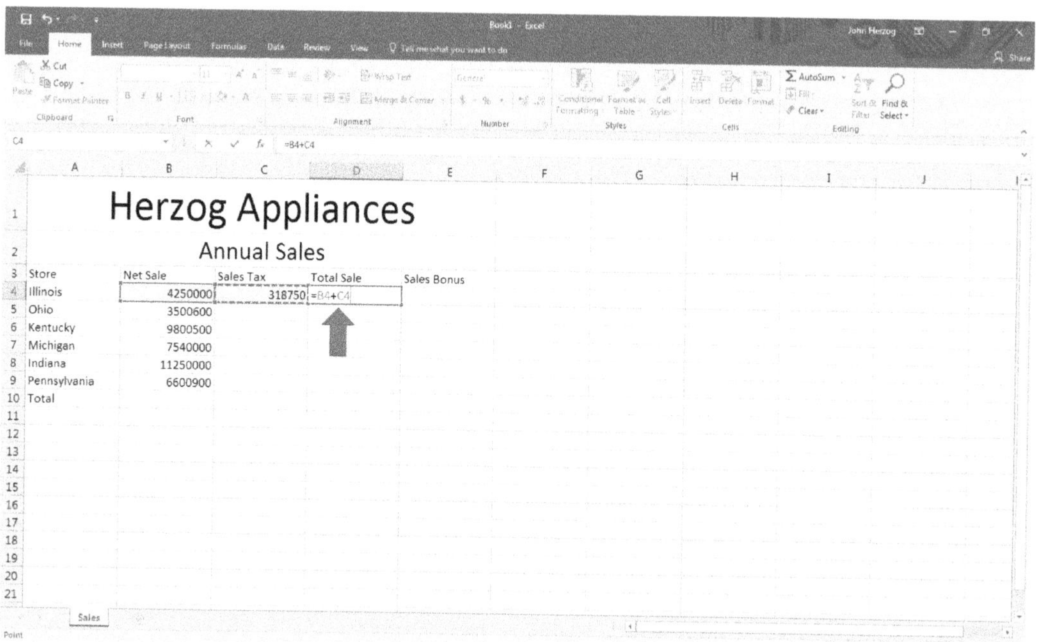

FIGURE 1-17

By doing so, you will have entered the formula that will read =*C4+B4*. When you press enter, it will calculate the total sales as *4568750* as seen in Figure 1-18.

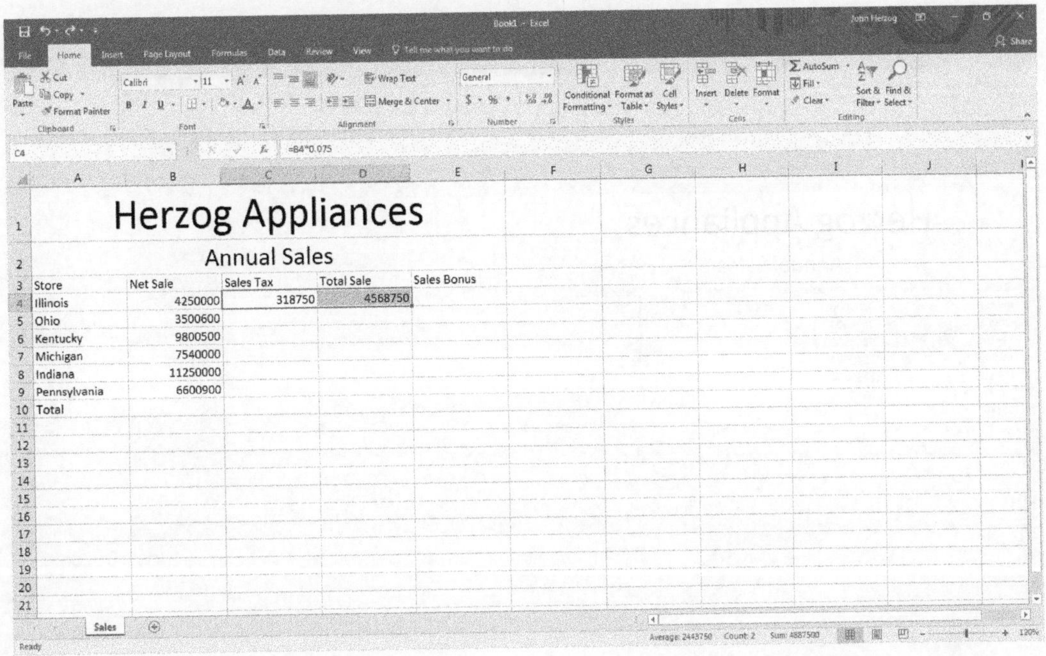

FIGURE 1-18

Filling a Formula

Some think that creating a formula takes more time than just using a calculator. Filling the formula to many cells, however, is one of the most powerful tools of Microsoft Excel or any spreadsheet. Once you fill the formulas to each cell below them, you will never again have to retype the formula. This is one of many things that makes a spreadsheet much faster than a calculator. To fill the formulas down, do the following:

Step 1: Select **both** cells (B4 and C4) by clicking on cell B4 and then clicking and dragging the mouse to cell C4 to the right (Figure 1-18). NOTE: Make sure you see a **white-block cross** before you do this. **If your mouse is pointing to a cell border you will not see the white-block cross, but rather you will see a thin black cross with arrows. If you click and drag your mouse at that point, you will wind up moving the cells rather than highlighting them.**

Step 2: Place your mouse on the small black box at the lower right of the area that you want to fill to the cells below. This is called the **Fill Handle**. When you do, your mouse will look like a thin black cross (Figure 1-19A, Arrow).

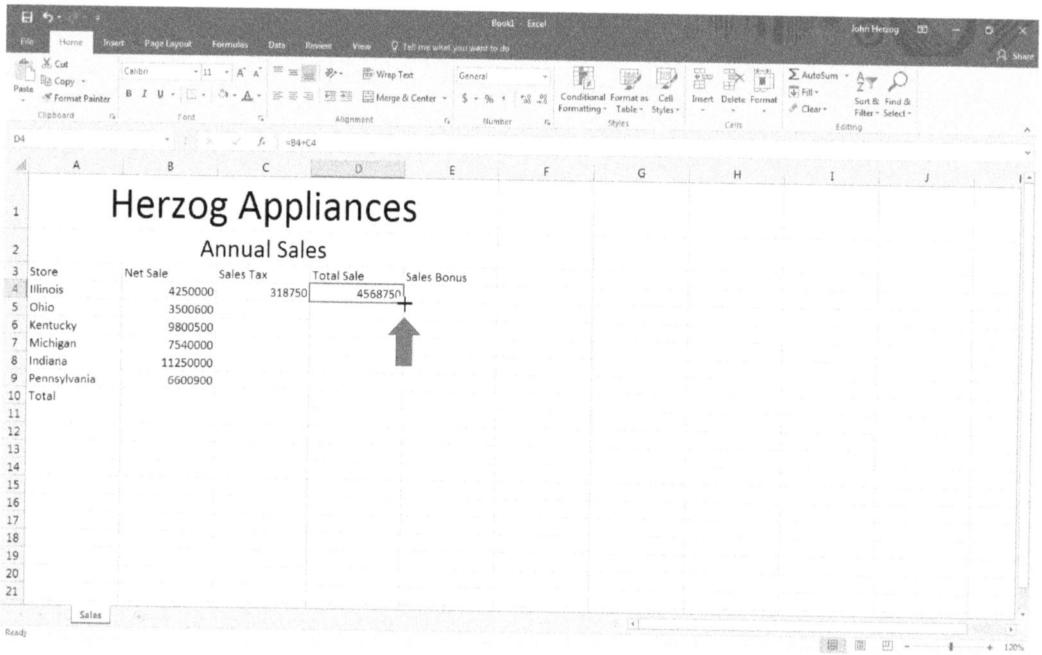

FIGURE 1-19A

Step 3: Click and drag that **Fill Handle** down to row 9. That will fill your formulas down to cells C9 and D9 inclusive. Your screen should look like Figure 1-19B.

NOTE: This is a major and important feature of spreadsheets! Filling formulas to other cells removes the need to type the formula over and over in each cell in which you want to show the calculations. Therefore, it makes it so that you no longer have to use a calculator for routine calculations as discussed earlier. Now that the formulas have been copied, each time you enter or change data in any of cells B4 through B9, the formulas in cells C4 through D9 will automatically **recalculate**!

GETTING TOTALS BY FUNCTIONS

Suppose we now want the total of the numbers in the column labeled as *Net Sale*. When you need totals or averages of a range of many cells, it is best **NOT** to enter a formula. For example, in this case it is best **NOT** to enter =B4+B5+B6+B7+B8+B9. That

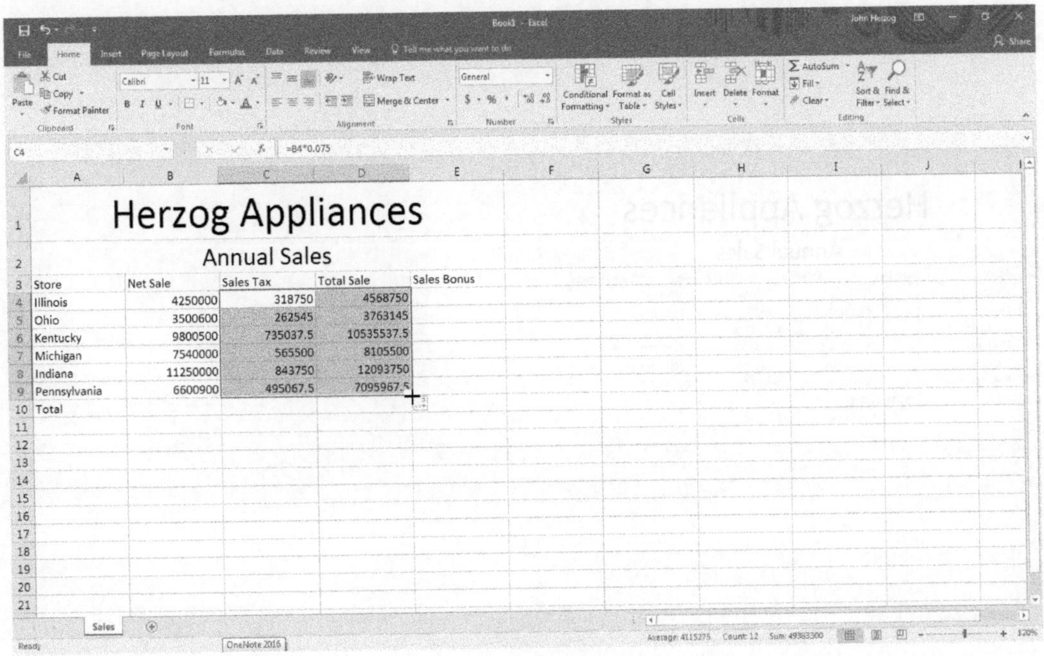

FIGURE 1-19B

would work for a few cells, but if you need to get totals of **many** cells, it would take a long time to make such a formula. That is why we use **functions**. As stated before, a function is an enhanced and easier-to-use formula. In this example, we now want to find the sum of the numbers in the columns labeled as *Net Sale, Sales Tax,* and *Total Sales* by using what is called the **SUM** function.

The SUM Function: Method One

Step 1: Place your cursor in cell B10.

Step 2: Type *=SUM(B4:B9)* as seen in Figure 1-20.

Step 3: Press **Enter**.

That would give you a total of 42942000, which is the sum of all of the numbers in cells B4 through B9 inclusive (Figure 1-21). NOTE: Anything inside the parentheses is called an **Argument**. In this case, there is only one argument, and it is the range of cells for which you want to find the sum.

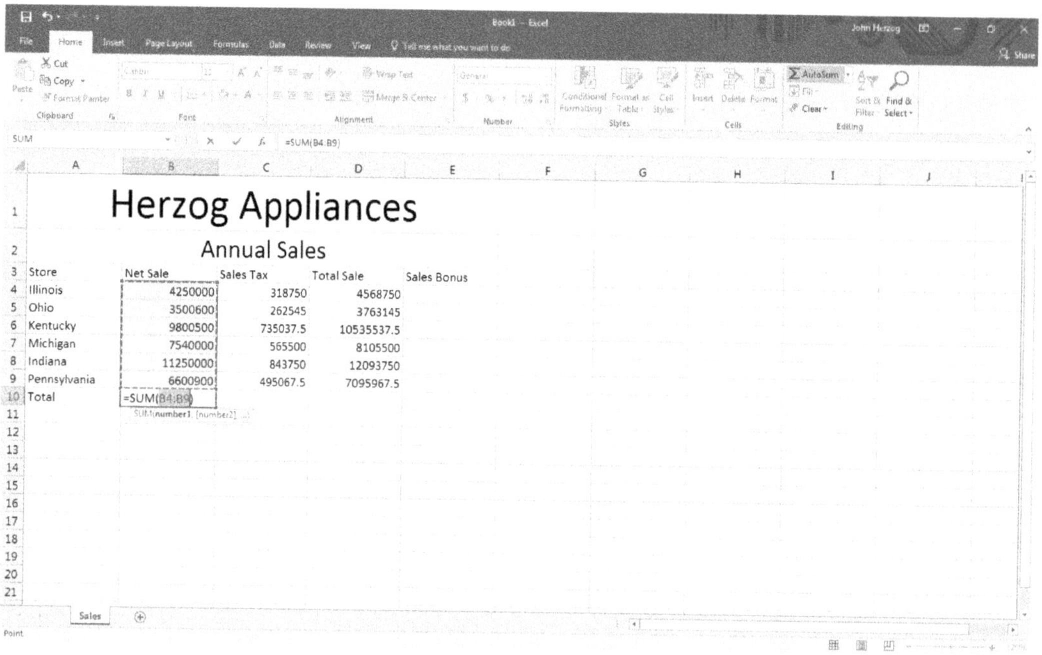

FIGURE 1-20

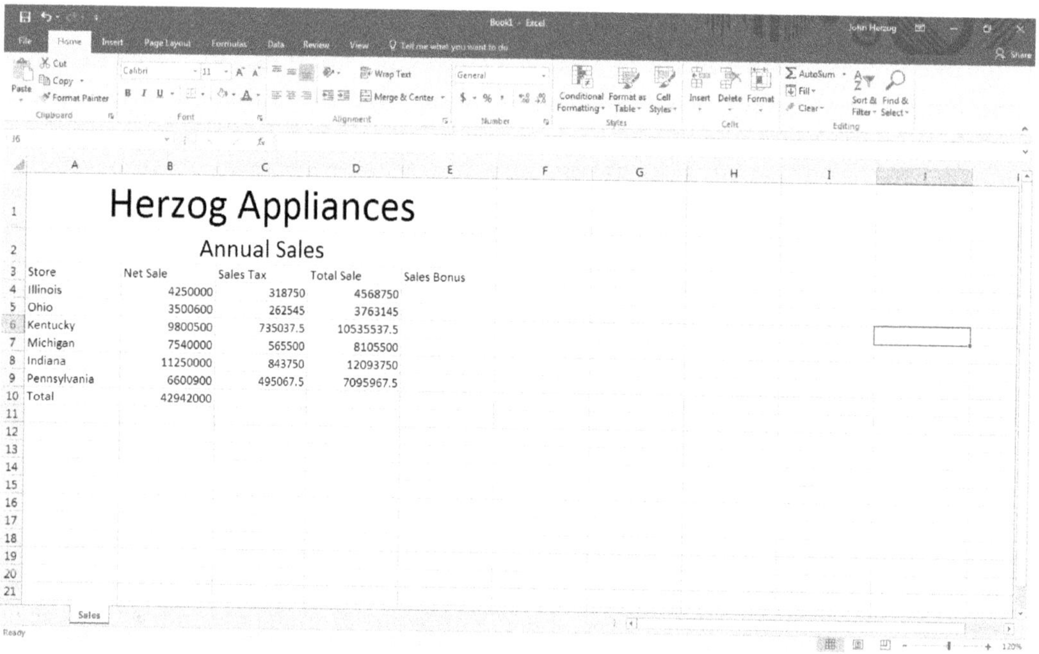

FIGURE 1-21

The SUM Function: Alternate Method Two

Click on cell B10, and then press the **Delete key** on your keyboard so we can remove the SUM function you just entered in order to use a different approach. NOTE: Selecting cells and then pressing **Delete** will always remove all content from any cell!

Step 1: After deleting the SUM function from cell B10, we will enter it a different way by typing =*SUM(*.

Step 2: Click and drag the cell range of B4 through B9 in the spreadsheet, and Microsoft will enter that cell range for you as seen in Figure 1-22.

Step 3: Press **Enter**. NOTE: All functions require a right parenthesis, but Excel will put it there for you in most cases if you forget to do so.

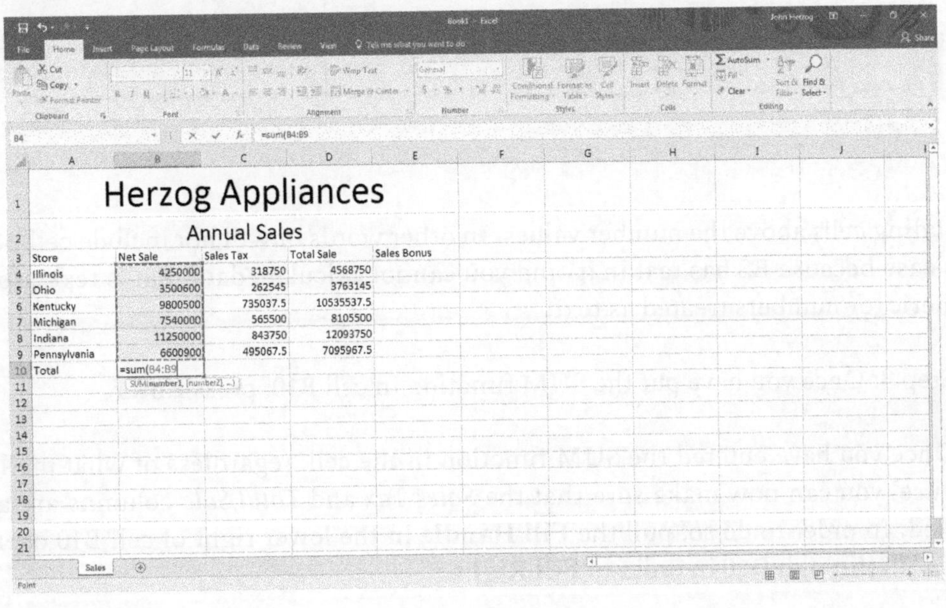

FIGURE 1-22

The SUM Function: Method Three (Using the AutoSum)

Once again, click on cell B10, and then press the **Delete key** on your keyboard.

Step 1: Make sure your cursor is still on cell B10.

Step 2: Click the **AutoSum** button in the **Home Ribbon** (Figure 1-23, Arrow).

When you do, it will automatically select all of the numbers above itself and it will enter that range as an argument in the SUM function. Notice that it will stop

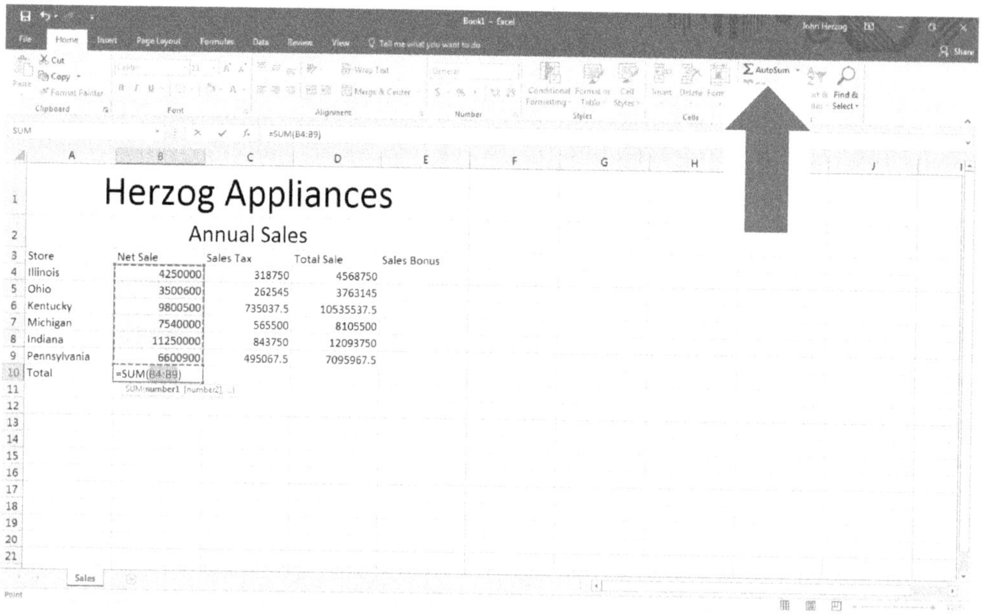

FIGURE 1-23

including cells above the number values. In other words, it will not include cell B3 in this case because B3 has text in it, and you cannot calculate data that is text, alpha-numeric, or numbers treated as text.

Step 3: Once you have put the SUM function in cell B10, press **Enter**.

Once you have entered the SUM function in the cell, regardless of what method you use, you can now make sure that the *Sales Tax* and *Total Sale* columns are also totaled. In order to do so, pull the **Fill Handle** in the lower right of cell B10 over to cells C10 through D10 in order to **Fill Right**.

TO USE THE AUTOSUM ON MULTIPLE CELLS

Let's return to our scenario. We are now going to get totals of all the numbers in cells B4 through D9 in a much easier and quicker way. Click on cell B10, and then press the **Delete key** on your keyboard.

Step 1: Select cells B10 through D10 as shown in Figure 1-24.

Step 2: Click the **AutoSum** command again (Figure 1-25, Arrow A). It will put the **SUM** function in all three selected cells simultaneously (Figure 1-25, Arrow B).

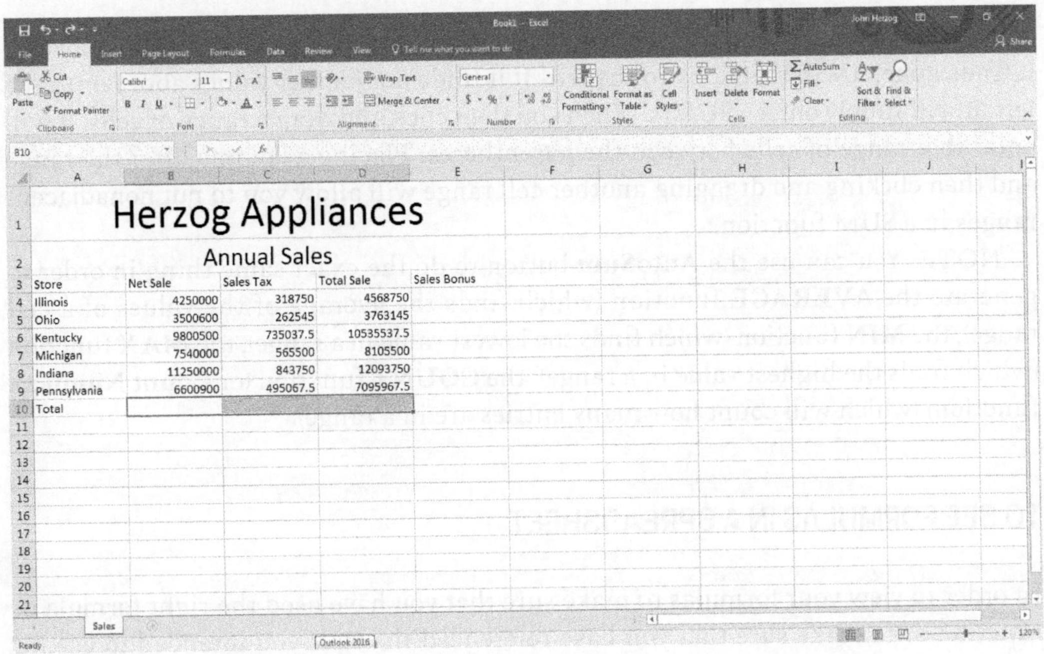

FIGURE 1-24

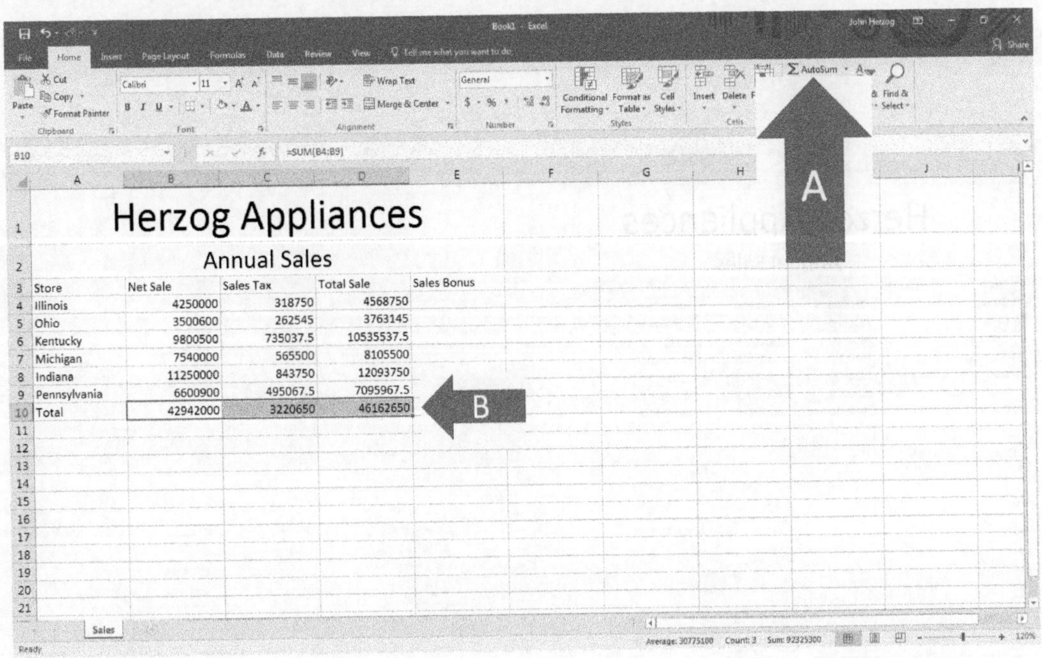

FIGURE 1-25

NOTE: Whenever you use the AutoSum, it will look for number values above. If it finds none, it will then look to the left. If it finds no number values above or to the left, it will simply enter =SUM(). When you then click and drag a range of cells, it will enter that range of cells between the parentheses. Placing a comma after the range and then clicking and dragging another cell range will allow you to put nonadjacent ranges in a SUM function.

NOTE: You can use the **AutoSum** button to do the exact same thing in order to generate the **AVERAGE** function (which finds the average of the values of a cell range), the **MIN** function (which finds the lowest value in a range), the **MAX** function (which finds the highest value in a range), the **COUNT** function (or **Count Numbers** function) (which will count how many entries are in a range).

TO SEE FORMULAS IN A SPREADSHEET

In order to view your formulas to make sure that you have used the right formula or function or to make sure that you have referenced the cells you intended to use for your calculations, you can look at them individually in the **Formula Bar** when you select the cell. For example, notice that cell D4 is selected in Figure 1-26. If you look

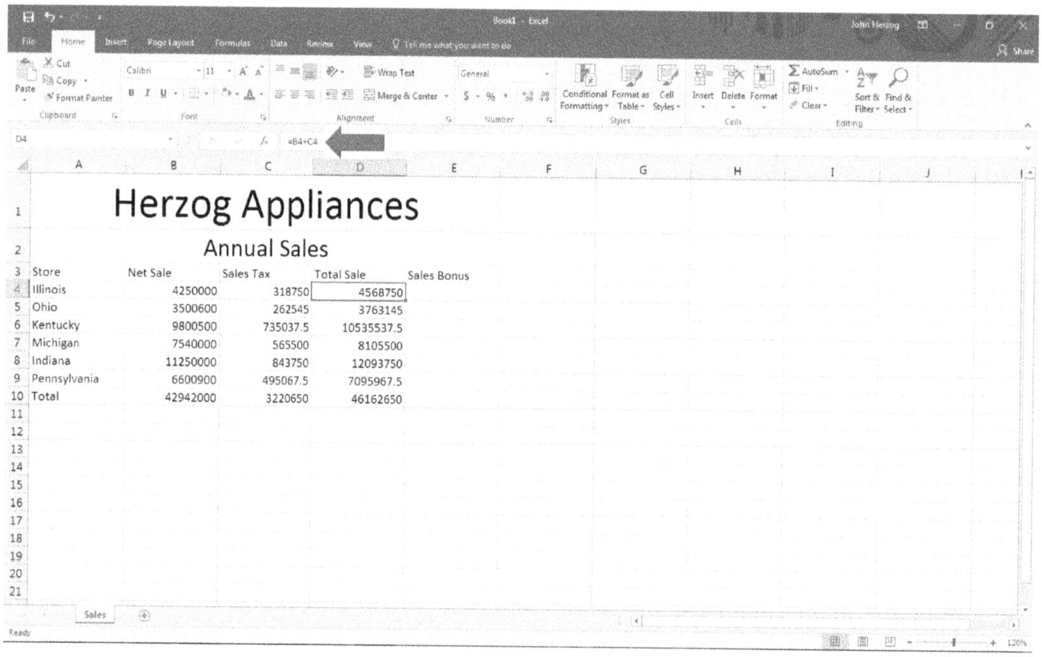

FIGURE 1-26

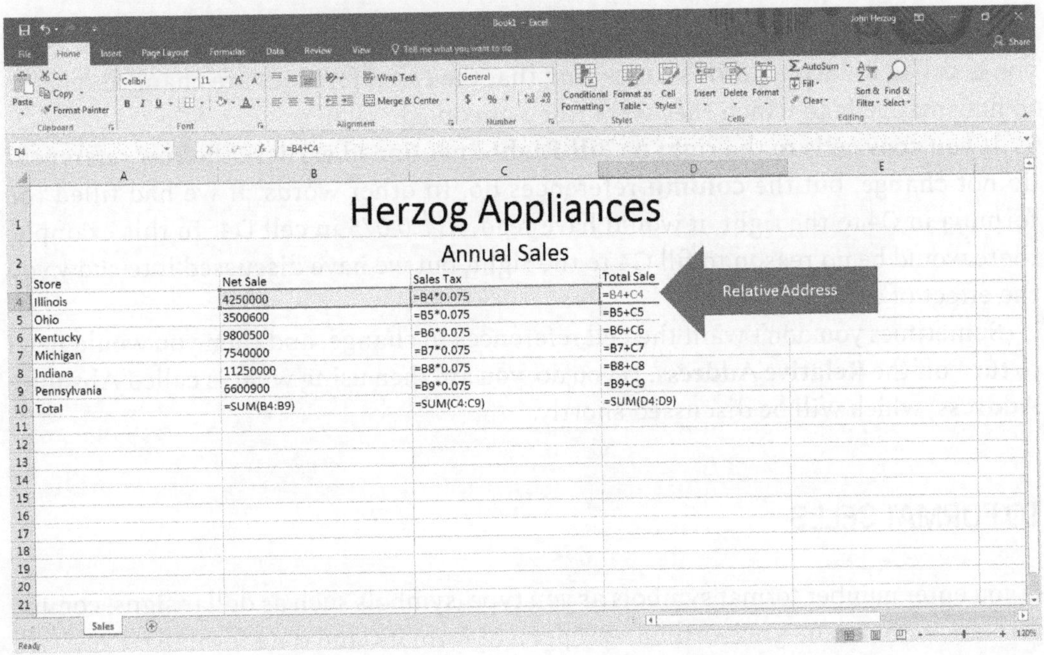

FIGURE 1-27

at the formula in the **formula bar** that is to the left of the arrow in Figure 1-26, you can see that the formula that actually generated the 4,568,750 in that cell is *=B4+C4*.

Another way to see your formulas is to press and hold **Ctrl** ~, and you will see what is shown in Figure 1-27.

It is actually better to see the formulas by using the **Ctrl** ~ command because you can see all of the formulas at one glance. Please notice that when you do, the formatting is temporarily removed. To hide the formulas and return to the previous view, simply press Ctrl~ again, and the formatting will be restored.

RELATIVE ADDRESS

Take another look at Figure 1-27. Notice that the formulas' cell references were automatically changed when the formulas were filled down. The original formula, for example, in cell C4 was *=B4*0.075*. But when it was filled down, Excel automatically changed those cell references by one row in each of the cells below. This is very useful because it removes the need for you to enter each individual formula. In our example, we do in fact want the formula to be *=B5*0.075* in the cell C5 and *=B6*0.075* in the cell C6 below that, and so on. Otherwise, if you had to calculate the sales tax for hundreds of stores, it would have taken a long time to type that formula into each cell individually. This process of Excel automatically changing the cell references for you is called **Relative Address**, meaning that the cell references are changed relative to how may cells to which they were filled or copied.

If you copy cells to the right or **Fill Right** (mentioned earlier), the **row** references do not change, but the **column** references do. In other words, if we had filled the formula in C4 to the right, it would have read *=C4*0.075* in cell D4. In this example, there would be no reason to fill C4 to the right, but we have discussed it to show you the effect of doing so.

Sometimes you don't want the cell references to change, and then you would need to turn off the **Relative Address**. If you do, you are then using what is called **Absolute Address**, which will be discussed shortly.

TO FORMAT CELLS

If you enter number format symbols as you type (symbols such as dollar signs, comma formats for numbers greater than 999, percent formats, and date formats), Excel will assume that format is what you want in that cell, and it will remain there unless you change it. You should type in symbols as you add data **ONLY** if the data is being entered only once or twice in a few cells.

When you enter a date into a spreadsheet or into most programs in a computer, the dates are treated as numbers. The numbers represent the number of days that have elapsed since January 1, 1900. For example, if you type *1/11/2018*, it will automatically set the format so that Excel will display it as a date. But if you change the format of that cell to a number format, *43,111* will be displayed because as of *1/11/2018*, *43,111* days had elapsed since January 1st, 1900.

Different date formats are shown in Figure 1-28.

Thursday, January 11, 2018	Long Date
1/11/2018	Short Date
1/11	Short Date Month/Day
11-Jan	Short Date Day, Month
11-Jan-18	Short Date Day, Month, Year
Jan-18	Short Date Month, Year
January 11, 2018	Full Month, Day, Year

FIGURE 1-28

There are also various formats for displaying money:

1. Accounting
2. Currency
3. Accounting with no dollar signs
4. Currency with no dollar signs

Accounting formats display zeros as a dash, and the dollar signs will be aligned to the left of each cell regardless of how large or small a number is (as shown in Figure 1-29). **Currency** is just the opposite. Zeros are shown as zeros, and the dollar signs float back and forth as the numbers get larger or smaller in value.

Thousands		Zeros
$ 1,000.00	Accounting	$ -
$1,000.00	Currency	$0.00
1,000.00	Accounting With NO Dollar Sign	-
1,000.00	Currency With NO Dollar Sign	0.00

FIGURE 1-29

In this example, we will use the **Currency with NO Dollar Sign**. Many often use that format because the dollar signs tend to clutter a spreadsheet, and it's really not necessary to use them if it is understood that all of the numbers in the spreadsheet represent dollar sales. This format will also give us the commas where they are needed, and it will display zeros as zeros. To use the **Currency with NO Dollar Sign** format, take the following steps:

Step 1: Select the cells you wish to format (cells B4 through D10 in this example) as seen in Figure 1-30.

Step 2: Click the **drop-down box** in the **number format box** (Figure 1-31, Arrow A) and choose the **More Number Formats** option (Figure 1-31, Arrow B).

Step 3: When the **pop-up menu** shown in Figure 1-32 is displayed, click **Currency** (Figure 1-32, Arrow A).

Step 4: Click **None** in the **symbols option box** (Figure 1-32, Arrow B).

That will give you what is seen in Figure 1-33 below.

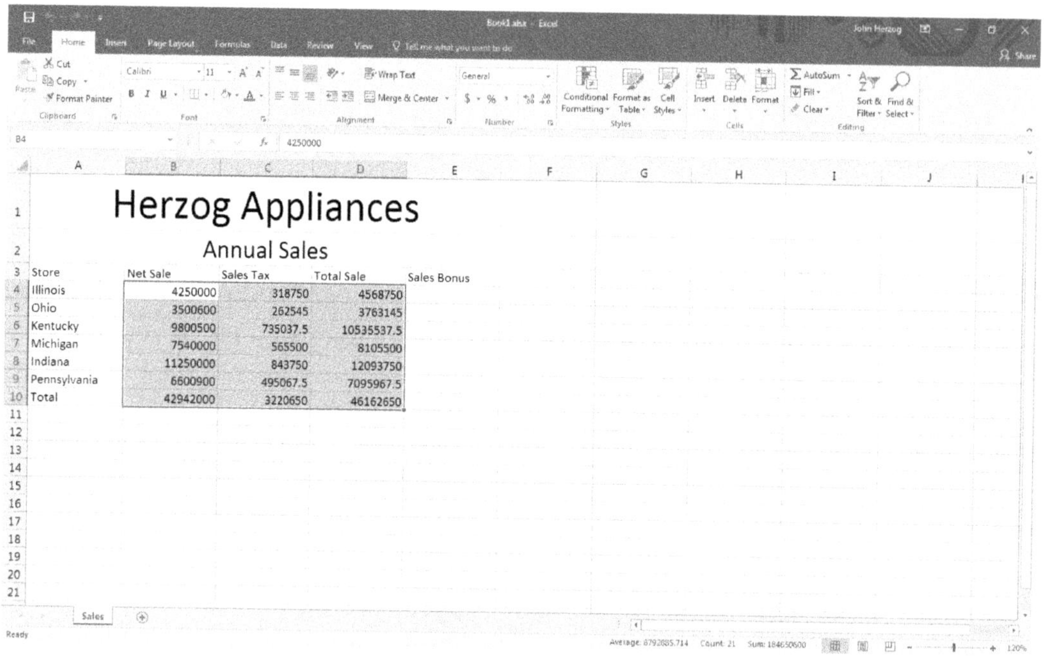

FIGURE 1-30

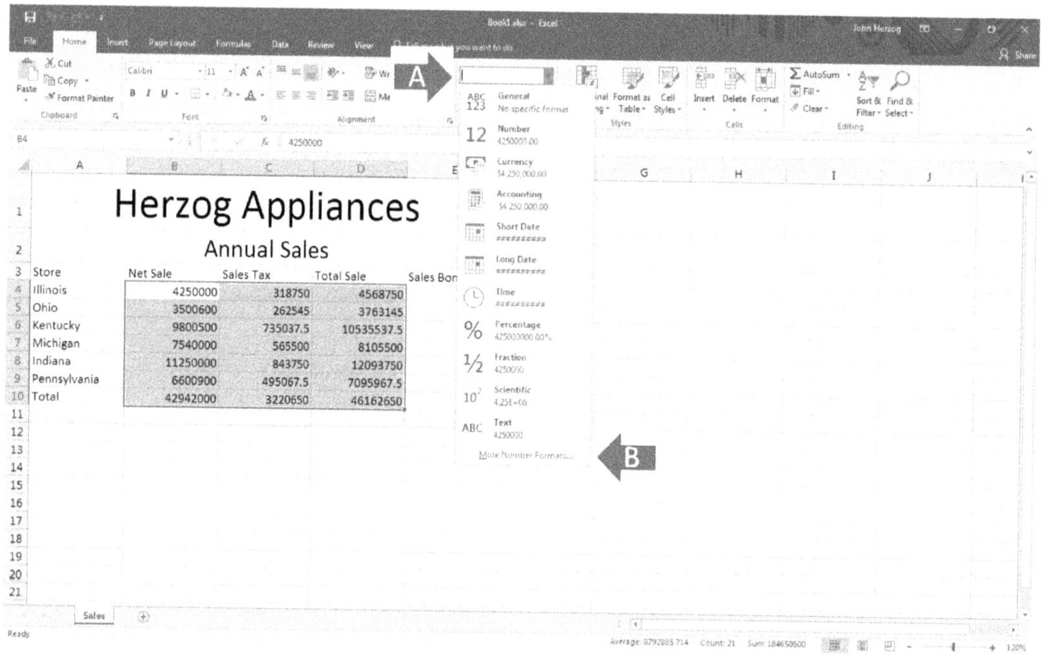

FIGURE 1-31

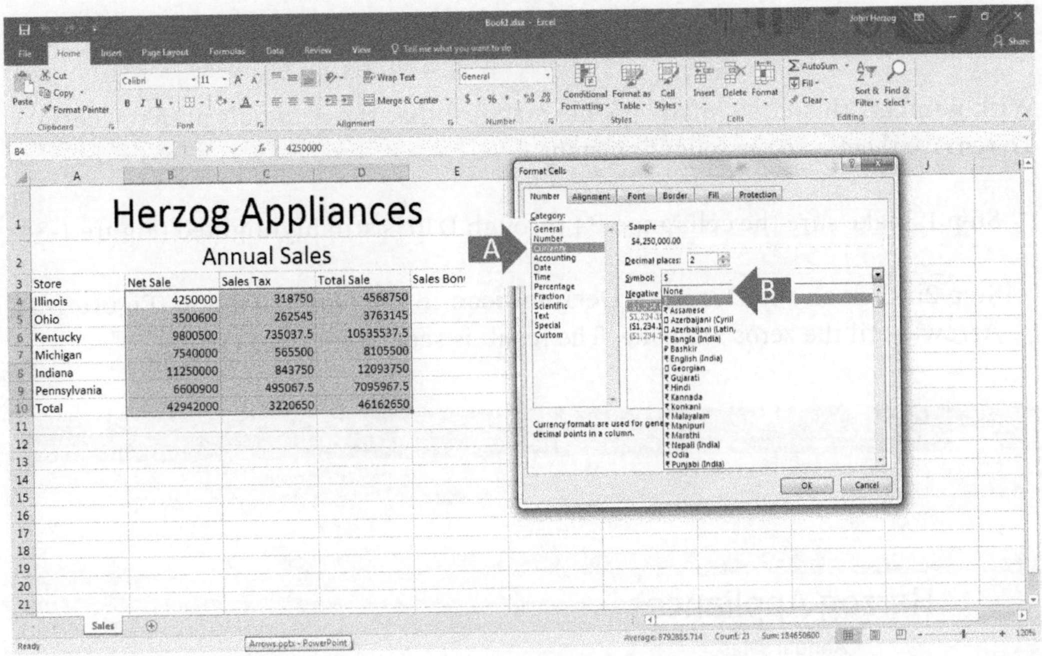

FIGURE 1-32

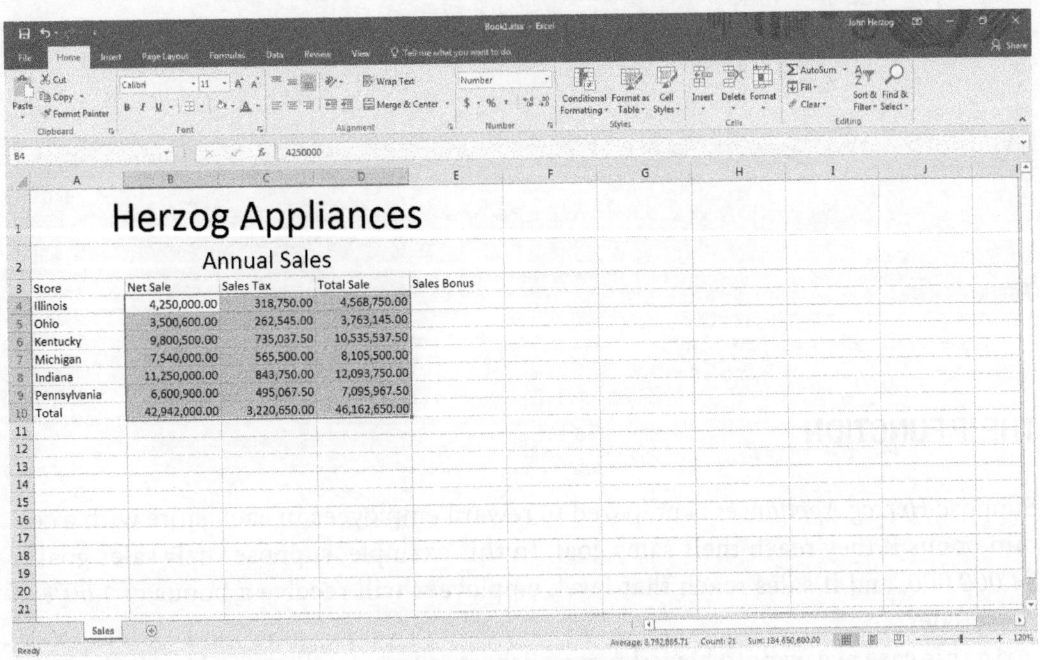

FIGURE 1-33

TO REMOVE THE DECIMALS

With numbers this large, many people may feel that the decimals aren't necessary. To remove them, use the following steps:

Step 1: Make sure the cell range B4 through D10 is still highlighted (Figure 1-34).

Step 2: Click on the **Decrease Decimal icon** in the **Home Ribbon** (Figure 1-34, Arrow) until the zeros are gone. The result is seen in Figure 1-34.

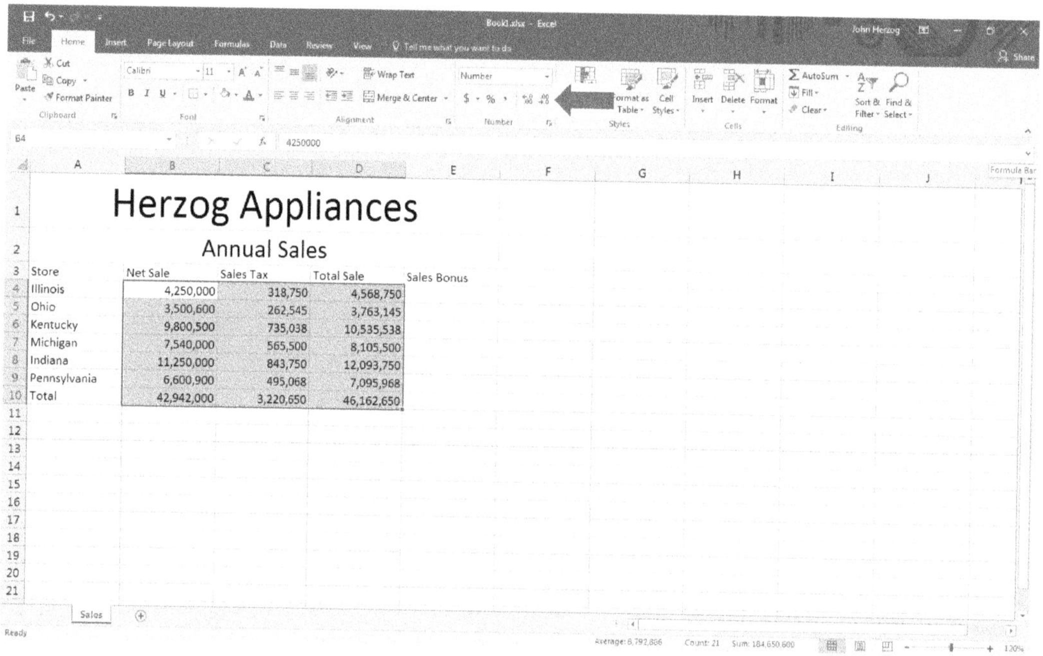

FIGURE 1-34

THE IF FUNCTION

Suppose *Herzog Appliances* is prepared to reward employees in each store with a certain bonus if they reach their sales goal. In this example, suppose their sales goal is *$4,000,000*, and if sales reach that level, employees will receive a bonus of *1.00%* of those sales.

In this case, we want to have the spreadsheet automatically calculate the bonuses as sales numbers are added. One way this can be done is by the use of the **IF** function. To prepare for the discussion of the **IF** function, it is important that you enter a few numbers in this spreadsheet that you can use for this example to understand it better. It is a rather complex function.

1. Enter the word *Goal* into cell C12 and enter the word and symbol *Bonus %* into cell C13 (Figure 1-35, Arrow A). Make sure they are **bold** and **right-aligned** by clicking the **Bold** and **Right-Align** icons in the **Home Ribbon** as seen in Figure 1-35, Arrow B.
2. Enter *$4,000,000* into cell D12 and enter *1%* into cell D13 (Figure 1-35, Arrow C). NOTE: Normally it is better to set the formats by using the **Home Ribbon** icons, but when you are formatting only one cell, you can type in the symbol as you type the number.

Here is how the **IF** function works.

As stated earlier, every function has an **argument**. Each argument is separated by a comma. When you used the **SUM** function earlier, there was only one argument, which was the range of cells to be totaled. In the **IF** function, there are three (3) arguments as show below:

=IF(Argument 1: Statement Made; Argument 2: The value displayed in the cell if the statement made in Argument 1 is true; Argument 3: The value displayed in the cell if the statement made in Argument 1 is false).

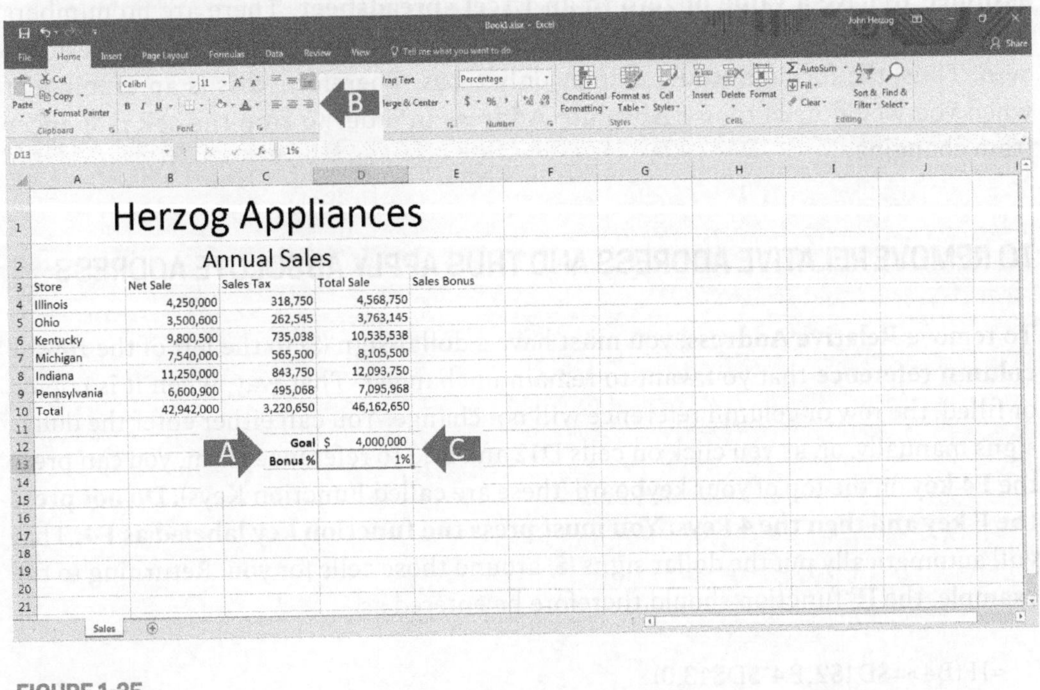

FIGURE 1-35

Therefore, in this example, we will put the **IF** function into cell E4, which would read:

=IF(B4>=D12,B4*D13,0)

If you want to read the logic of the **IF** function in this example, it would read, "If B4 is greater than or equal to D12, then display the product of B4 times D13 in this cell; otherwise, display a zero in this cell."

There is just one problem. After you place that function into cell E4 and fill it down to cell E9, the **Relative Address** would change it to read as follows:

Cell E5 would read =IF(B5>=D13,B5*D14,0)

Cell E6 would read =IF(B6>=D14,B6*D15,0)

Cell E7 would read =IF(B7>=D15,B7*D16,0)

Cell E8 would read =IF(B8>=D16,B8*D17,0)

Cell E9 would read =IF(B9>=D17,B9*D18,0)

If that happens, only the first cell would be correct. **Blank cells are always assumed to have a value of zero in an Excel spreadsheet**. There are no numbers in cells D14 through D18, and when you multiply anything times zero, the result is zero. Thus, the formula would display only zeros. Therefore, this is an example of when you would need to use **Absolute Address** in order to stop the cell references from changing.

TO REMOVE RELATIVE ADDRESS AND THUS APPLY ABSOLUTE ADDRESS

To remove **Relative Address**, you must have a dollar sign ($) to the left of the **row or column reference** that you want to remain unchanged. That way, when it is copied or filled, the row or column reference will not change. You can either enter the dollar signs manually, or, as you click on cells D12 and D13 to reference them, you can press the **F4 key** on the top of your keyboard (these are called Function Keys). **Do not press the F key and then the 4 keys. You must press the function key labeled as F4**. That will automatically put the dollar signs ($) around those cells for you. Returning to the example, the **IF** function should therefore be entered as:

=IF(B4>=$D1$2,B4*$D$13,0)

When you do, the subsequent formulas would then read as follows:

Cell E5 would read =IF(B5>=$D$12,B5*$D$13,0)

Cell E6 would read =IF(B6>=$D$12,B6*$D$13,0)

Cell E7 would read =IF(B7>=$D$12,B7*$D$13,0)

Cell E8 would read =IF(B8>=$D$12,B8*$D$13,0)

Cell E9 would read =IF(B9>=$D$12,B9*$D$13,0)

Please notice that you **DO** want the B4 to change to B5, B6, B7, B8, and B9 in each cell in order to read each different line of sales from each store; therefore, no **Absolute Address** is required in those cell references.

After adding the formula and filling it down, the results will be as seen in Figure 1-36. Please notice that cell E5 has a value of 0 because the sales for the *Ohio* store are below the *Goal*. If you change some of the sales, the bonuses will change accordingly if they go above or below the sales *Goal* in cell D12.

In this scenario, in order to add up the total store bonus, place the **SUM** function in cell E10 that will sum cells E4 through E9.

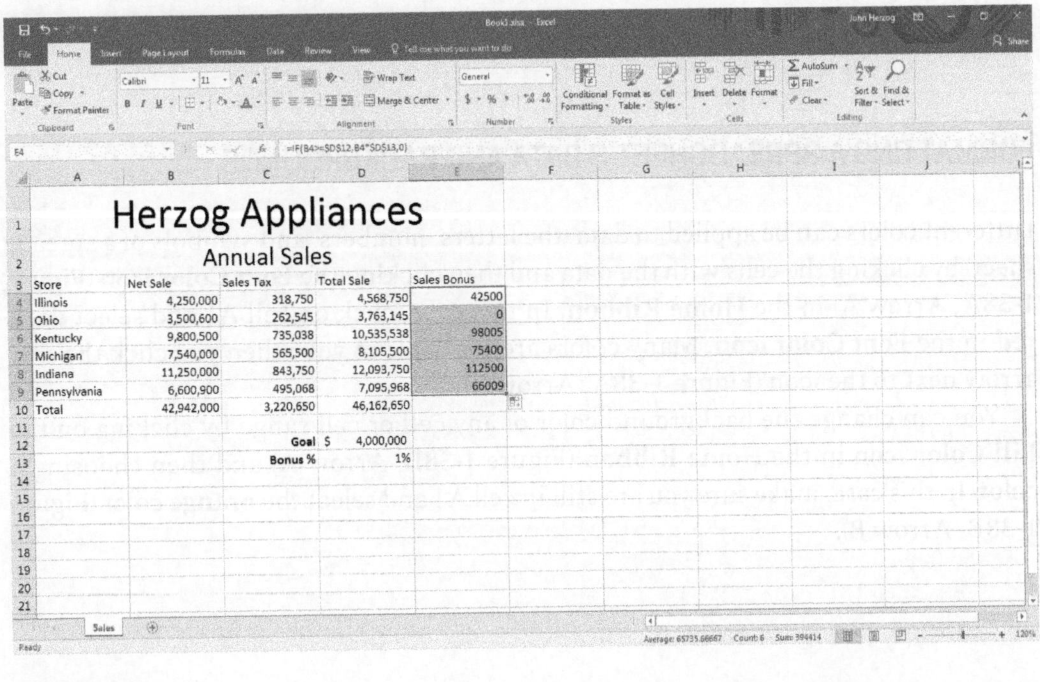

FIGURE 1-36

That will give you what you see in Figure 1-37 below.

After doing this, apply the same currency with no symbols format to cells E4 through E10.

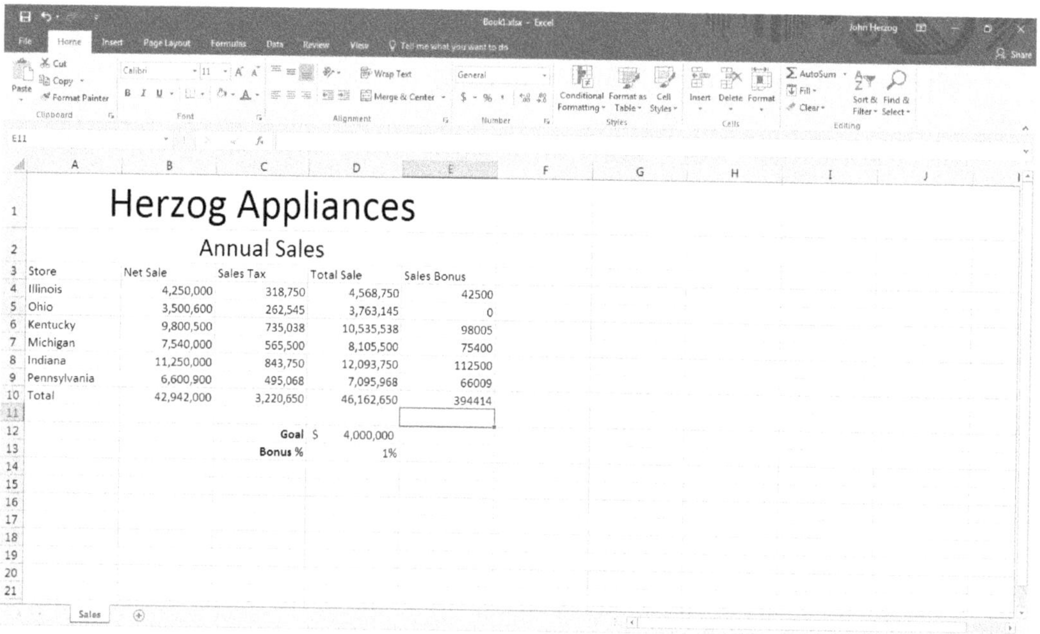

FIGURE 1-37

FORMATTING A SPREADSHEET'S DATA AND BACKGROUNDS

Different colors can be applied to data (the letters, numbers, and symbols of a spreadsheet) by clicking the cells with the data and then clicking the **Font Color icon** (Figure 1-38A, Arrow A) in the **Home Ribbon**. In this case, click on cell A1 and select **dark red** in the **Font Color icon**. Many colors are available to you when you click the tiny arrow next to the icon (Figure 1-38A, Arrow B).

You can change the background color of any cell or cell range by clicking on the **Fill Color icon** in the **Home Ribbon** (Figure 1-38B, Arrow A) and then choosing a color. In this case, make sure you are still in cell A1 and select the **orange** color (Figure 1-38B, Arrow B).

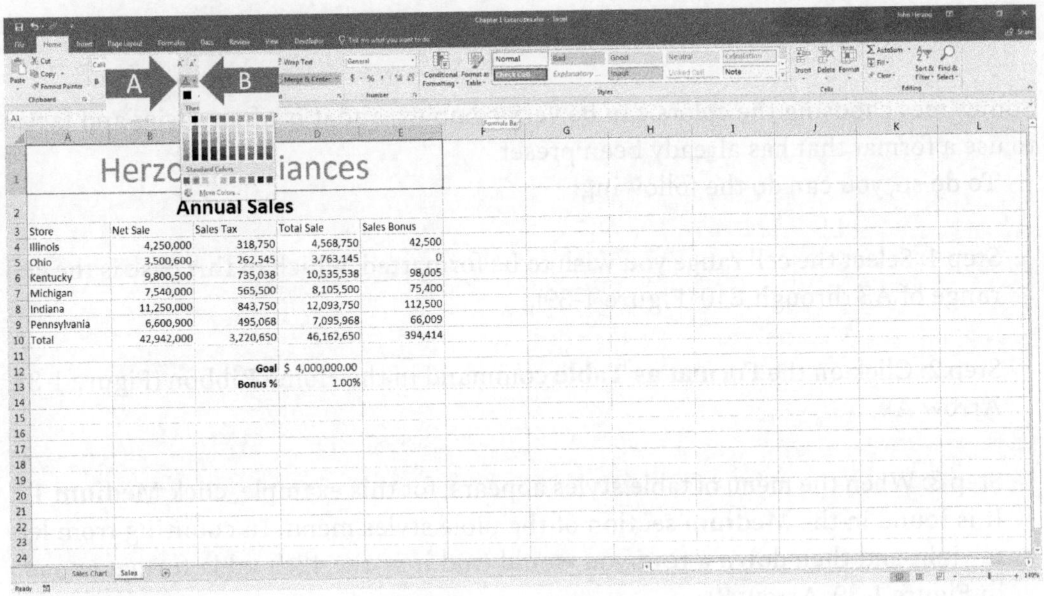

FIGURE 1-38A

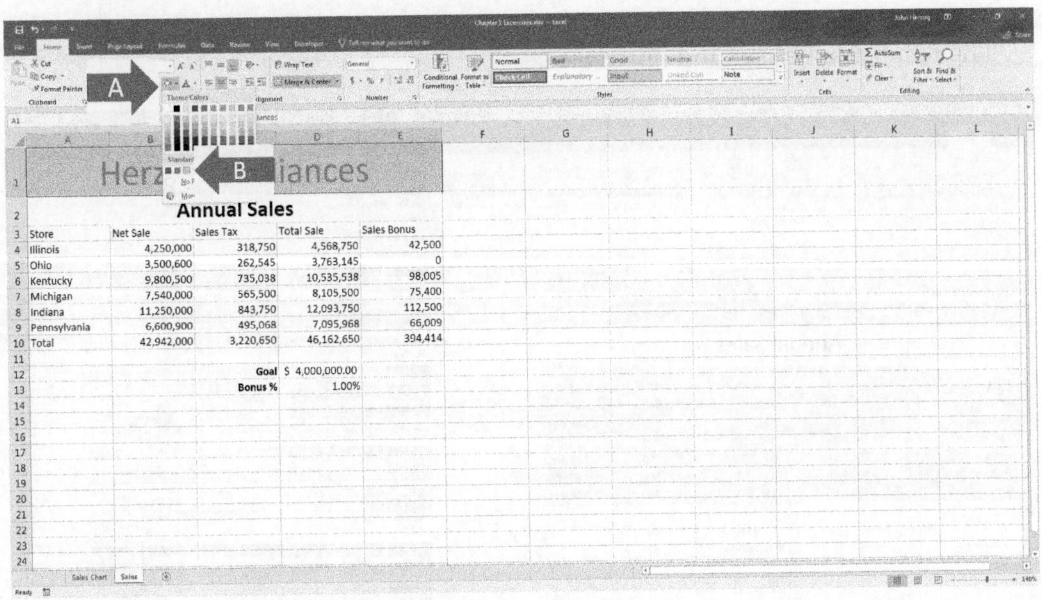

FIGURE 1-38B

FORMATTING A SPREADSHEET AS A TABLE

Sometimes choosing the colors can be very complex, and it is often easier and faster to use a format that has already been preset.

To do so, you can do the following:

Step 1: Select the cell range you wish to be formatted, which in this case is the cell range of A3 through E10 (Figure 1-39).

Step 2: Click on the **Format as Table** command in the Home Ribbon (Figure 1-39, Arrow A).

Step 3: When the menu of table styles appears, for this example, click **Medium 12.** It is found in the **Medium** section of the table styles menu. In counting from left to right and then down a row, you would find it as the 12th table format as seen in Figure 1-39, Arrow B).

Step 4: When the **confirmation pop-up menu** is displayed, make sure you have chosen the range you want (which is shown in Figure 1-40, Arrow A), then click **OK** (Figure 1-40, Arrow B).

The result can be seen in Figure 1-41.

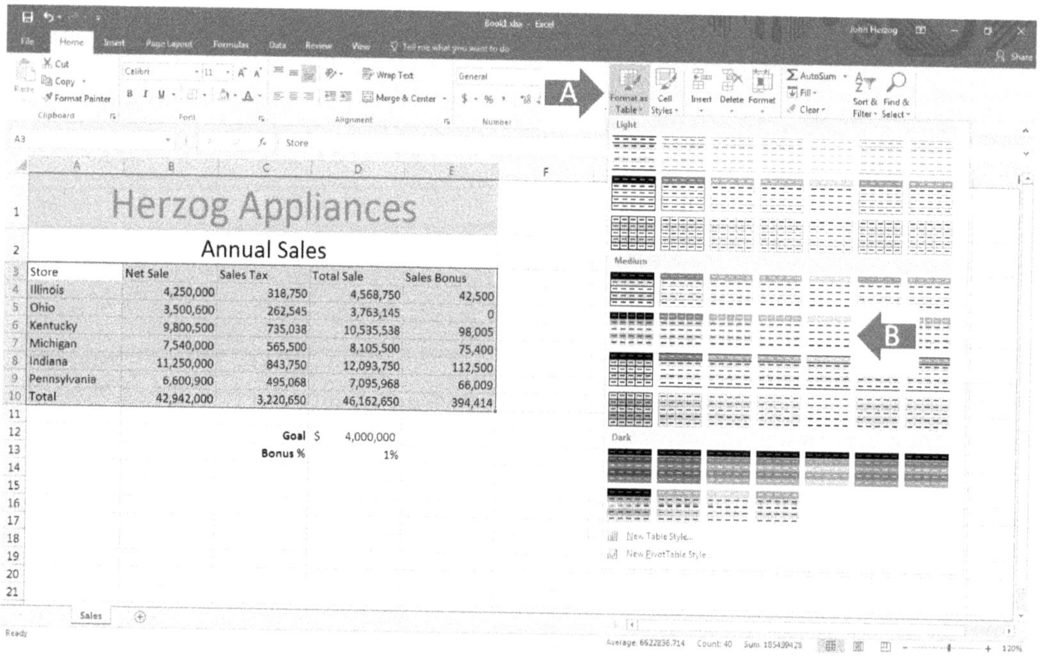

FIGURE 1-39

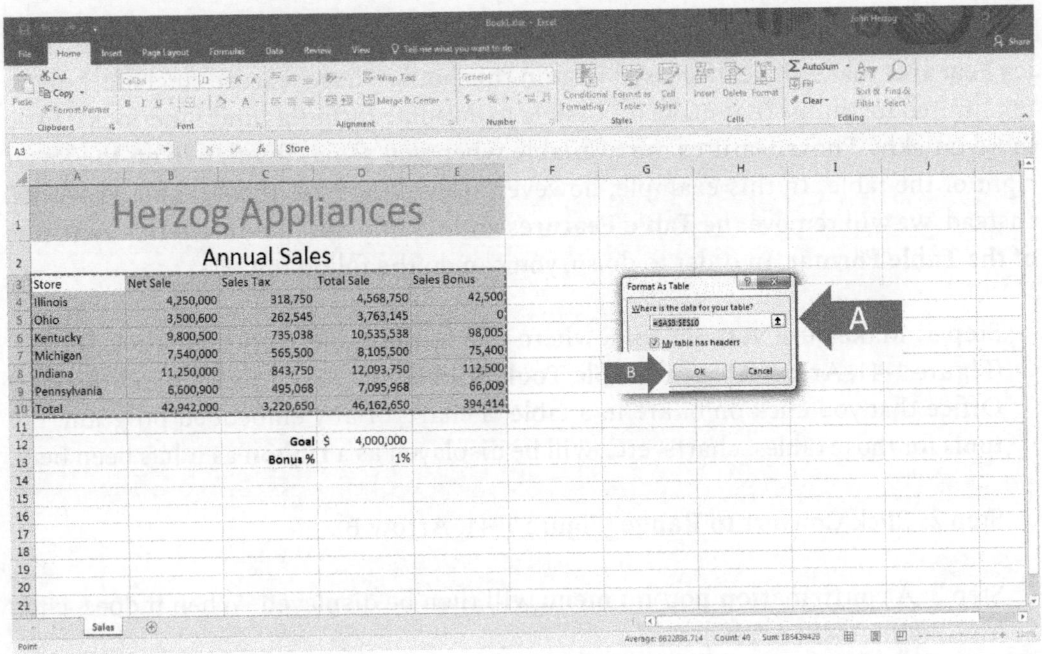

FIGURE 1-40

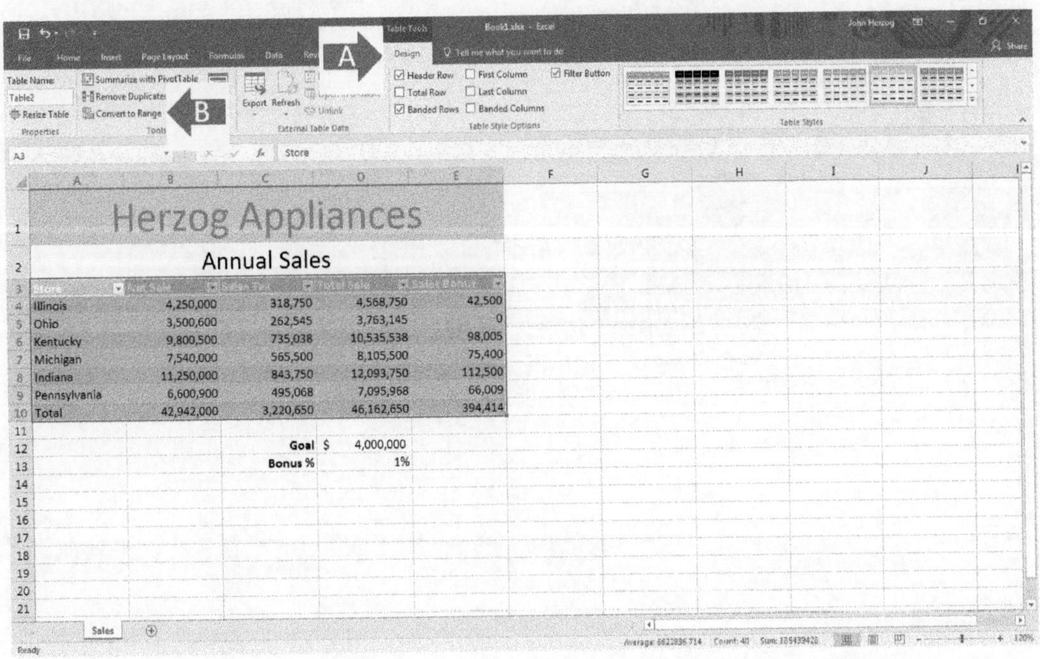

FIGURE 1-41

When you format a spreadsheet as a table, there are **Table Features** and **Aesthetics Features** that will be discussed later in this book. Those features include, but are not limited to, things that make it possible to do some formulas and functions a little more quickly. Those features are available when you see a tiny flange on the lower right of the table. In this example, however, we will not use those **Table Features**. Instead, we will remove the **Table Features** and strictly use the **Aesthetics Features** of the **Table Format**. In order to do so, you can do the following:

Step 1: Make sure you are somewhere inside the table, and then click **Design** (Figure 1-41, Arrow A) in the **Table Tools Ribbon**. NOTE: Any time in Microsoft Office that you click on or are in a table, a chart, or any embedded program, the **tools** for those tables, charts, etc., will be displayed as a **ribbon** as it has been here.

Step 2: Click **Convert to Range** (Figure 1-41, Arrow B).

Step 3: A **confirmation pop-up menu** will then be displayed. When it does, click **Yes** (Figure 1-42, Arrow).

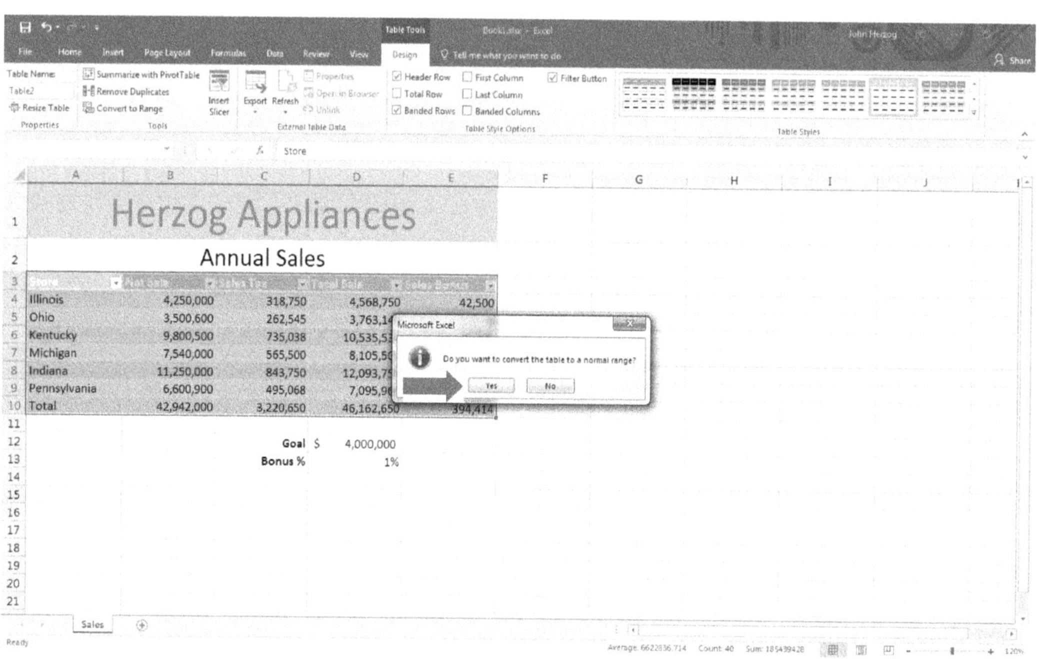

FIGURE 1-42

AUTOCOMPLETE/PICK FROM LIST

If for some reason you want to type the state of *Pennsylvania* additional times in the spreadsheet (in this spreadsheet, it would not be practical), **AutoComplete** will usually display a previously entered word (not numbers) when you type the first letter of that word, as you can see in Figure 1-43.

However, if you type the "*I*" key in this example, you will not get the state of *Illinois* because there are two text entries on column A that begin with "*I*" (*Illinois* and *Indiana*). If you typed *In*, then *Indiana* would be seen, only because there are no **other** entries in that column that begin with *In*. If you want to find what states you have entered (or other text entered in the column), you can do the following:

Step 1: Right-click on the cell and in the **shortcut menu**, click **Pick from Dropdown List**. (Figure 1-44, Arrow).

Step 2: That will take you to the screen shown in Figure 1-45. Simply click the text item you want to enter, and Excel will enter it into the cell for you (Figure 1-45, Arrow). This will maintain consistency in your data entry.

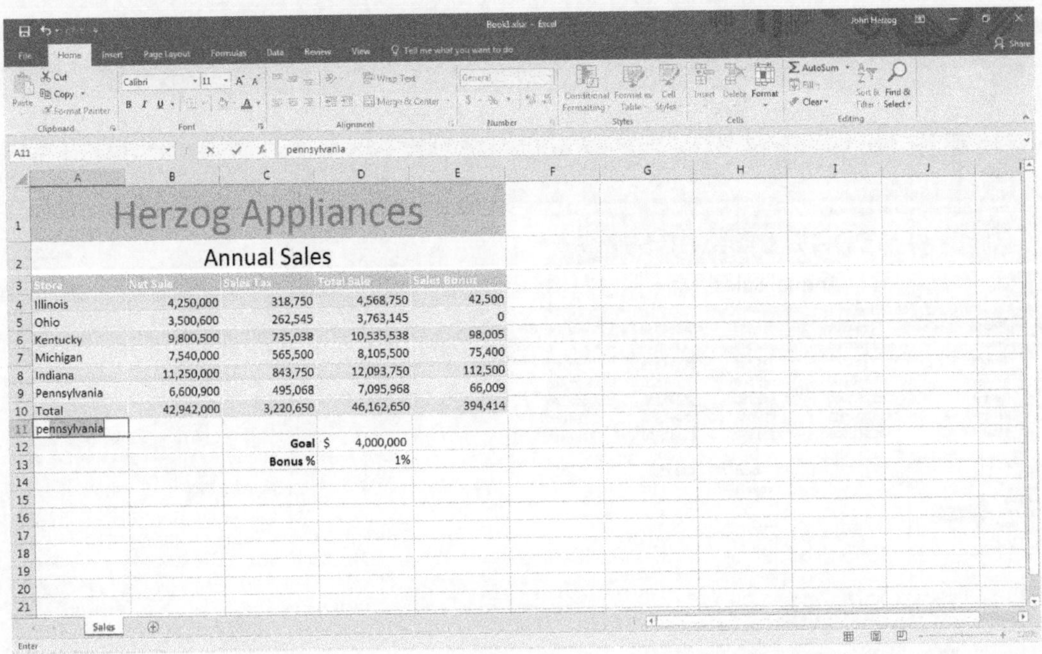

FIGURE 1-43

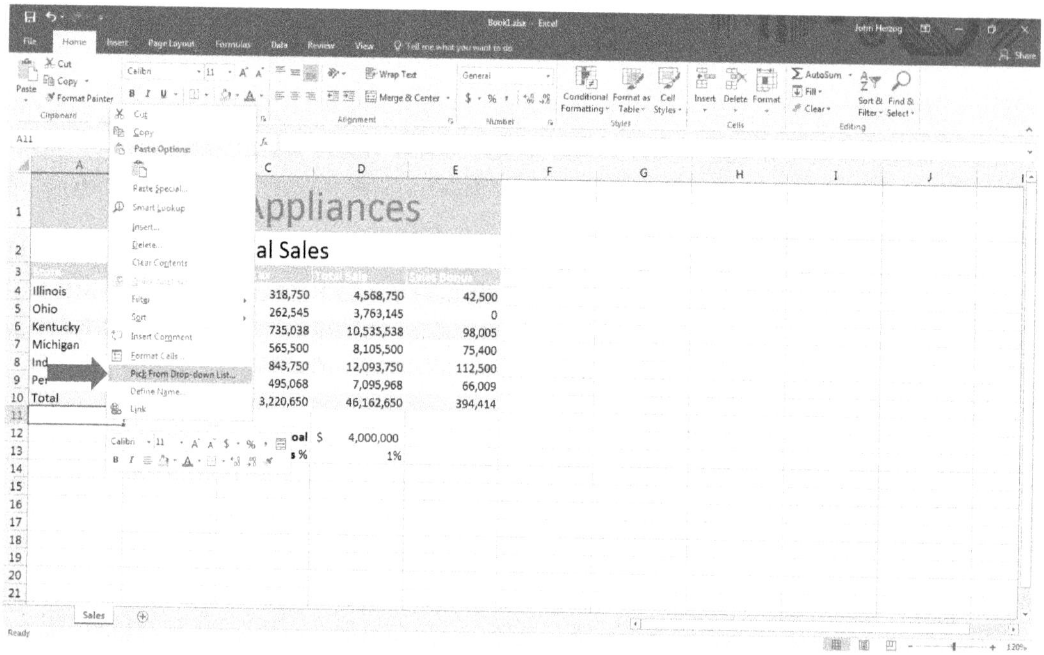

FIGURE 1-44

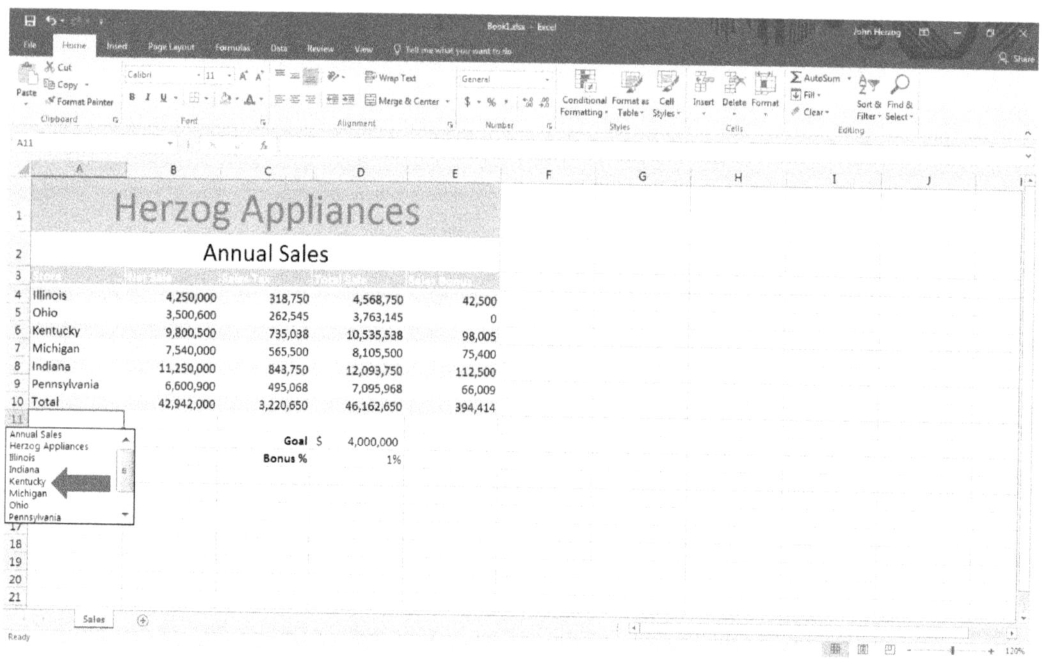

FIGURE 1-45

TO CREATE A CHART

A picture is worth a thousand words. That is also true for spreadsheets. Data is much easier to convey in an instant when you use a chart.

In order to graph your data, you must first select the range of data to be graphed. In this example, we would like to graph the *Store* names with the *Net Sale* of each store (as well as their labels), which are found in cells A3 through B9 (Figure 1-46). To make this happen:

Step 1: Select cells A3 through B9. **Never include a Total row in a selection for chart data** as you can see in Figure 1-46. Doing so would cause a chart to have one large column (or bar or pie slice, depending upon what kind of chart you are creating) and the remaining data would be made very small—in some cases, it could not be read at all. NOTE: When you select **nonadjacent cells** for creating a chart, you must remember to press and hold the **Ctrl key** as you click and drag the ranges of cells. When you do, please remember two additional important requirements for selecting the data for creating charts:

a. The ranges must be equal in length.
b. You cannot stop in the middle of a cell range while selecting it and then resume the selecting.

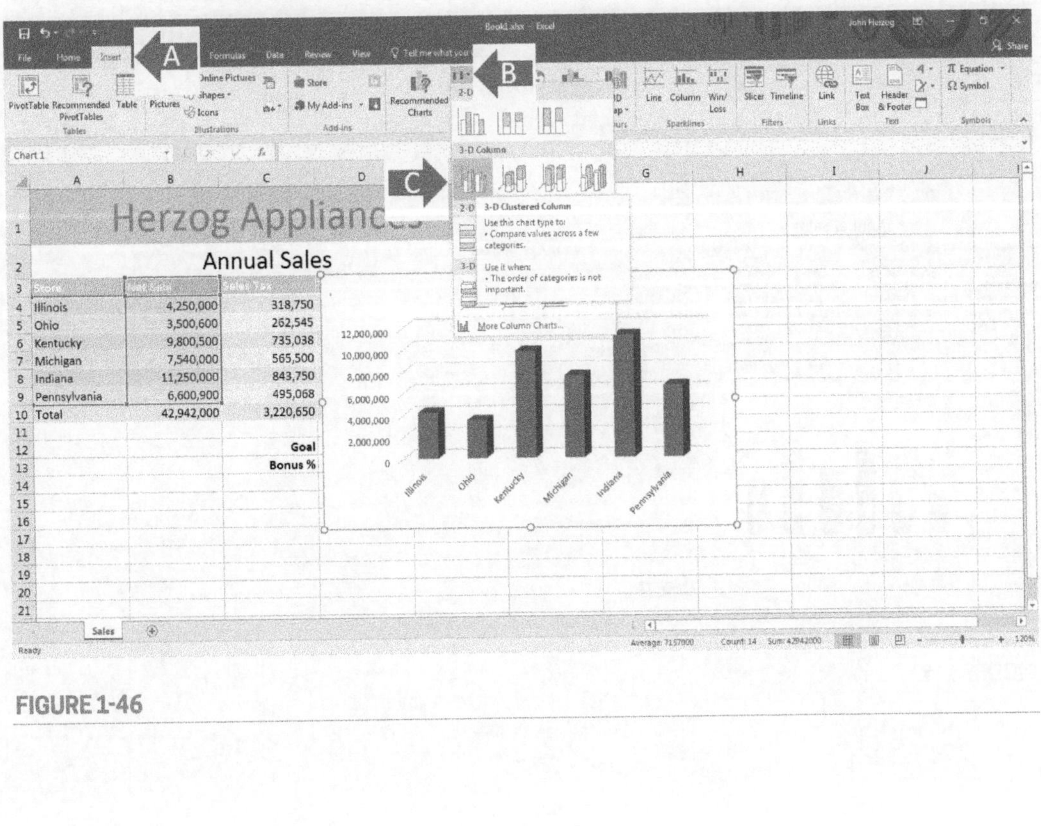

FIGURE 1-46

Failure to remember these two requirements will make your graph completely unusable.

Step 2: Click **Insert** in the **Home Ribbon** (Figure 1-46, Arrow A).

Step 3: Click the tiny box to the right of the **Column Chart icon** in the **Insert Ribbon** (Figure 1-46, Arrow B),

Step 4: You can choose from various chart types, but in this case, we will create a **3-D Column Chart**. Click the first **3-D Column** option (Figure 1-46, Arrow C), which is called the **3-D Clustered Column** chart.

That will give you the screen you see in Figure 1-46. This is called an **Embedded Chart** when the chart remains inside the same sheet that holds your data.

Suppose you are asked to place the chart in a specific cell range. In this example, we will put it in the cell range of cell A15 through half of cell E28. You can do so by doing the following:

Step 1: Click and drag the border of the chart until its upper left corner is in cell A15 (Figure 1-47).

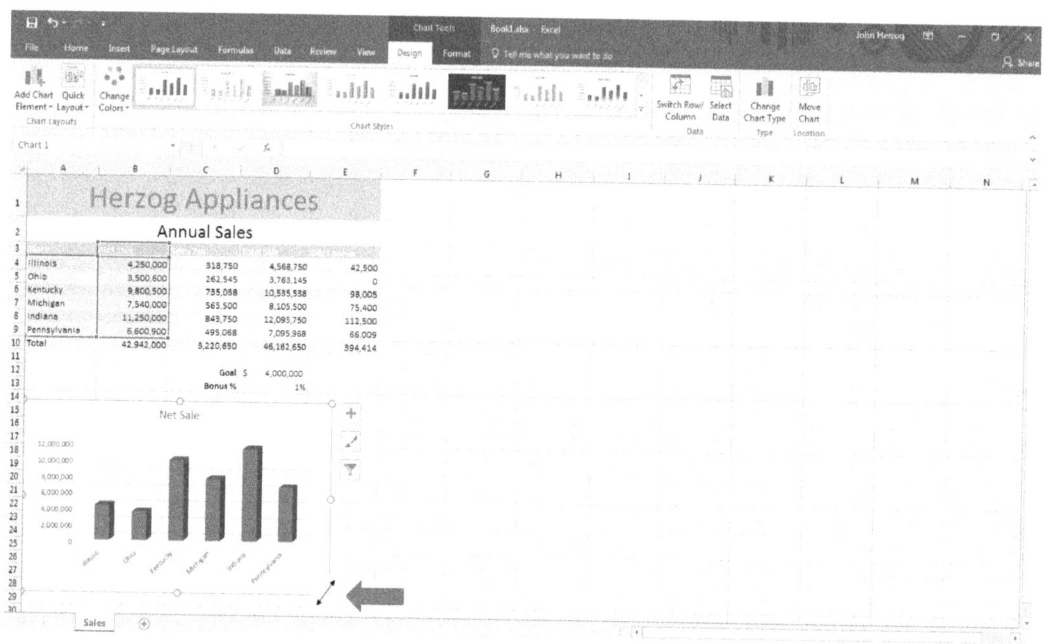

FIGURE 1-47

Step 2: Point your mouse to the lower right corner of the chart until you get a **double arrow** in order to resize the chart to the size of your liking. In this example, resize the chart until its lower right corner is in the middle of cell E28 (Figure 1-47, Arrow).

TO MAKE A CHART ITS OWN SHEET

If you make the chart its own sheet, it can be easier to see and to read. If you want to have the chart become its own sheet rather than an **embedded chart**, you can do the following:

Step 1: Click on the chart and then click **Design** in the **Chart Tools** (Figure 1-48, Arrow A).

Step 2: Click **Move Chart** (Figure 1-48, Arrow B).

Step 3: At the next menu, click the **New Sheet** option. When you do, it has a place to name the sheet you are about to create that will become this chart. In this example, name the new sheet as **Sales Chart** (Figure 1-49, Arrow).

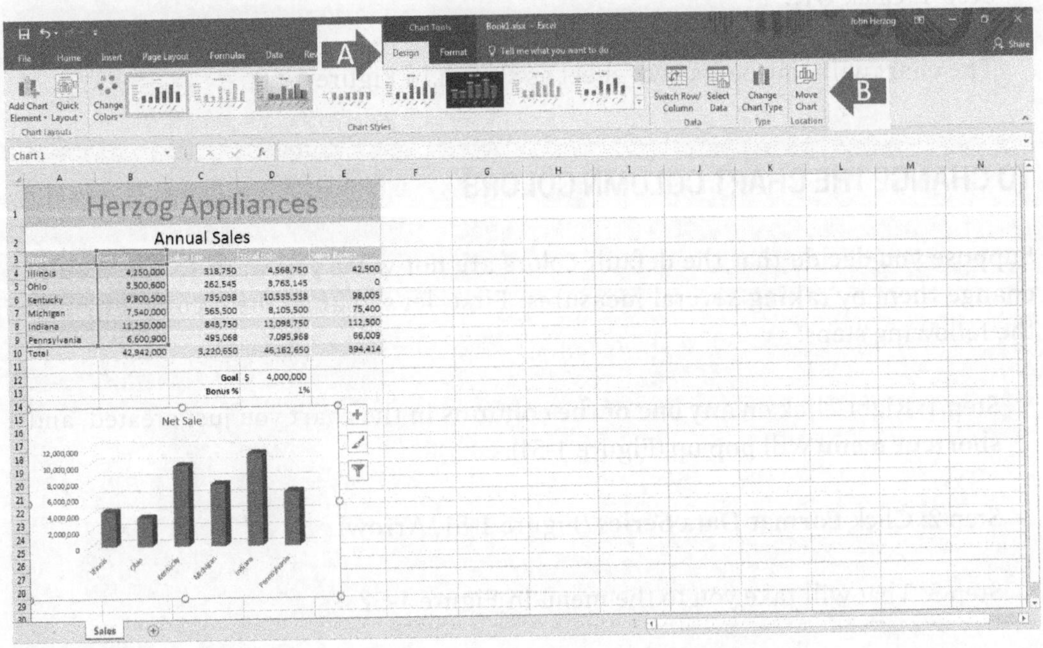

FIGURE 1-48

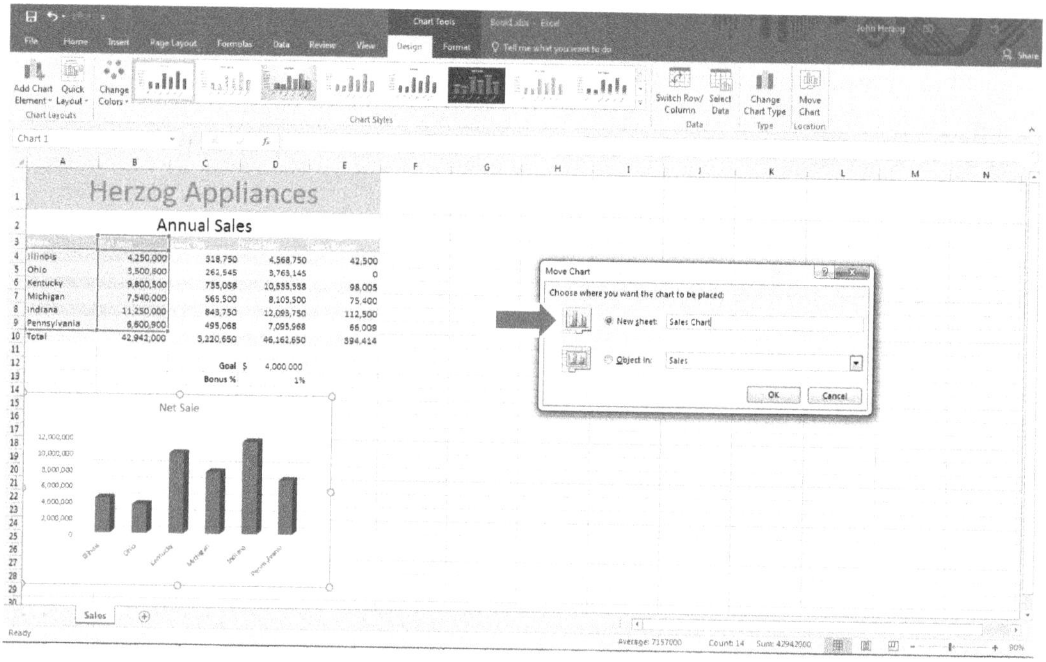

FIGURE 1-49

Step 4: Click **OK**.

The chart will now be its own sheet as shown in Figure 1-50.

TO CHANGE THE CHART COLUMN COLORS

Suppose you decide that the default colors are not what you want to use. You can change them by taking several measures. First, let's change the columns by using the following steps:

Step 1: Right-click on **any one of the columns** in the chart you just created, and a **shortcut menu** will pop up (Figure 1-51).

Step 2: Click **Format Data Series** (Figure 1-51, Arrow).

Step 3: That will take you to the menu in Figure 1-52.

Step 4: Click the **Fill** option (beneath **Series Options**, Figure 1-52, Arrow A).

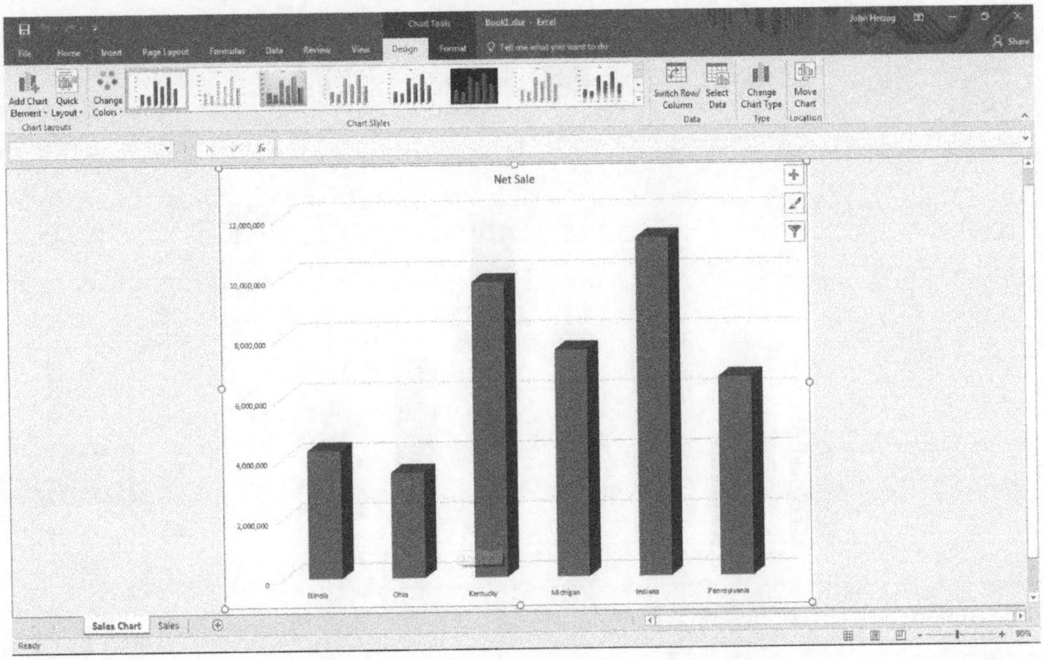

FIGURE 1-50

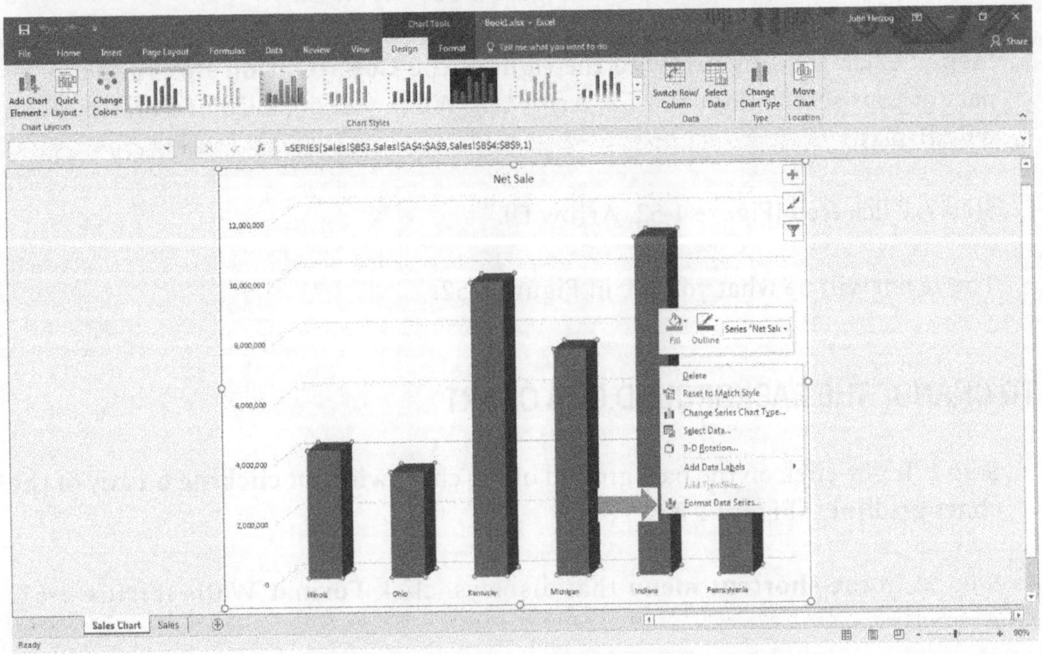

FIGURE 1-51

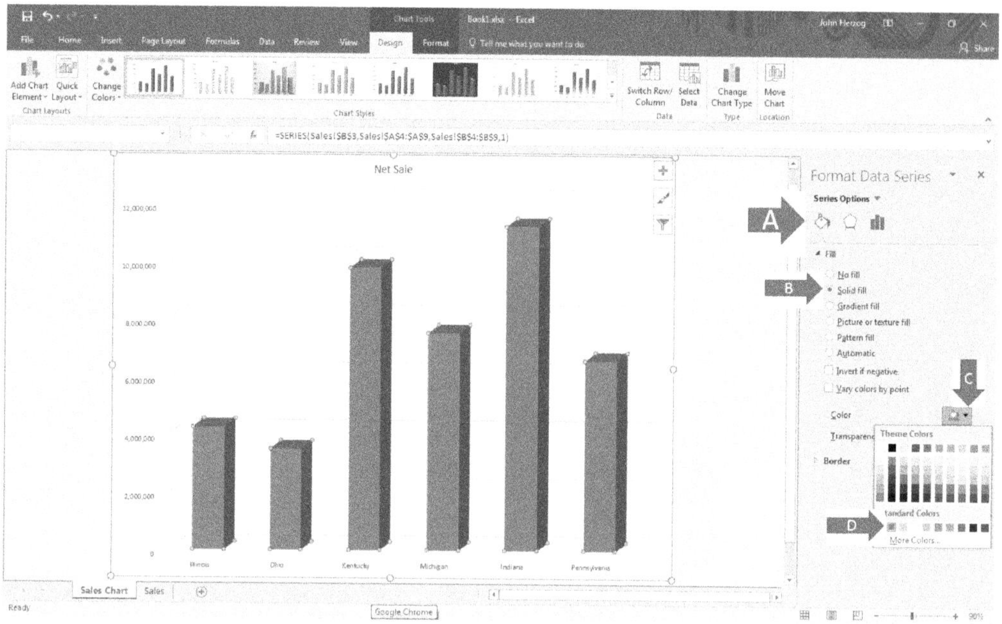

FIGURE 1-52

Step 5: Click the **Solid Fill** option (Figure 1-52, Arrow B).

Step 6: Click the tiny arrow to the right of the **Color Option Box** in the lower part of your screen (Figure 1-52, Arrow C). When you do, a **Color Palette Menu** will display.

Step 7: Click **Red** (Figure 1-52, Arrow D).

The result will be what you see in Figure 1-52.

TO CHANGE THE BACKGROUND OF A CHART

Step 1: Right-click on the background of the chart **without clicking on any of the chart gridlines** (Figure 1-53).

Step 2: In the **shortcut menu** that displays, click **Format Walls** (Figure 1-53, Arrow).

Step 3: That will take you to the menu you see in Figure 1-54. Click the **Fill** option (Figure 1-54, Arrow A).

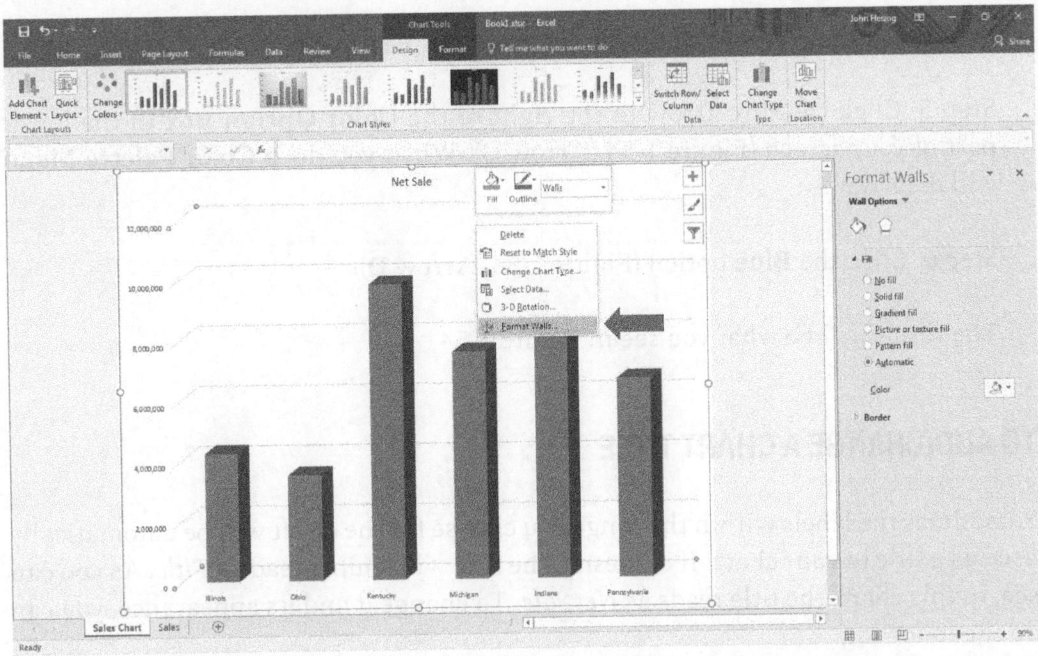

FIGURE 1-53

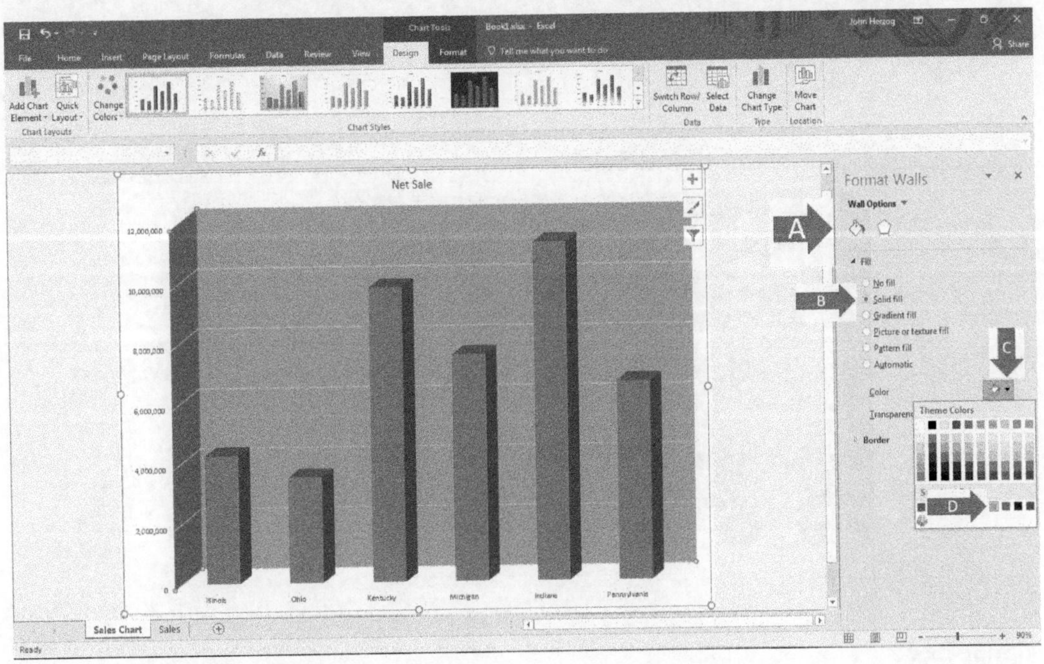

FIGURE 1-54

Step 4: Click the **Solid Fill** option (Figure 1-54, Arrow B).

Step 5: Click the tiny arrow to the right of the **Color Option Box** in the lower part of your screen (Figure 1-54, Arrow C). When you do, a **Color Palette Menu** will display.

Step 6: Click the **Blue** option (Figure 1-54, Arrow D).

The result will be what you see in Figure 1-54.

TO ADD/CHANGE A CHART TITLE

Sometimes the labels within the range you choose for the chart will be automatically used as a title to your chart. If it doesn't, the title will simply read as *Title*. As you can see, in this chart, the title reads as *Net Sale*. To change it and its appearance, you can do the following:

Step 1: Click on the chart title (Figure 1-55).

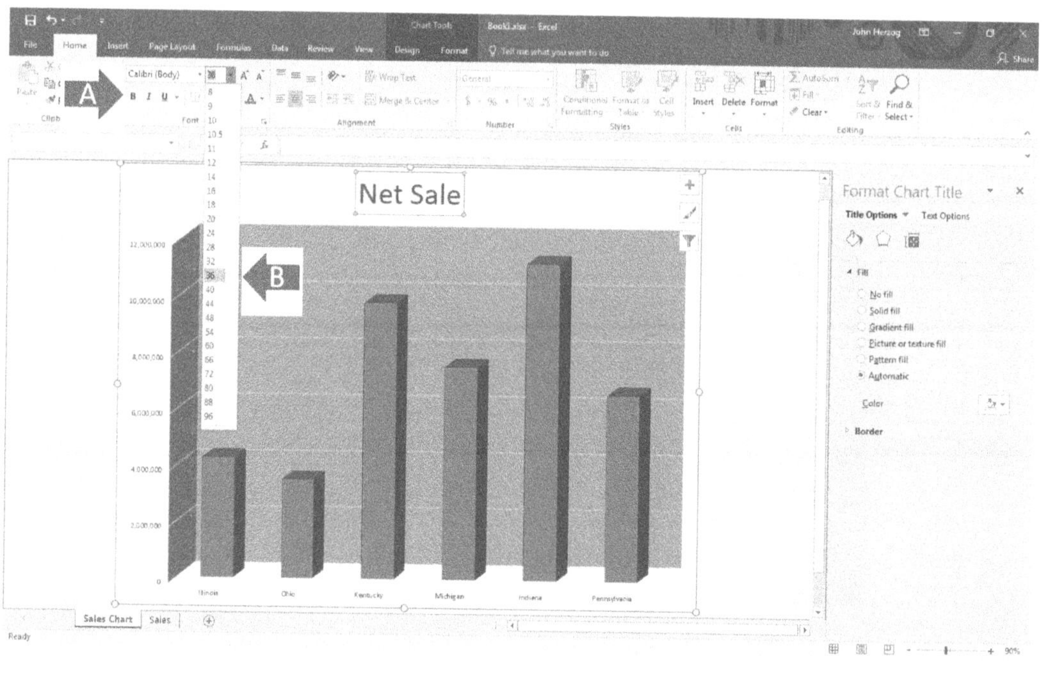

FIGURE 1-55

Step 2: Click on the **Home Ribbon**.

Step 3: Click the buttons labeled **B, I,** and **U** (Figure 1-55, Arrow A), which will give your chart title the format settings of **Bold**, **Italics**, and **Underlining**.

Step 4: Choose the font size of *36* (Figure 1-55, Arrow B).

You can change other formats in the title if you wish, such as the color of the font, the background, and much more by using the other formatting options in the **Home Ribbon** (although we will not be doing so in this example).

TO CHANGE NUMBER FORMATS IN CHARTS

In some cases, the default format for your numbers may not be the same as what you want. Suppose, for example, that you want your sales figures to be displayed with **currency**. To make that happen, you can do the following:

Step 1: Right-click anywhere on the **axis** with the numbers that need to have a different format. In this example, that would be the Y-axis (Figure 1-56).

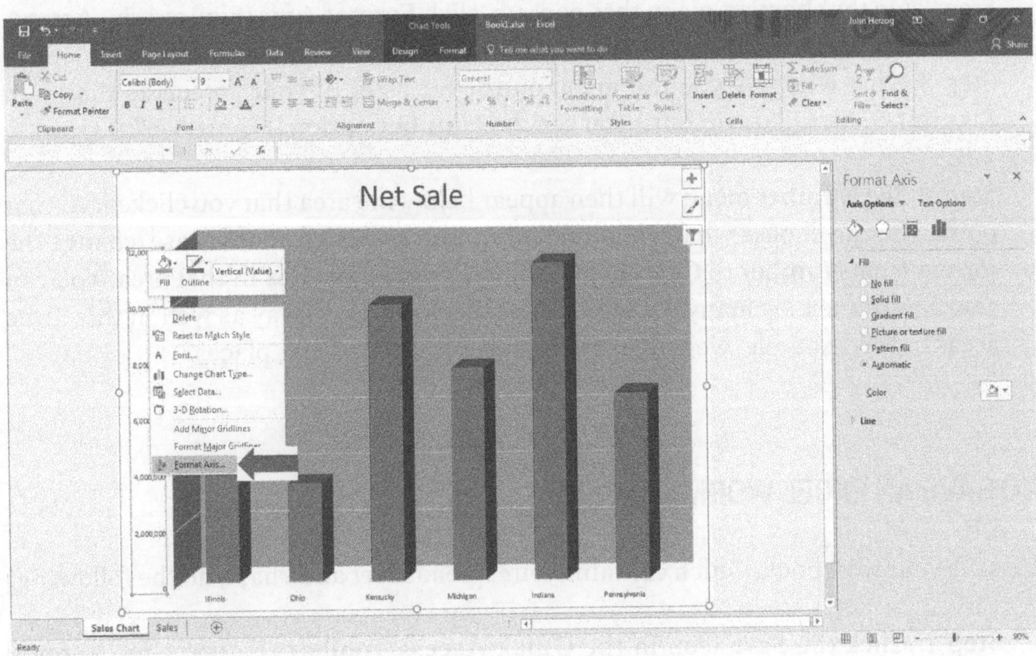

FIGURE 1-56

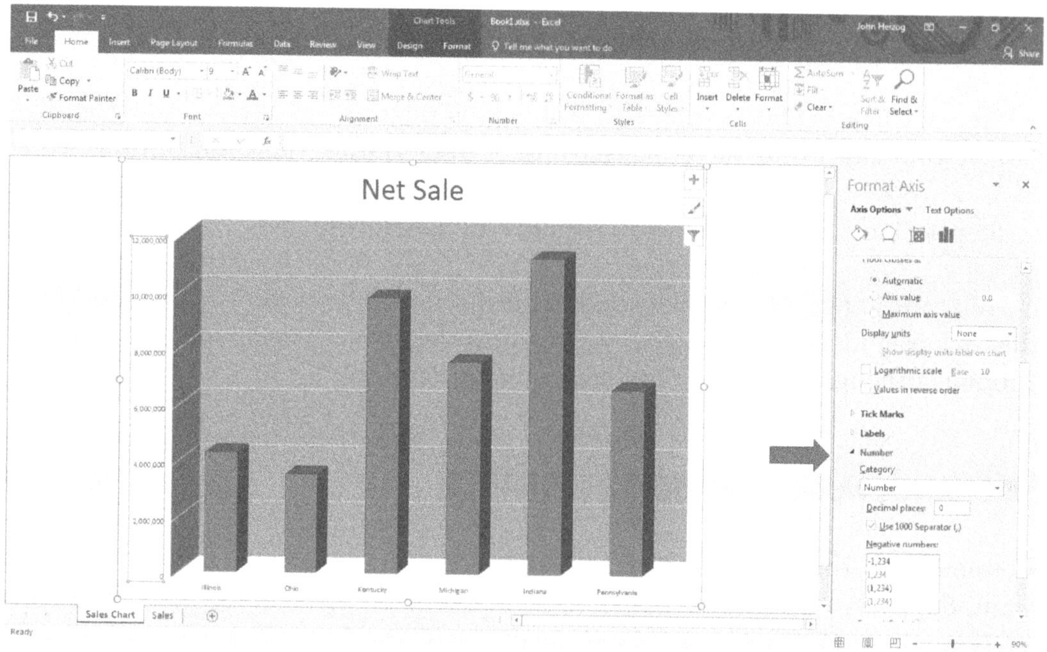

FIGURE 1-57

Step 2: On the **shortcut** menu that pops up, click **Format Axis** (Figure 1-56, Arrow).

Step 3: That will give you the menu on the far right of your screen as shown in Figure 1-57. Scroll down until you see the **Number** option (Figure 1-57, Arrow).

Step 4: The **Number** menu will then appear below the area that you clicked. At that point, you can choose whatever number format you desire. In this case, change the format from **Number** to **Currency** in the drop-down list box in that area. You can also determine how many decimal places you want to display as well in this same area. In this example, we will keep the setting at 0 decimal places.

TO SAVE AN EXCEL WORKBOOK FILE

To save your workbook, which contains your spreadsheet and chart, do the following:

Step 1: Click the **Save** icon in the **Quick Access Toolbar** (Figure 1-58, Arrow). NOTE: You can also click the **File Menu** shown below the **Quick Access Toolbar**;

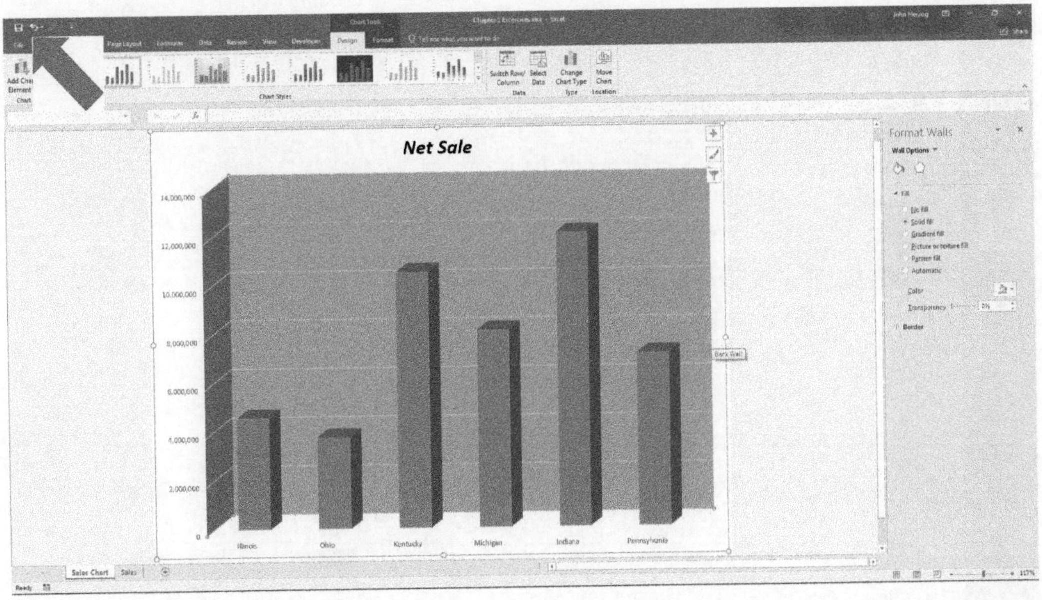

FIGURE 1-58

then click **Save As** in that menu. In either case, that will take you to the menu shown in Figure 1-59.

Step 2: Click **Save As** (Figure 1-59, Arrow A).

Step 3: Click **Browse** (Figure 1-59, Arrow B).

Step 4: Choose a destination folder in a given drive that is accessible by your computer (Figure 1-59, Arrow C). In this example, a folder is chosen on the computer's USB drive.

Step 5: Name the file as **Homework** in the **File Name: text box** (Figure 1-59, Arrow D). NOTE: Regarding the **File Type: text box**, Excel will always save the file as an Excel Workbook with the most recent version's extension. Later, you will be shown how save a workbook as a different file type.

Step 6: Click the **Save** command to the right of the **Tools** option (Figure 1-59, Arrow E).

Your workbook will now be saved. Congratulations! You have just created your first workbook with a spreadsheet and chart!

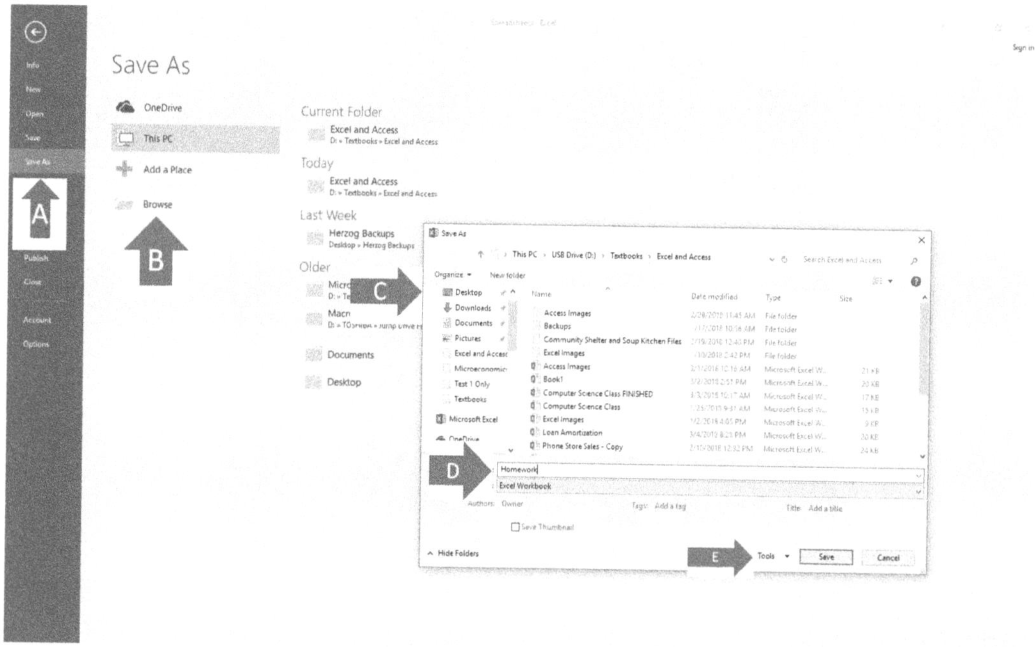

FIGURE 1-59

CHAPTER 1—ASSIGNMENT (SUMMARY OF TASKS)

Herzog Appliances wants to create a spreadsheet that will show the sales breakdown for all of their appliance stores from a given year. Your first task is to create a spreadsheet that tallies the numbers and then displays them in an understandable manner. It then involves the formatting of the data and then adding a graph. Create the spreadsheet shown in Figure 1-60, using the following specifications:

1. Start a new spreadsheet.
2. Make sure that all of the column widths in the spreadsheet are each *15* characters wide.
3. Type the words *Herzog Appliances* in cell A1.
4. Make the words *Herzog Appliances 36* points high.
5. **Merge and Center** cells A1 through E1.
6. Type the words *Annual Sales* in A2.
7. Make the words *Annual Sales* 22 points high.
8. **Merge and Center** cells A2 through E2.
9. Give the newly merged cell A1 a background of *orange* and a font of *red*.

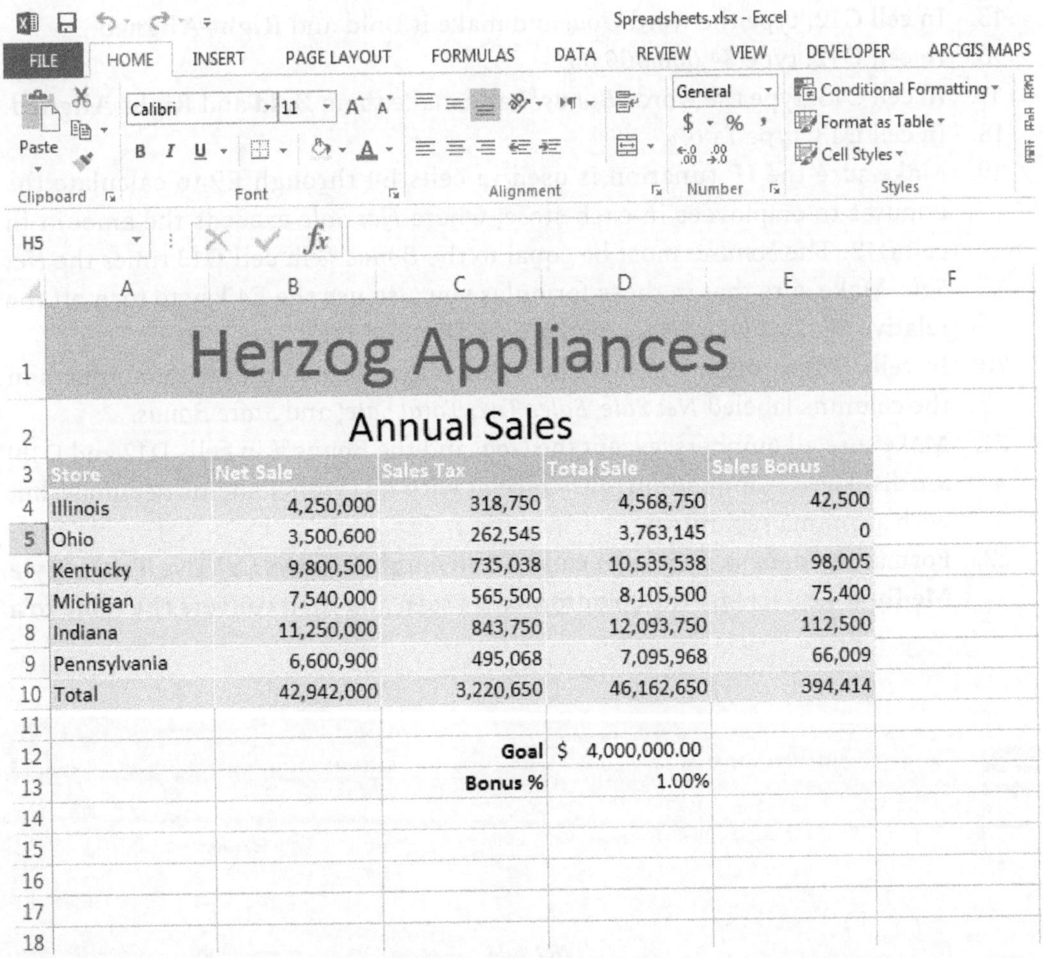

FIGURE 1-60

10. Enter all of the labels (*Store*, *Illinois*, *Ohio*, etc.) in cells A3 through A10 as shown in Figure 1-60.
11. Enter all of the labels (*Net Sale*, *Sales Tax*, *Total Sale*, *etc.*) in cells B3 through E3 as shown in Figure 1-60.
12. In column B, enter the net sales for all of the store locations, also shown in Figure 1-60. No other data is to be entered after this (except for the goal and bonus %). All other numbers shown here must be calculated by using formulas and functions.
13. In cell C4, calculate the sales tax for the *Illinois* store by multiplying their *Net Sale* in cell B4 by .075. Fill that formula down to cells C5 through C9.
14. In cell D4, calculate the *Total Sale* for the *Illinois*, which is equal to the *Net Sale* plus the *Sales Tax*. Fill that formula down to cells D4 through D9.

15. In cell C12, type the word *Goal* and make it **Bold** and **Right-Aligned**.
16. In cell D12, type *$4,000,000.00*.
17. In cell C13, type the words *Bonus %* and make them **Bold** and **Right-Aligned**.
18. In cell D13, type *1.00%*.
19. Make sure the **IF** function is used in cells E4 through E9 to calculate the bonuses to employees in each store, where *Net Sale* exceeds the amount in cell D12. The *bonuses* must be equal to the *Bonus %* in cell D13 times the *Net Sale*. Make sure that in these formulas you also use the F4 key to turn off the relative address when referencing cells D12 and D13.
20. In cells B10 through E10, use the **SUM** function to get totals of numbers in the columns labeled *Net Sale*, *Sales Tax*, *Total Sale*, and *Store Bonus*.
21. Make sure all numbers (except the *Goal* and the *Bonus %* in cells D12 and D13) are displayed using the currency format with no symbol and no decimals, but with a comma separator.
22. Format the data as a table **in cells A3 through E10 ONLY!** Use **Table Style Medium 12** to create the color format as seen and then **convert the table to a**

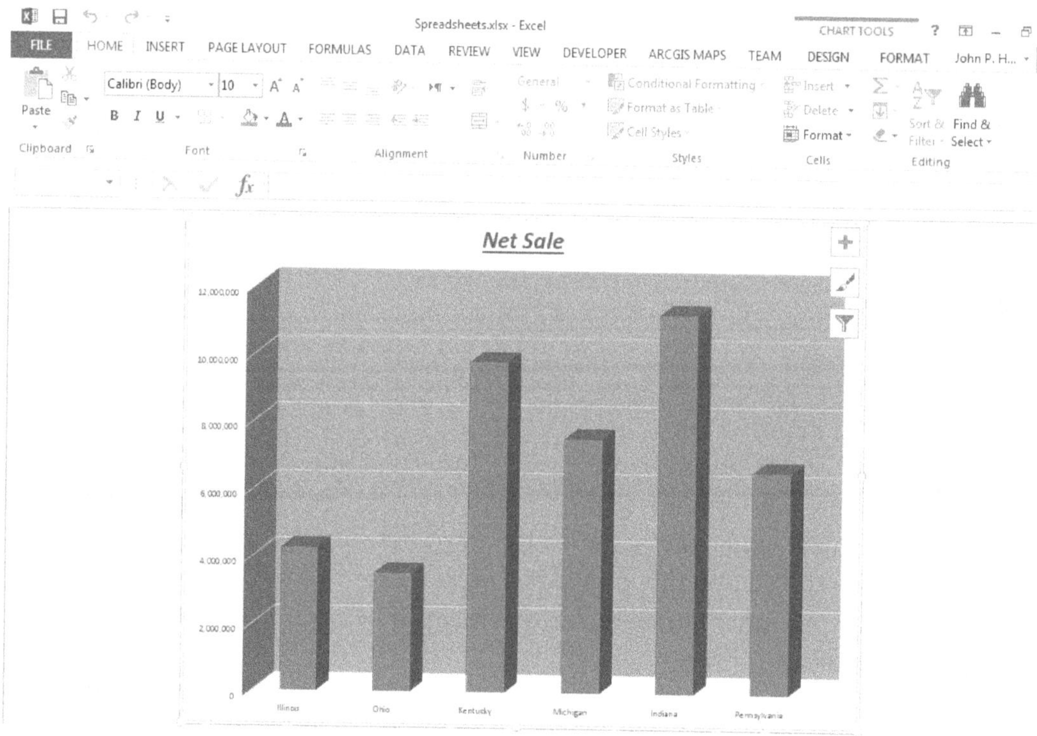

FIGURE 1-61

range. NOTE: Some versions or installations of Excel may give you a different look with **Table Style Medium 12** than what is shown here.

23. Name the sheet *Sales.*
24. Make a **3-D Clustered Column** chart of the stores and their total sales, using cells A3 through B9 as the data source for the chart. Make sure it is moved to become its own sheet (as shown in Figure 1-61).
    a. Make a title that reads *Net Sale* that will have a font size of *24* points high and formatting that is **Bold, Italics**, and **Underlined**.
    b. Make the wall light blue and the series red.
    c. Name the sheet tab with the chart *Sales Chart.*
25. Save the whole workbook as *Homework.*

CHAPTER 1—ALTERNATE ASSIGNMENT (SUMMARY OF TASKS)

Acme Computer Store wants to create a spreadsheet that will show the sales breakdown for a given year. Your task is to create a spreadsheet that tallies the numbers and then displays them in an understandable manner. It also involves the formatting of the data and then adding a graph. Create the spreadsheet shown in Figure 1-62, using the following specifications:

1. Start a new spreadsheet.
2. Make sure that all of the column widths of the spreadsheet are each *15* characters wide.
3. Type the words *Acme Computer Store* in cell A1.
4. Make the words *Acme Computer Store 36* points high.
5. **Merge and Center** cells A1 through E1.
6. Type the words *Annual Sales* in cell A2.
7. Make the words *Annual Sales 22* points high.
8. **Merge and Center** cells A2 through E2.
9. Give the newly merged cell A1 a background of **light blue** and a font of **dark blue.**
10. Enter all of the labels (*Product Type, CPUs, Keyboards,* etc.) in cells A3 through A10 as shown in Figure 1-62.
11. Enter all of the labels (*Net Sale, Service/Shipping, Total Sale, etc.*) in cells B3 through E3 as shown in Figure 1-62.
12. In cells B4 through B9, enter the *Net Sale* for all of the product types, also shown in Figure 1-62. No other data is to be entered after this (except for the goal). All other numbers must be calculated using formulas and functions.

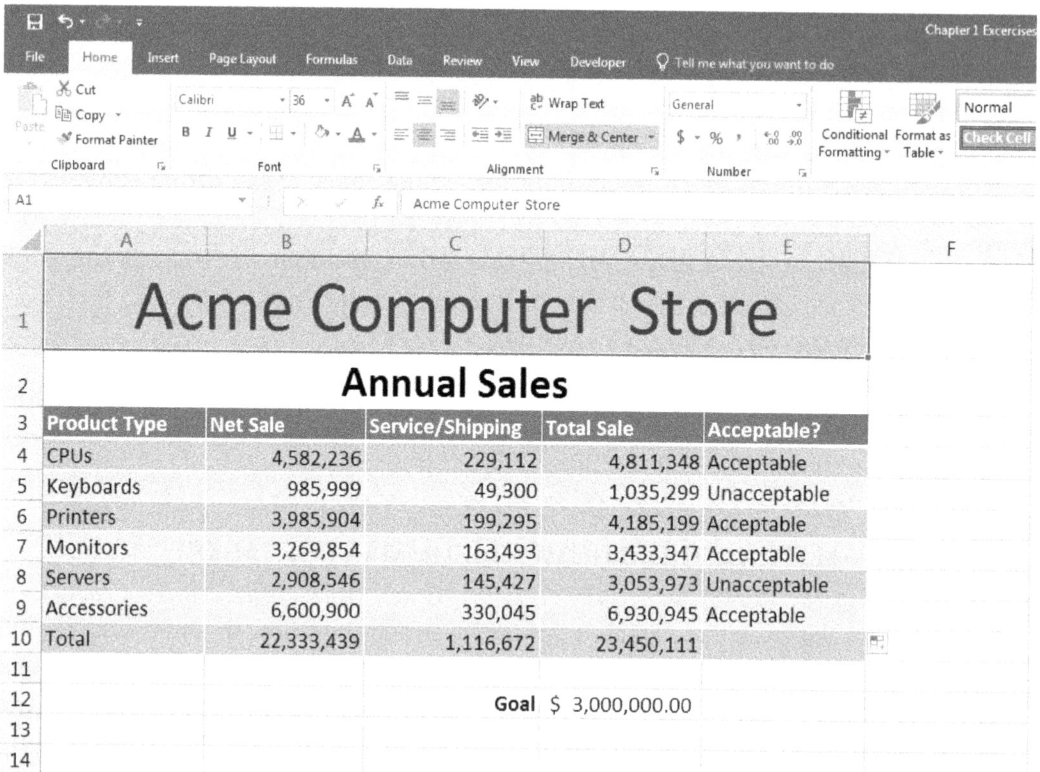

FIGURE 1-62

13. Assuming that the store charges 5% for a service charge and for shipping of all of their goods, in cell C4, calculate the *Service/Shipping* for *CPUs* by multiplying their *Net Sale* in cell B4 by *.05*. Fill that formula down to cells C4 through C9.
14. In cell D4, calculate the *Total Sale* for each *Product Type*, which is equal to the *Net Sale* plus the *Service/Shipping*. Fill that formula down to cells D4 through D9.
15. In cell C12, type the word *Goal* and make it **Bold**.
16. In cell D12, type *$3,000,000.00*.
17. Use the **IF** function in cells E4 through E9, which will declare that *Net Sales* for each product type are either *Acceptable or Unacceptable*. They are acceptable if they exceed the value in cell D12, but they are unacceptable if they do not. Make sure that in these formulas you also use the F4 key to turn off the relative address when referencing cell D12.
18. In cells B10 through D10, use the **SUM** function to get totals of numbers in the columns labeled *Net Sale*, *Service/Shipping*, and, *Total Sale*.

19. Make sure all numbers (except the Goal in cell D12) are displayed using the currency format with no symbol and no decimals, but with a comma separator.
20. Format the data as a table (cells A3 through E10 ONLY!). Use Table Style Medium 9 to create the color format as seen, and then convert the table to a range.
21. Name the sheet *Annual Sales.*
22. Make a **3-D Clustered Bar** chart of the *Product Type* and *Total Sale* columns only. Make sure it is moved to a new sheet and will thus become its own sheet (as shown in Figure 1-63).
    a. Make a title that reads *Net Sale* that will have a font size of *24* points high that is **Bold, Italics**, and **Underlined**.
    b. Make the wall **yellow** and the series **green**.
    c. Name the sheet tab with the chart *Sales Bar Chart.*
23. Save the whole workbook as *Homework II* and submit it.

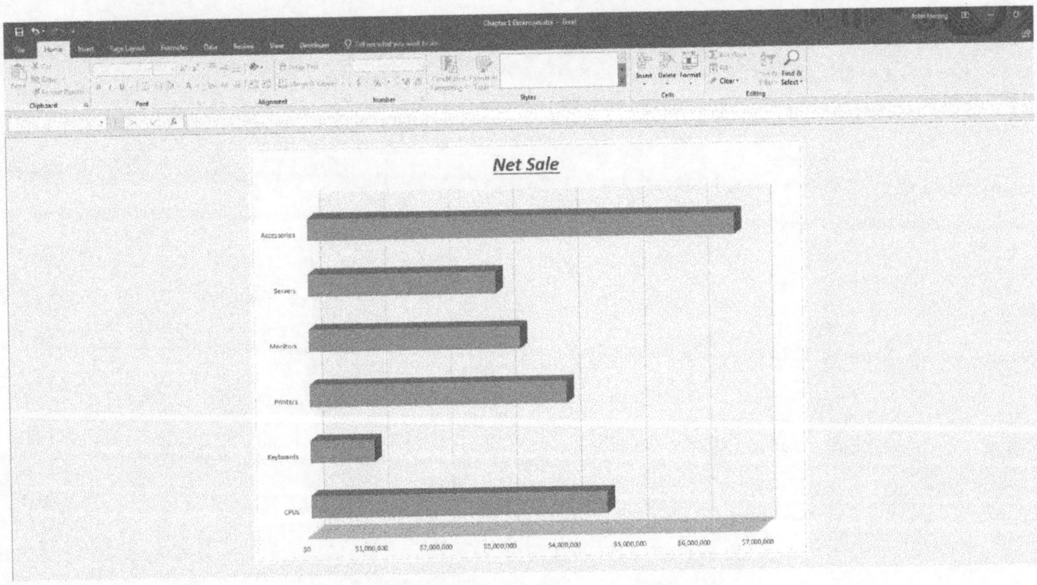

FIGURE 1-63

CHAPTER TWO

Using a Spreadsheet to Find Stats About Given Numbers

BY THE END of this chapter, it is our objective that the user will be able to:

1. Use the AVERAGE, MIN, MAX, COUNT, VLOOKUP, VARIANCE, STDEVA, and MEDIAN functions.
2. Rotate text.
3. Use Conditional Formatting.
4. Use the **IF** function in more ways.
5. Use the **SQRT** function.
6. Use **Absolute Address** in more ways.
7. Use Cell Styles.
8. Use Themes.
9. Wrap text.
10. Create a Weighted Average.

Any instructions/specifics given without a step-by-step guide are those already covered, and it is assumed that you will recall how to do them from before.

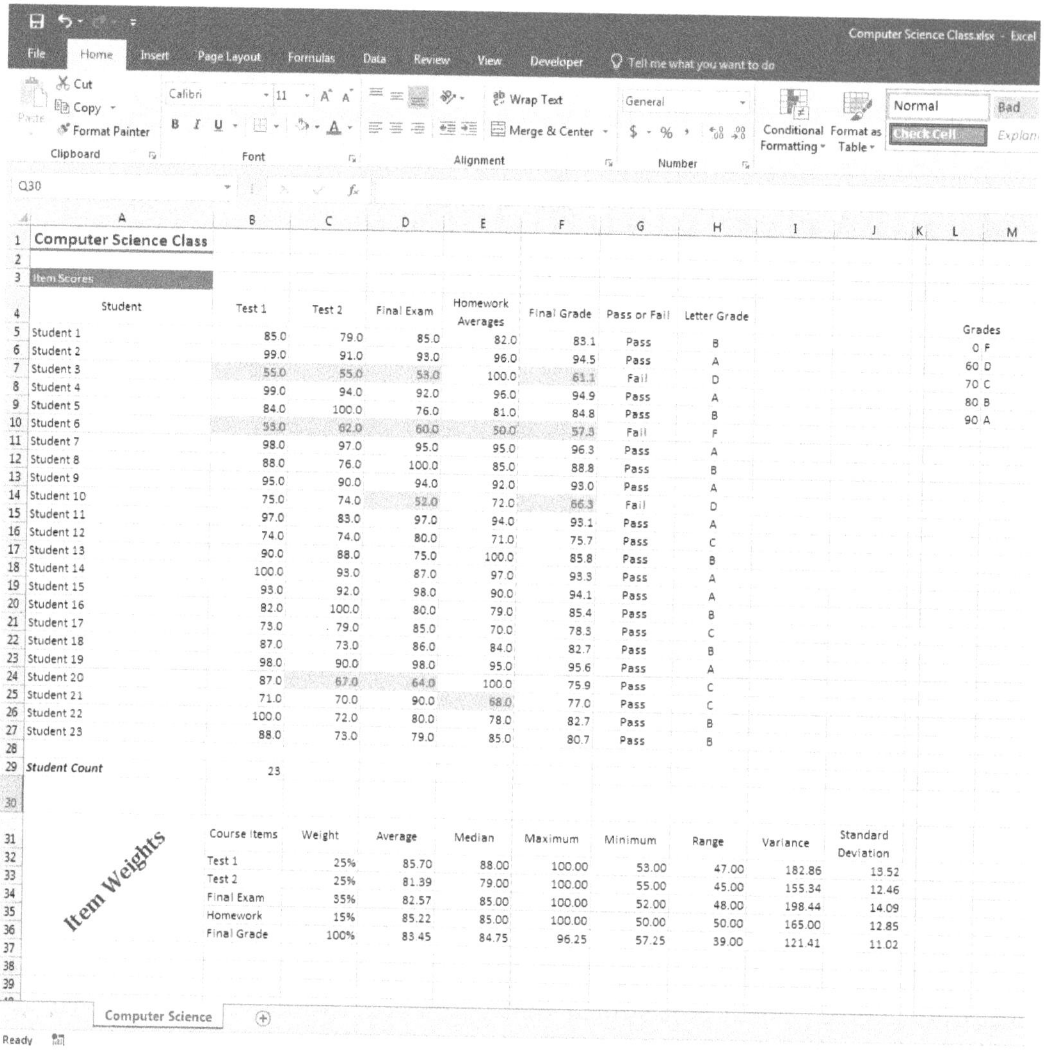

Computer Science Class

Item Scores

Student	Test 1	Test 2	Final Exam	Homework Averages	Final Grade	Pass or Fail	Letter Grade
Student 1	85.0	79.0	85.0	82.0	83.1	Pass	B
Student 2	99.0	91.0	93.0	96.0	94.5	Pass	A
Student 3	55.0	55.0	53.0	100.0	61.1	Fail	D
Student 4	99.0	94.0	92.0	96.0	94.9	Pass	A
Student 5	84.0	100.0	76.0	81.0	84.8	Pass	B
Student 6	53.0	62.0	60.0	50.0	57.3	Fail	F
Student 7	98.0	97.0	95.0	95.0	96.3	Pass	A
Student 8	88.0	76.0	100.0	85.0	88.8	Pass	B
Student 9	95.0	90.0	94.0	92.0	93.0	Pass	A
Student 10	75.0	74.0	52.0	72.0	66.3	Fail	D
Student 11	97.0	83.0	97.0	94.0	93.1	Pass	A
Student 12	74.0	74.0	80.0	71.0	75.7	Pass	C
Student 13	90.0	88.0	75.0	100.0	85.8	Pass	B
Student 14	100.0	93.0	87.0	97.0	93.3	Pass	A
Student 15	93.0	92.0	98.0	90.0	94.1	Pass	A
Student 16	82.0	100.0	80.0	79.0	85.4	Pass	B
Student 17	73.0	79.0	85.0	70.0	78.3	Pass	C
Student 18	87.0	73.0	86.0	84.0	82.7	Pass	B
Student 19	98.0	90.0	98.0	95.0	95.6	Pass	A
Student 20	87.0	67.0	64.0	100.0	75.9	Pass	C
Student 21	71.0	70.0	90.0	68.0	77.0	Pass	C
Student 22	100.0	72.0	80.0	78.0	82.7	Pass	B
Student 23	88.0	73.0	79.0	85.0	80.7	Pass	B
Student Count	23						

Grades	
0	F
60	D
70	C
80	B
90	A

Item Weights

Course Items	Weight	Average	Median	Maximum	Minimum	Range	Variance	Standard Deviation
Test 1	25%	85.70	88.00	100.00	53.00	47.00	182.86	13.52
Test 2	25%	81.39	79.00	100.00	55.00	45.00	155.34	12.46
Final Exam	35%	82.57	85.00	100.00	52.00	48.00	198.44	14.09
Homework	15%	85.22	85.00	100.00	50.00	50.00	165.00	12.85
Final Grade	100%	83.45	84.75	96.25	57.25	39.00	121.41	11.02

FIGURE 2-1

CHAPTER 2 TASK: CALCULATING CLASS GRADES

Suppose you are a teacher and you want to keep track of your students' test and homework grades, and then calculate their actual course grade at the end of the semester. Many teachers give different weights to different tests and assignments toward a final grade, either because one test or assignment may be larger than another, or because a final exam might be comprehensive. When that happens, a weighted average must be calculated. In addition to that, suppose you want to keep other stats regarding your students' performance, such as the average test scores, the median homework scores, and more. This assignment will show you how you can accomplish all of these things while using various formulas, functions, and formats.

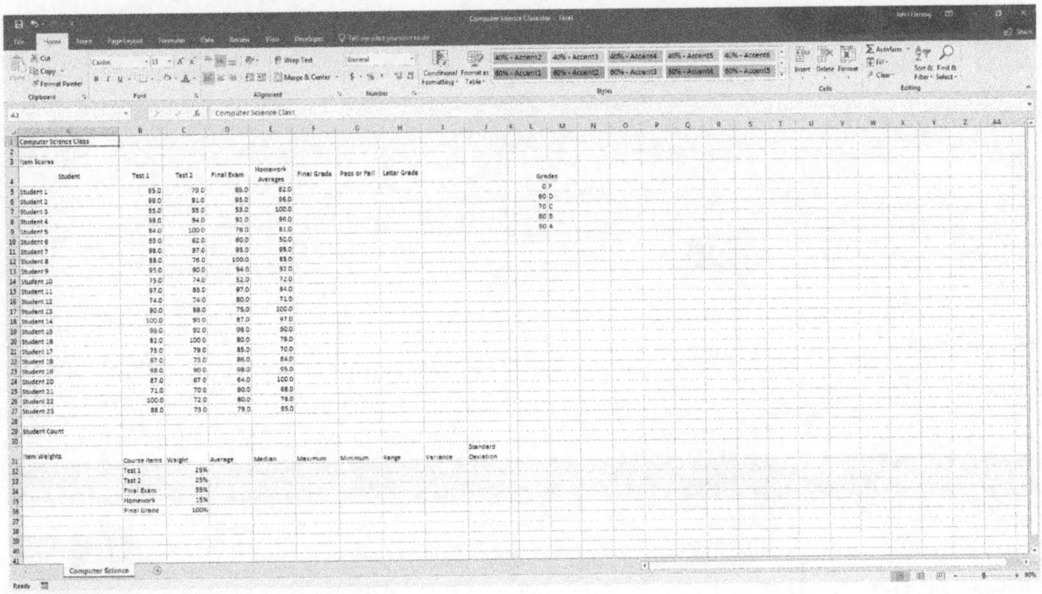

FIGURE 2-2

Open the file named *Computer Science Class* that is associated with this text. It will appear as it does in Figure 2-2.

APPLYING CELL STYLES

When you want to use a preset format for given cells, you can use the **Cell Styles** that Excel has already created. Doing so saves time, and you can benefit from the creativity of Microsoft. Suppose you want to apply **Cell Styles** to cells A1, A3, and A10. You can do so by taking the following steps:

Step 1: Go to cell **A1** (Figure 2-3A, Arrow A).

Step 2: Click the **Cell Styles** icon (Figure 2-3A, Arrow B) in the **Home Ribbon**. NOTE: Sometimes the **Cell Style** area in the **Home Ribbon** will look like what you see in Figure 2-3A, depending upon the resolution of your screen or the way Excel was installed on your particular computer. In that case, you need only to click on the scroll bar to the right of the area to see the styles (Figure 2-3B, Arrow).

After opening the **Cell Styles**, you will see a menu that looks like what you see in Figure 2-4.

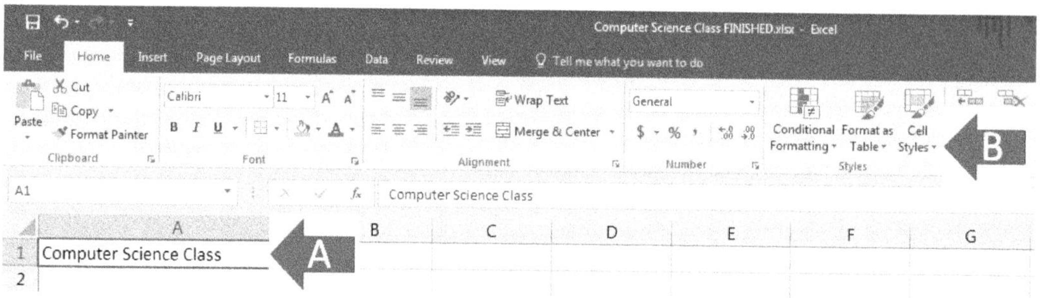

FIGURE 2-3A

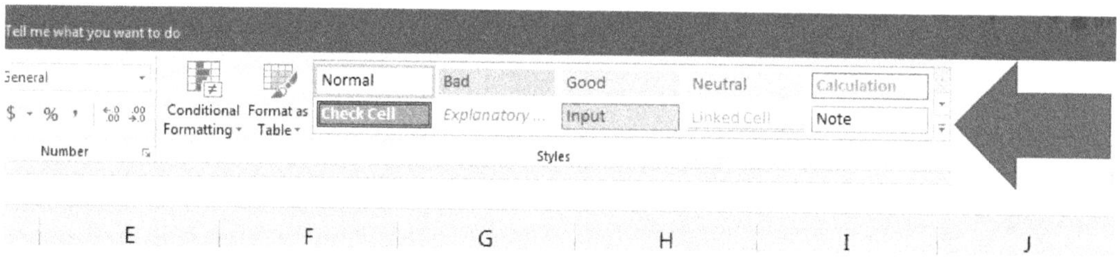

FIGURE 2-3B

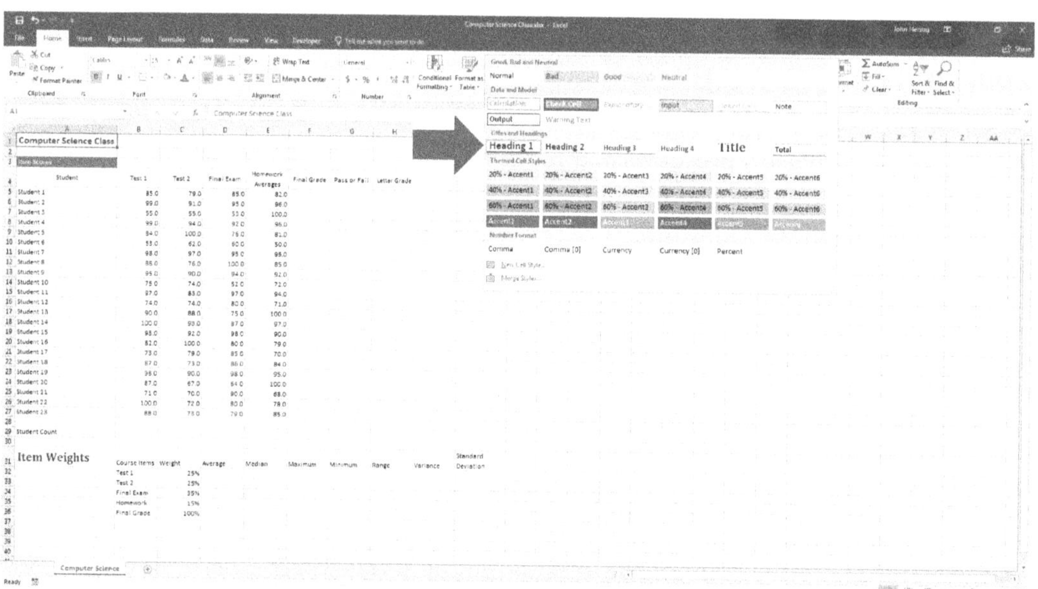

FIGURE 2-4

Step 3: Click **Heading 1** (Figure 2-4, Arrow). Repeat this process in order to apply the **Accent 1** style to cell **A3** and the cell style of **Title** to cell **A31**. NOTE: Your computer may not have the same color tone or styles shown in this text. If so, choose the color scheme closest to what is requested.

Make sure that column A is set at a width of *29* characters. After doing these things, your screen should look like Figure 2-5.

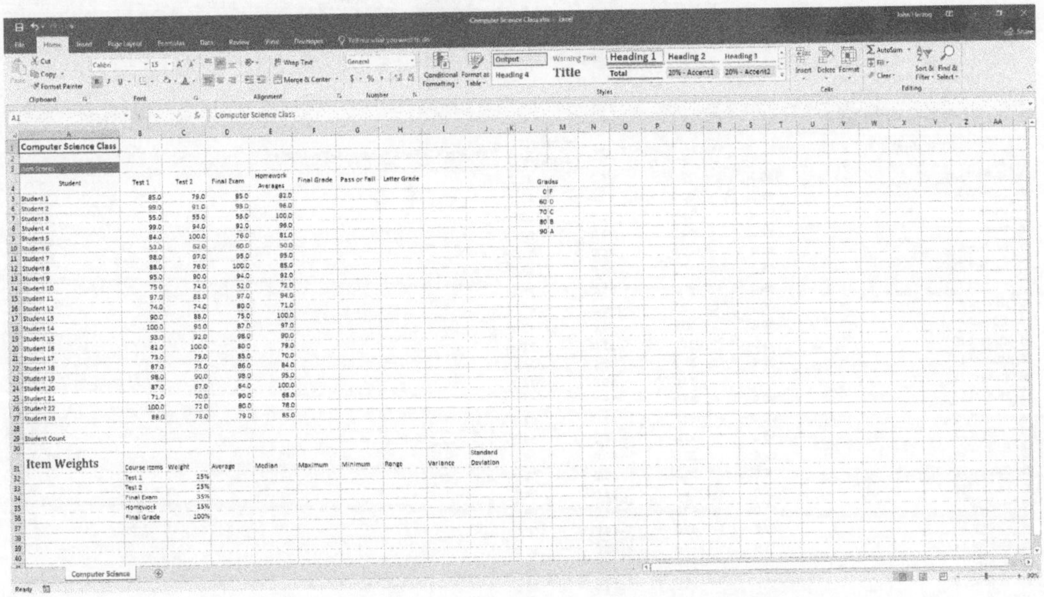

FIGURE 2-5

TO ROTATE TEXT

In some cases, you may want to rotate or tilt labels for a special effect or to make more room for them. Before you do, in this example you will want to first **merge and center** cells A31 through A36 as you learned to do in Chapter 1.

You can rotate the text in cell A31 by doing the following:

Step 1: Select the cell or cells needing to be rotated. In this example, it will be cell A31 that you just merged and centered.

Step 2: Click the tiny arrow next to the **Orientation** icon (Figure 2-6A, Arrow A) in the **Home Ribbon**.

Step 3: Click **Angle Counterclockwise** (Figure 2-6A, Arrow B). Your screen should then look like Figure 2-6B.

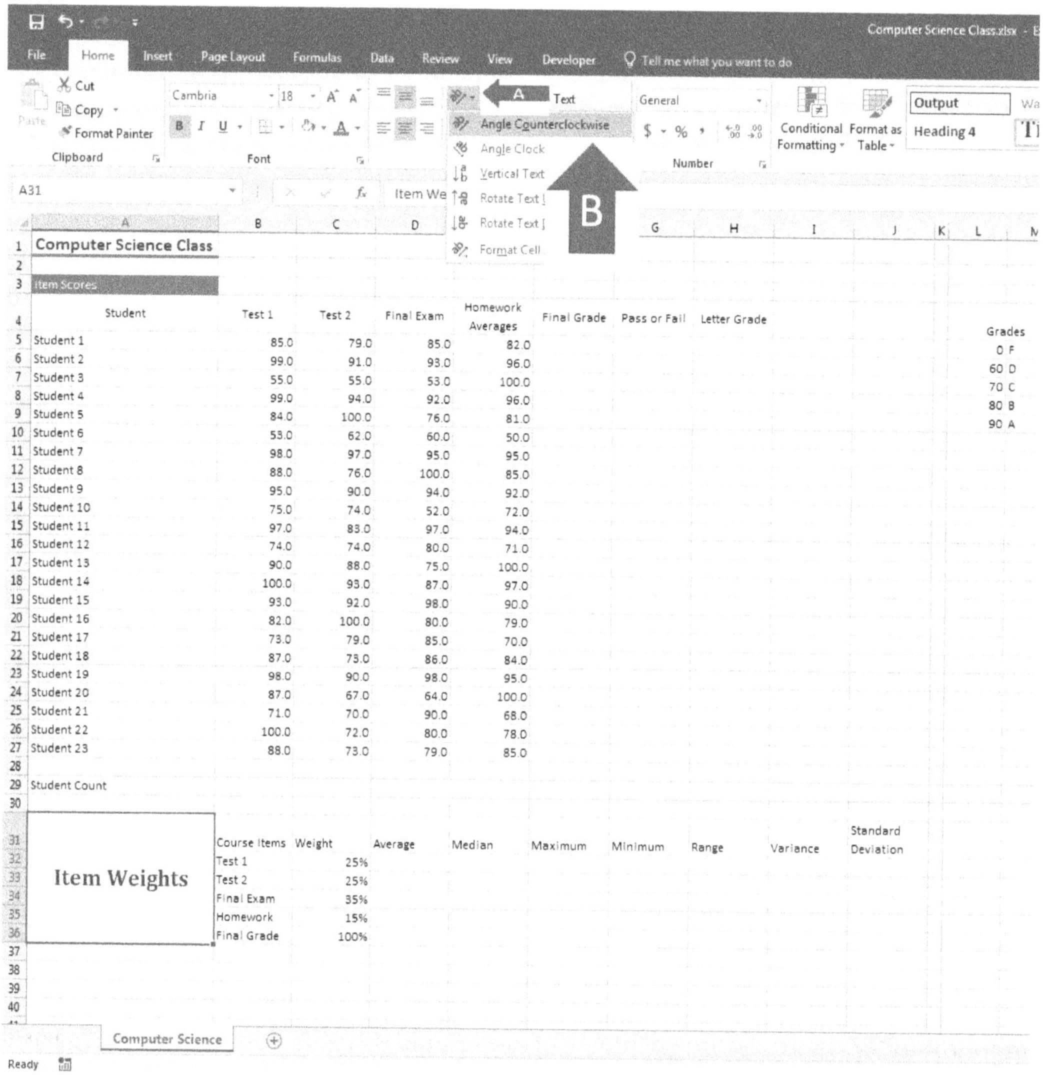

FIGURE 2-6A

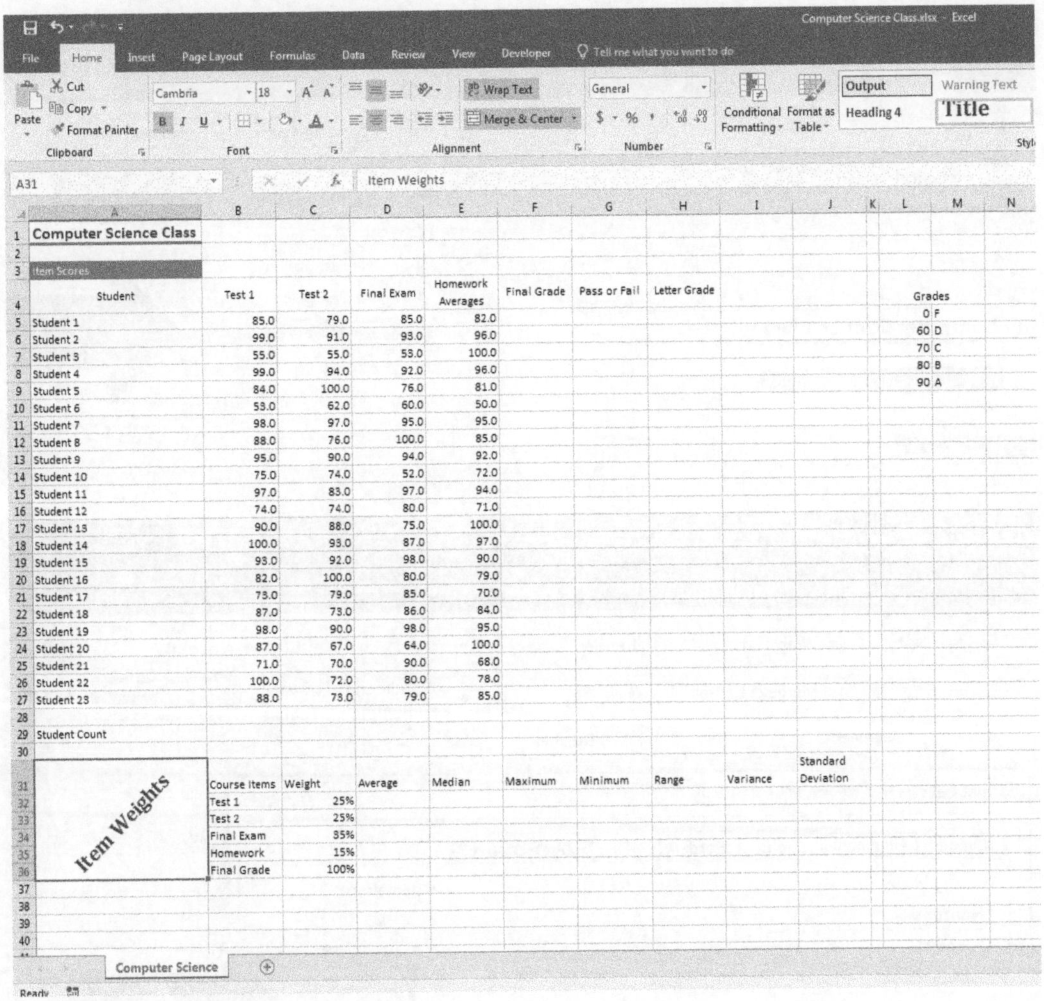

FIGURE 2-6B

NOTE: If you want to rotate text in any other direction or at an angle that is not necessarily 90 degrees, you can do so with the following steps.

Step 1: In the same **Orientation** box, click **Format Cell Alignment** (Figure 2-6C, Arrow). That will give you the menu you see in Figure 2-6D.

Step 2: Click and drag the **red diamond** (Figure 2-6D, Arrow A) up or down to the exact angle you would like. As you do, the **Degrees box** will change accordingly (Figure 2-6D, Arrow B). You can also just bypass the **red diamond** and type the degrees of the angle you want in the **Degrees box** (Figure 2-6D, Arrow B).

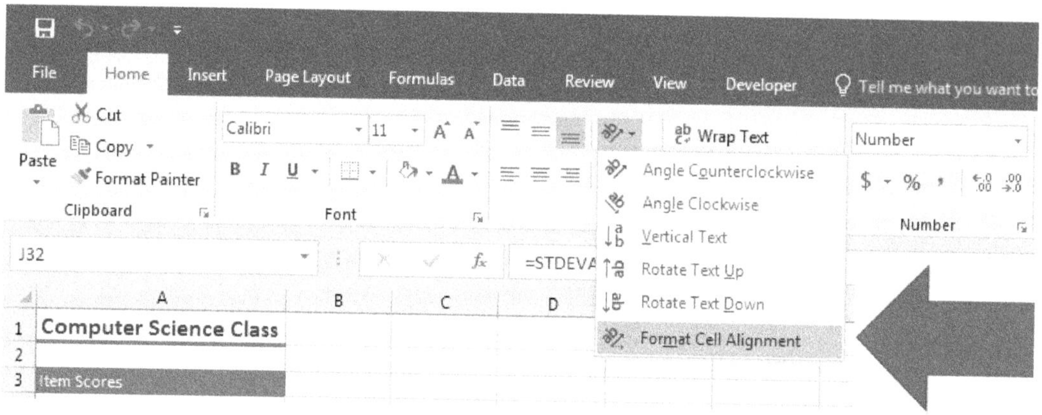

FIGURE 2-6C

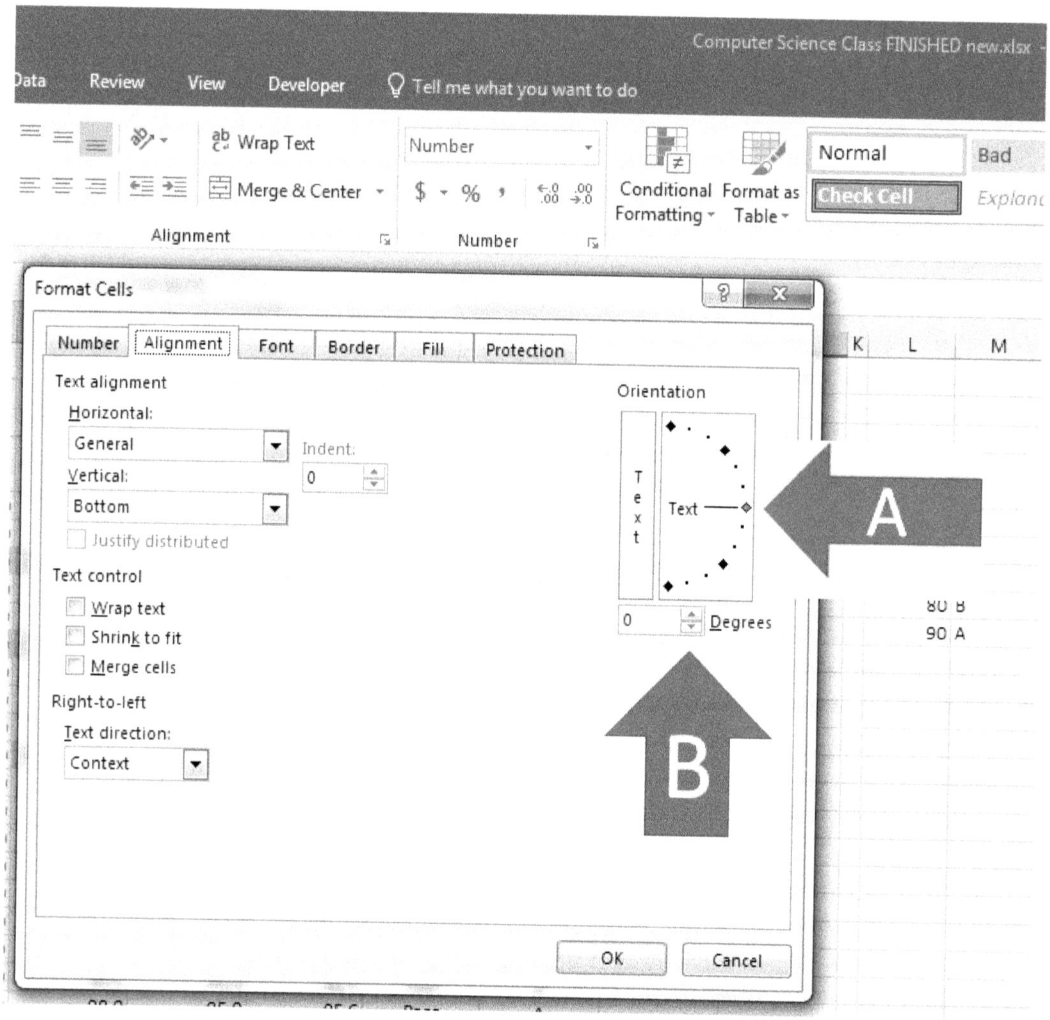

FIGURE 2-6D

TO FIND THE AVERAGE

Suppose you would like to display the **Average** of all of the scores for each course item. Let's start with placing the average score of *Test 1* in cell D32. This can be done by doing the following:

Step 1: Select cell D32.

Step 2: Click on the tiny arrow to the right of the **Autosum** button in the **Home Ribbon** (Figure 2-7A, Arrow A).

Step 3: Click **Average** (Figure 2-7A, Arrow B).

Notice that when you do, Excel will begin the **Average** function in cell D32 (Figure 2-7A, Arrow C). However, Excel will look for values in the cells above cell D32 that they assume you want to average (just as mentioned before when you click the **Auto-sum**). If it finds none, it will search for values in the cells to the left of D32. In this example, cell C32 has a number or value, but it's not what you want to average. Therefore, you can click and drag the range of cells that you **DO** want to average. In this case, that would be cells B5 through B27 that hold the test scores from *Test 1*.

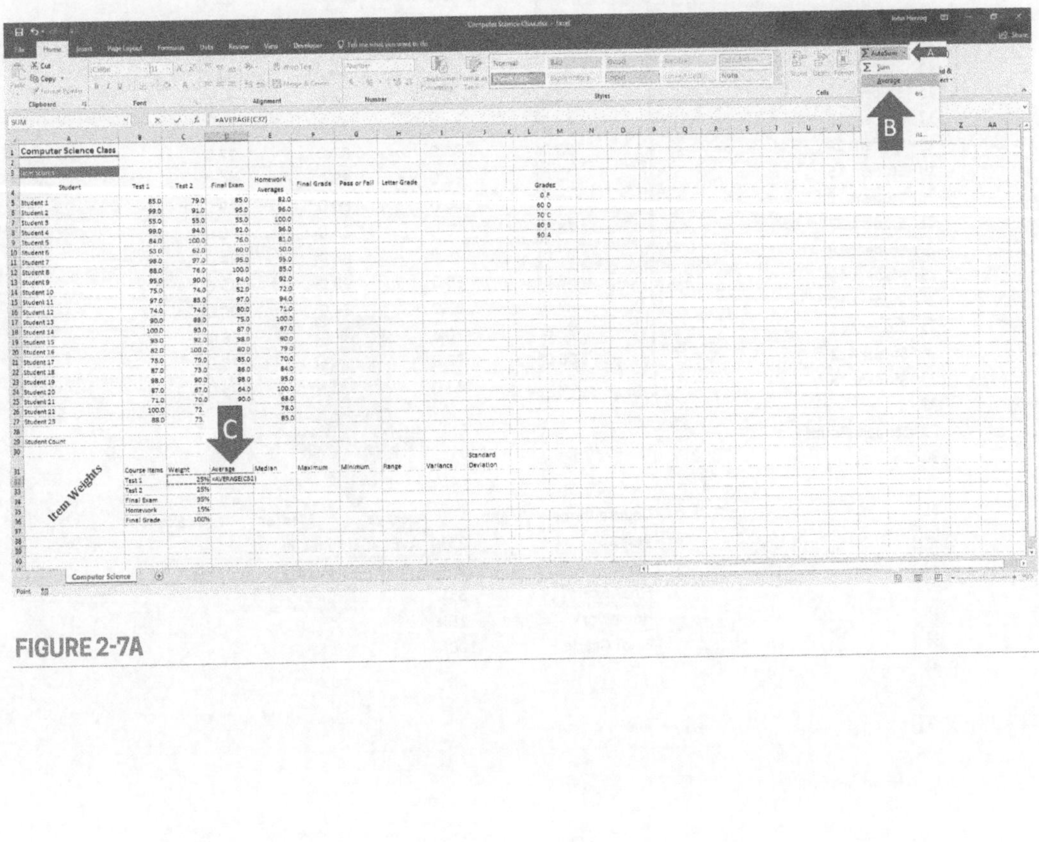

FIGURE 2-7A

Step 3: Click and drag the cell range B5 through B27, and you will see that Excel will automatically enter in that range as the argument to your **Average** function in cell D32 (Figure 2-7B, Arrow).

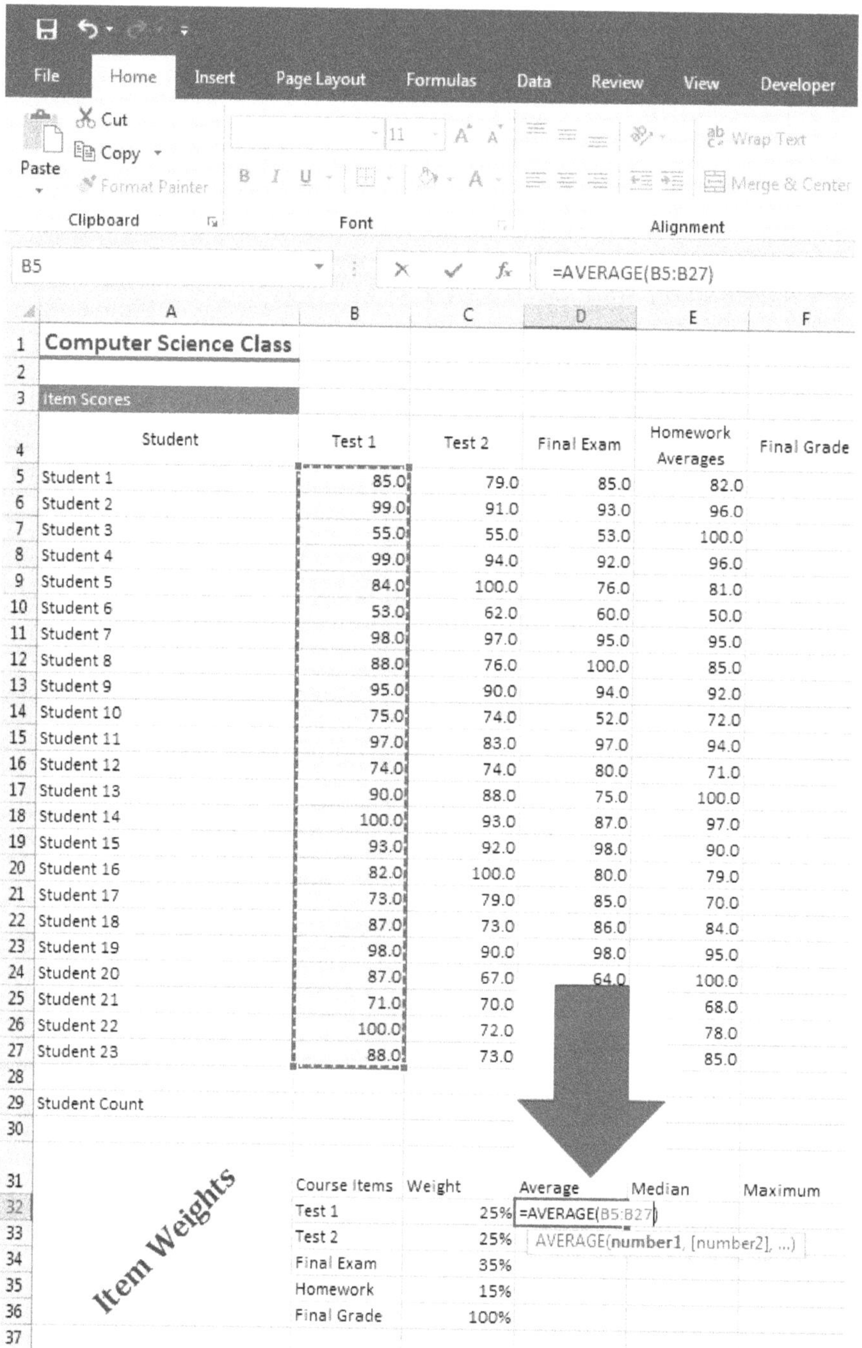

FIGURE 2-7B

Step 4: Press **Enter**. You will see in Figure 2-7C that the average score of *Test 1* will be calculated in cell D32.

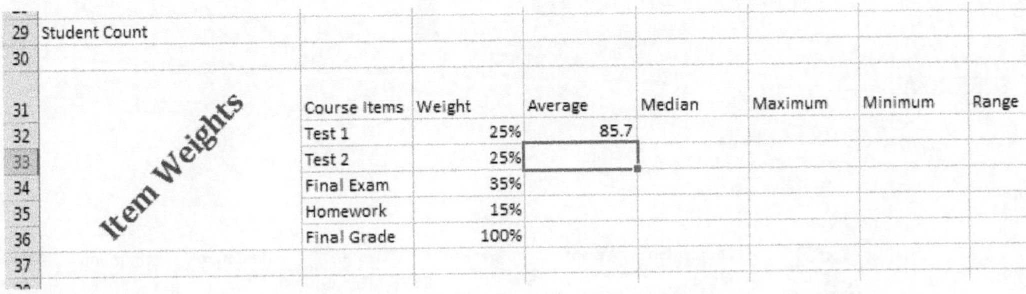

	A	B	C	D	E	F	G	H
29	Student Count							
30								
31	Item Weights	Course Items	Weight	Average	Median	Maximum	Minimum	Range
32		Test 1	25%	85.7				
33		Test 2	25%					
34		Final Exam	35%					
35		Homework	15%					
36		Final Grade	100%					
37								

FIGURE 2-7C

We will get the **Average** for *Test 2*, the *Final Exam*, and the *Homework* shortly.

TO FIND THE MEDIAN

Suppose you would like to display the **Median** of all of the scores for each course item. Let's start with placing the median score of *Test 1* in cell E32. The **Median** is the middle value (or score, in this case) after the scores are ranked from the lowest to the highest value. If you have an even number of items of observations (or tests), the **Median** would then be equal to the average of the middle two values (or test scores) after they have been ranked. This can be done by doing the following:

Step 1: Select cell E32.

Step 2: Type =MEDIAN(.

NOTE: When you begin to type anything in a cell that resembles a **function**, a **named cell** (discussed later), or a **named range** (also discussed later), you will see a small prompt below the cell in which you are working (Figure 2-8A). Excel will try to guess what you are entering, and it will give you some information about what it does. For example, when you began to type the **MEDIAN** function, Excel gave a prompt, anticipating that the **MEDIAN** function was what you wanted to enter (Figure 2-8A, Arrow). If you are in fact entering the **MEDIAN** function, all you need to do is double-click the prompt, and it will automatically finish typing the function as well as the left parenthesis (Figure 2-8B). If you are not entering one of the functions or named cells on their list, you can just ignore it and keep typing.

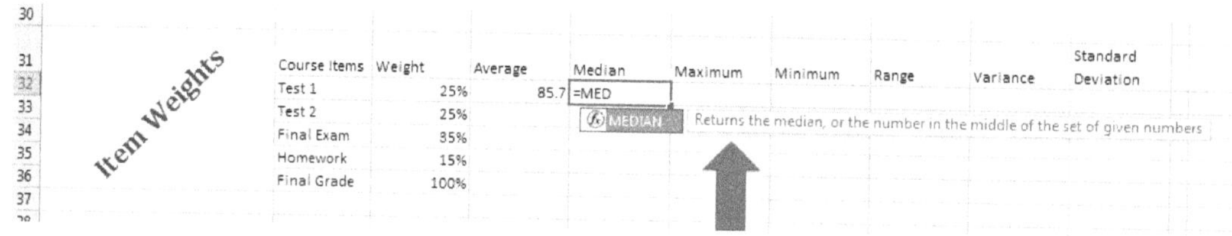

FIGURE 2-8A

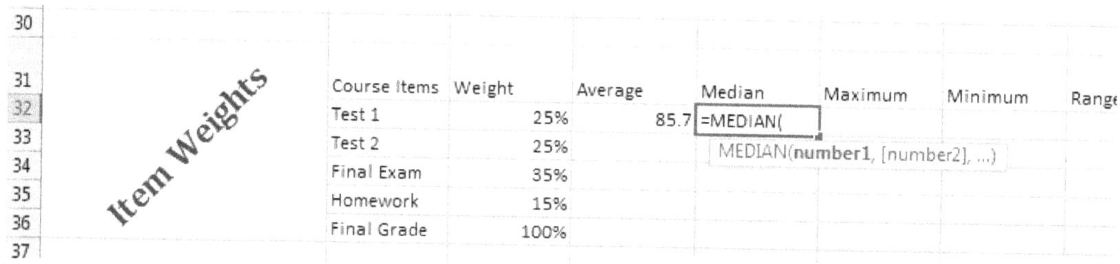

FIGURE 2-8B

Step 3: Click and drag the cell range B5 through B27 again (Figure 2-8C, Arrow).

Step 4: Press **Enter**, and the **median** will be calculated in cell E32.

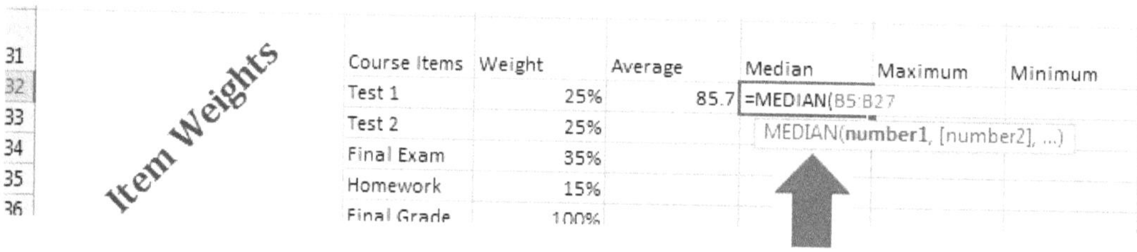

FIGURE 2-8C

USING THE FUNCTION WIZARD

If you want Excel to help you to create a function, you can use the **Function Wizard.** When you click on a cell where the function is to be placed, you can click the **Function Wizard** on the **Formula Bar** (Figure 2-9, Arrow A). When you do, a menu will pop up. If you don't know what the category is for the function you need, you can search for a function in the first box labeled **Search for a function**: and then click **Go.** If you know the category, you have the option to choose it at the box labeled **Or select a**

category: (Figure 2-9, Arrow B). In this example, we are creating the **MEDIAN**, so we want to look under the **Statistical** category (Figure 2-9, Arrow B). After choosing the **Statistical** category, the program will give you a list of statistical functions on the left that you can scroll through in order to find the one you are looking for. In this case, you want the **Function Wizard** to help you with the **Median** function. Once you find it, click it (Figure 2-9, Arrow C), and then click **OK**.

Once you do, the program will take you to what you see in Figure 2-10. At this menu, Excel will explain how the function works (Figure 2-10, Arrow A), and you can then add the first range of cells to be used in the function in the box labeled **Number1**

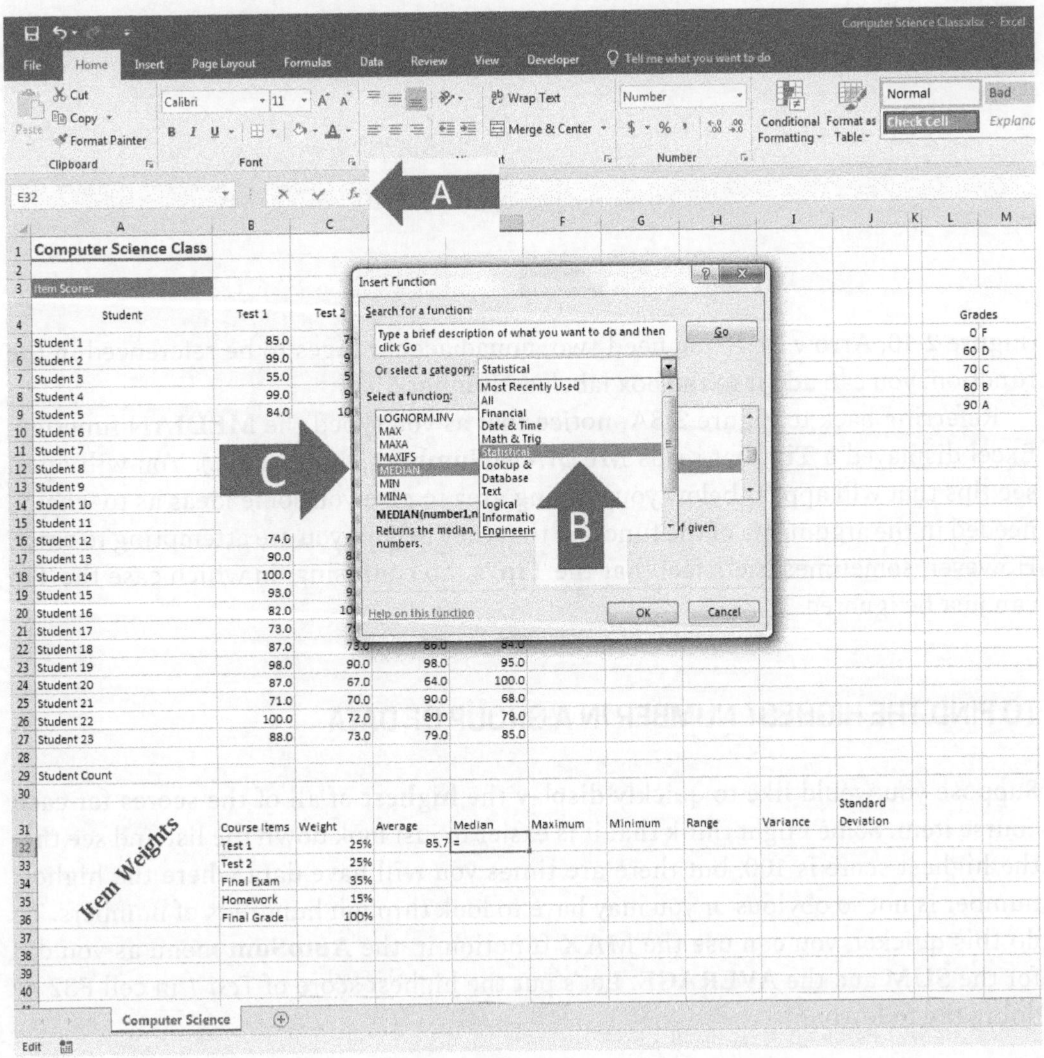

FIGURE 2-9

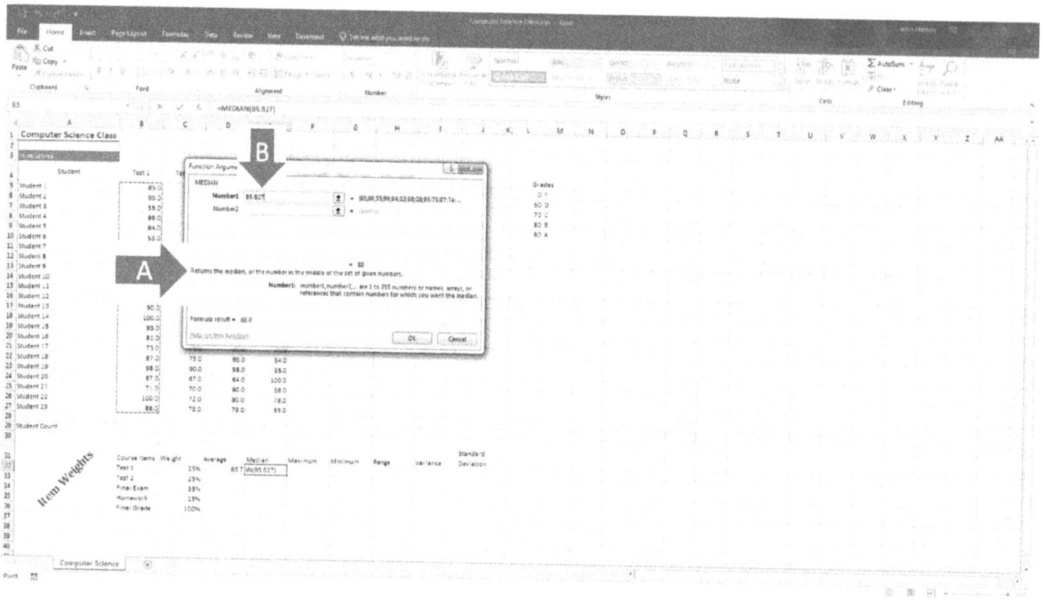

FIGURE 2-10

(Figure 2-10, Arrow B). If you need two, nonadjacent ranges to be referenced in the function, you can add it to the box labeled **Number2**.

Referring back to Figure 2-8A, notice that as you typed the **MEDIAN** function, Excel displayed a **Tip** that reads **MEDIAN(Number1,Number2,...)**. You will often see tips that will appear below your typing area to give you some ideas as to what is needed in the arguments of the function that they believe you are attempting to enter. However, sometimes users feel that the **Tip** is too confusing, in which case the **Tip** can just be ignored.

TO FIND THE HIGHEST NUMBER IN A GROUP OF DATA

Suppose you would like to quickly display the **highest** of all of the scores for each course item. Some might think that it is easier to just look down the list and see that the highest score is 100, but there are times you will have data where the highest number is not so obvious or you may have to look through hundreds of numbers. To do this quicker, you can use the **MAX** function in the **AutoSum** menu as you did for the **SUM** and the **AVERAGE**. Let's put the highest score of *Test 1* in cell F32 by doing the following:

Step 1: Select cell F32.

Step 2: Type =MAX(.

Step 3: Click and drag the cell range B5 through B27. Once again, Excel will put that cell range inside the parentheses of the **MAX** function for you (Figure 2-11).

Step 4: Press **Enter,** and you will see your result.

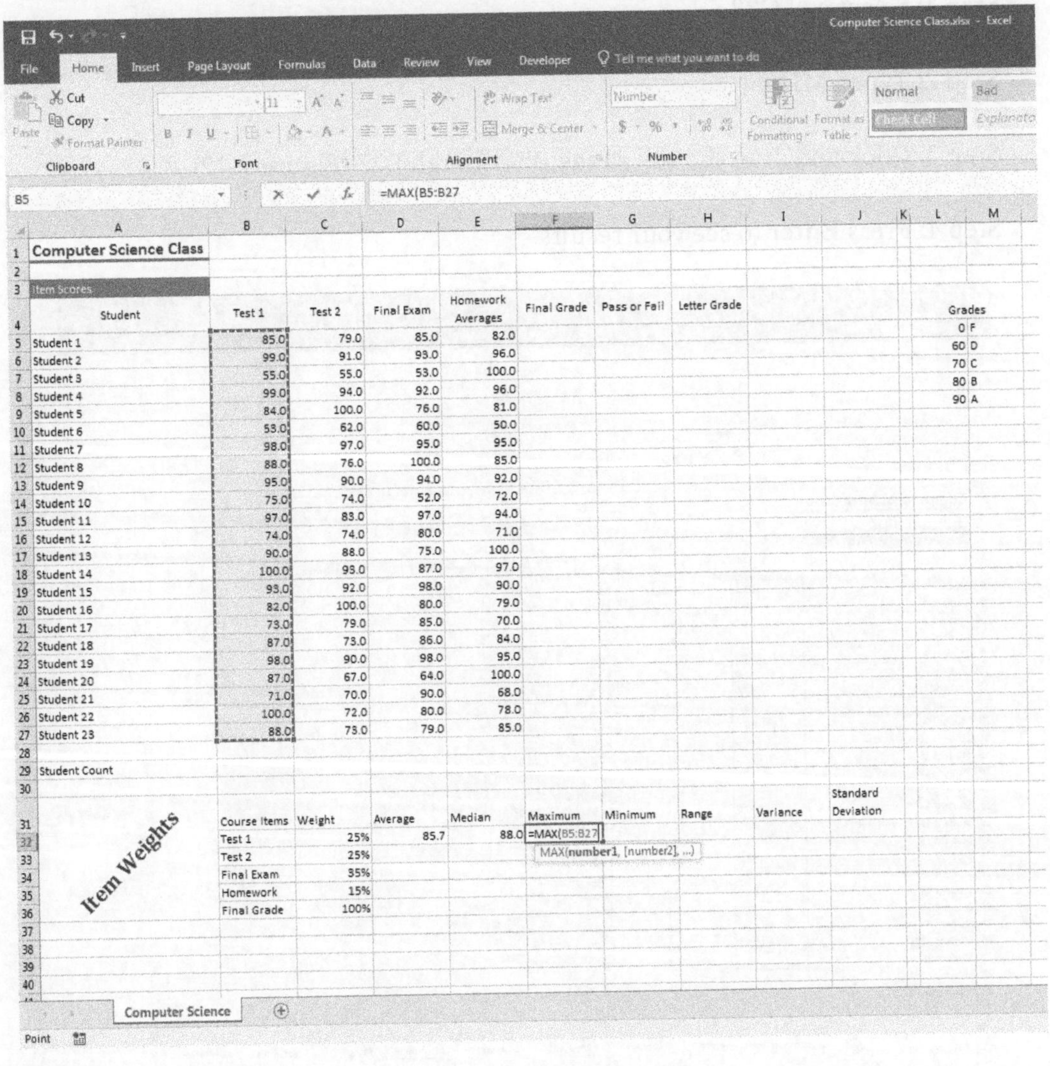

FIGURE 2-11

TO FIND THE LOWEST NUMBER IN A GROUP OF DATA

Suppose you would like to quickly display the **lowest** of all of the scores for each course item in cell G32. Once again, you can use the **MIN** function in the **AutoSum** menu as you did for the previous functions, or you can do the following:

Step 1: Select cell G32.

Step 2: Type =MIN(.

Step 3: Click and drag the cell range B5 through B27 (Figure 2-12).

Step 4: Press **Enter** to see your results.

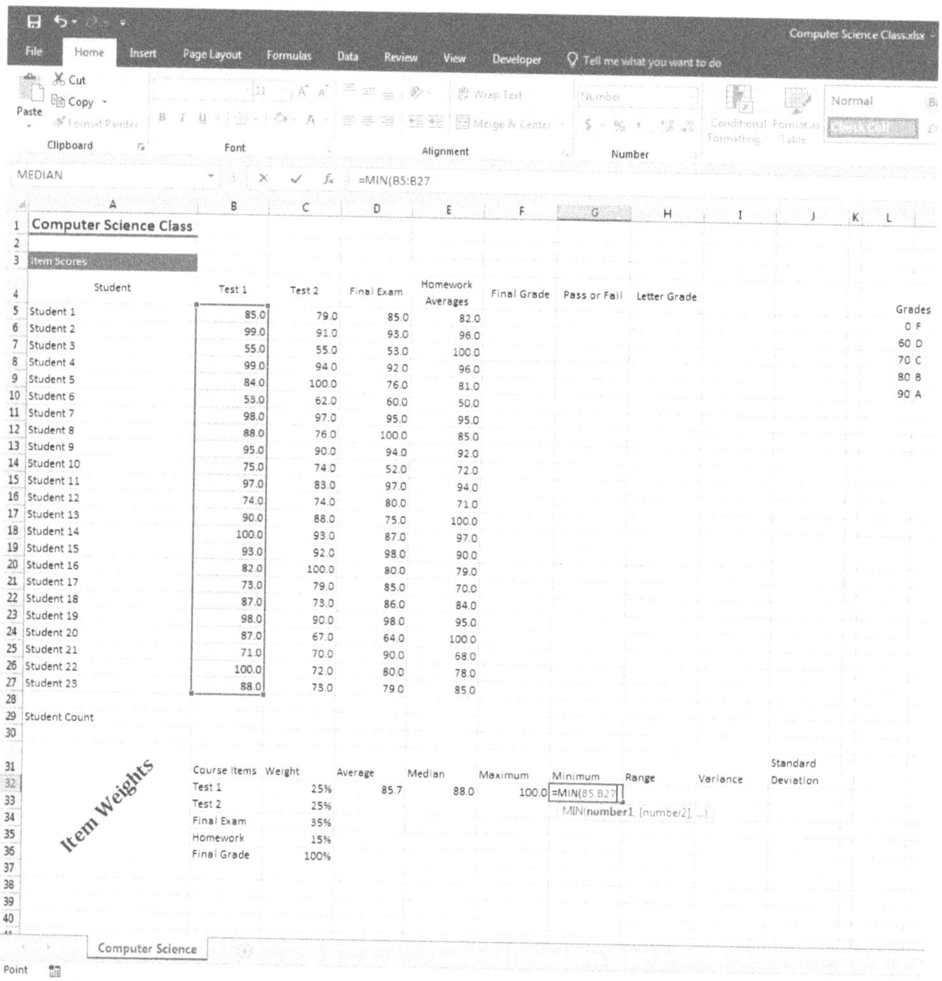

FIGURE 2-12

TO FIND THE RANGE OF DATA

Suppose you now want to know the grade range in cell H32. For example, if the lowest of a test score is 50 and the highest is 98, the range would be 48. This gives you an idea of how well and how poorly the students did on a test or on homework assignments. **There is no function in Excel called the Range.** Instead, you simply subtract the lowest score (the **MIN**) from cell G32 from the highest score (the **MAX**) in cell F32 by doing the following:

Step 1: Select cell H32.

Step 2: Type = and then click on cell F32.

Step 3: Press the **minus sign** (-) on the **numeric keypad** of your computer.

Step 4: Click on Cell G32. The formula you just created should read *=F32-G32* (Figure 2-13).

Step 5: Press **Enter** to see your results.

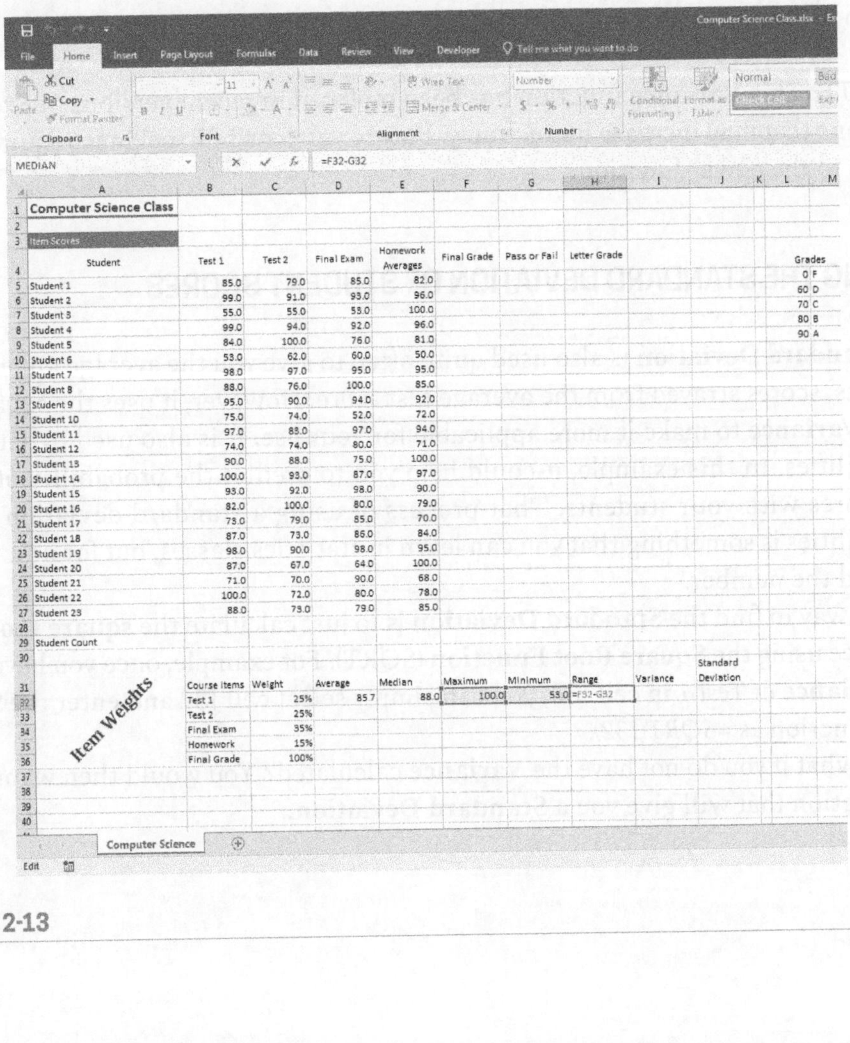

FIGURE 2-13

FINDING THE VARIANCE OF STUDENT SCORES

Sometimes the averages and the ranges of numbers aren't very telling. One student can do poorly while the rest do very well. That low score from the one student will make the average of all scores lower and the range higher, indicating that the overall performance of the class was not good, which may not be the case. The **Variance** will give you the squared average amount that students' scores strayed from the average test score. If this number is high, it may indicate that there may be one test score that is dragging your score average down and your range upward. Suppose you want to put the **Variance** of *Test 1* scores into cell I32. You can do so using the following steps:

Step 1: Select cell I32.

Step 2: Type *=VAR(*.

Step 3: Click and drag the cell range B5 through B27 (Figure 2-14A).

Step 4: Press **Enter** to see your results.

NOTE: This **Variance** function has been replaced in recent versions of Excel with functions that may increase its accuracy. Future versions of Excel may require you to change to the newer function.

FINDING THE STANDARD DEVIATION OF STUDENT SCORES

The **Standard Deviation** is also used quite often to give you the average amount that students' scores strayed from the average test score. However, it uses the square root of the **Variance** to make it more applicable for your use. It is also used to determine probabilities. In this example, it could help you to predict the probability of future test scores with your students. That process of using a standard deviation to find probabilities is something that you can learn in statistics classes, but for now we will just find the number.

One way to find the **Standard Deviation** is to just calculate the **square root** of the **Variance** using the **Square Root Function** (SQRT). For example, once you have found the **Variance** of *Test 1* in cell I32, you can simply select cell J32 and enter the **Square Root** function as *=SQRT(I32)*.

But what if you do not have the **Variance** calculated? You would then want to use the function that will give you a **Standard Deviation**.

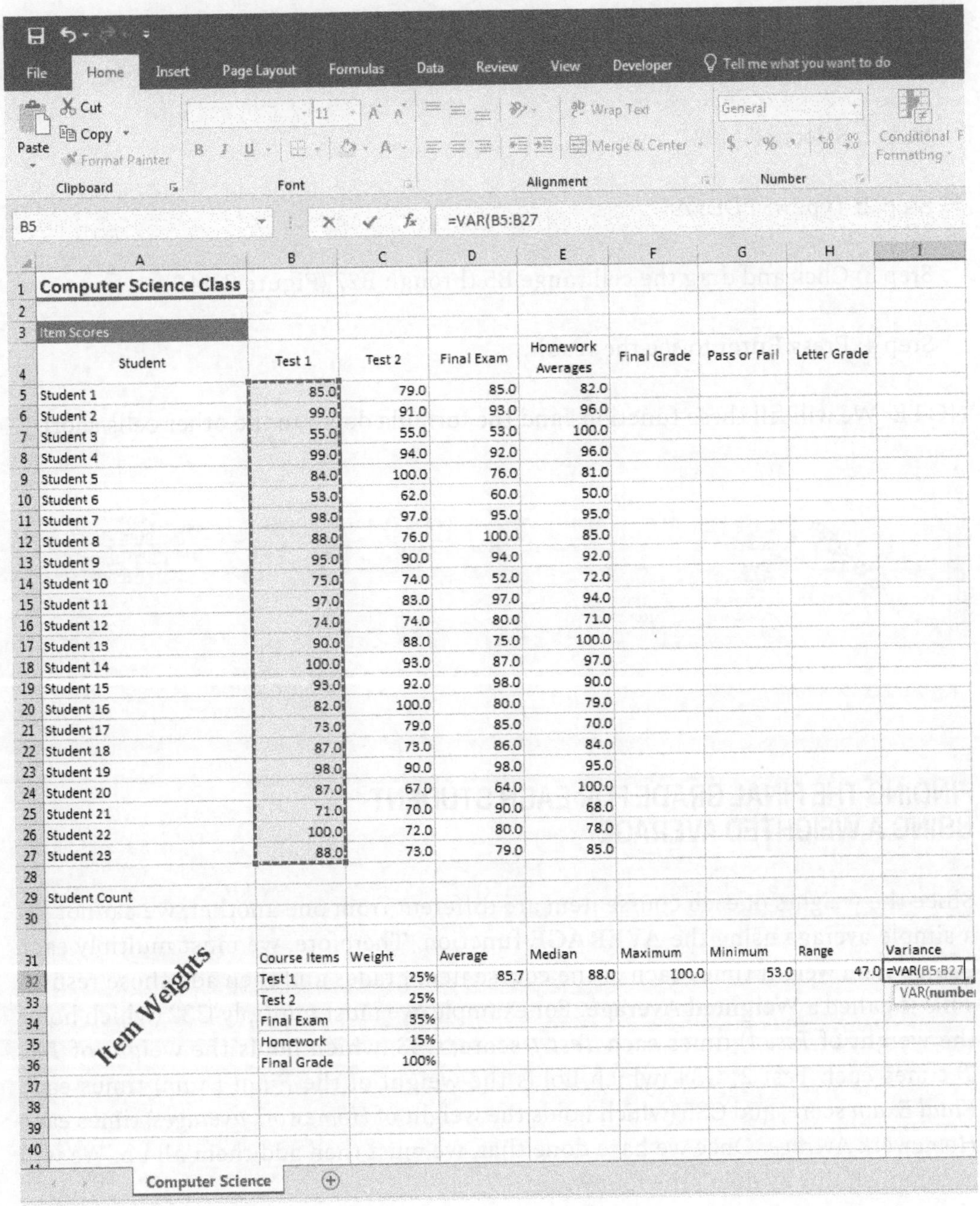

Computer Science Class

Item Scores

Student	Test 1	Test 2	Final Exam	Homework Averages	Final Grade	Pass or Fail	Letter Grade
Student 1	85.0	79.0	85.0	82.0			
Student 2	99.0	91.0	93.0	96.0			
Student 3	55.0	55.0	53.0	100.0			
Student 4	99.0	94.0	92.0	96.0			
Student 5	84.0	100.0	76.0	81.0			
Student 6	53.0	62.0	60.0	50.0			
Student 7	98.0	97.0	95.0	95.0			
Student 8	88.0	76.0	100.0	85.0			
Student 9	95.0	90.0	94.0	92.0			
Student 10	75.0	74.0	52.0	72.0			
Student 11	97.0	83.0	97.0	94.0			
Student 12	74.0	74.0	80.0	71.0			
Student 13	90.0	88.0	75.0	100.0			
Student 14	100.0	93.0	87.0	97.0			
Student 15	93.0	92.0	98.0	90.0			
Student 16	82.0	100.0	80.0	79.0			
Student 17	73.0	79.0	85.0	70.0			
Student 18	87.0	73.0	86.0	84.0			
Student 19	98.0	90.0	98.0	95.0			
Student 20	87.0	67.0	64.0	100.0			
Student 21	71.0	70.0	90.0	68.0			
Student 22	100.0	72.0	80.0	78.0			
Student 23	88.0	73.0	79.0	85.0			
Student Count							

Course Items	Weight	Average	Median	Maximum	Minimum	Range	Variance
Test 1	25%	85.7	88.0	100.0	53.0	47.0	=VAR(B5:B27
Test 2	25%						
Final Exam	35%						
Homework	15%						
Final Grade	100%						

FIGURE 2-14A

If you want to display the standard deviation in cell J32 using the **Standard Deviation** function, you can do so by taking the following steps:

Step 1: Select cell J32.

Step 2: Type =*STDEVA(*

Step 3: Click and drag the cell range B5 through B27 (Figure 2-14B).

Step 4: Press **Enter** to see the result.

NOTE: We will fill these functions and the formula down to the other cells shortly.

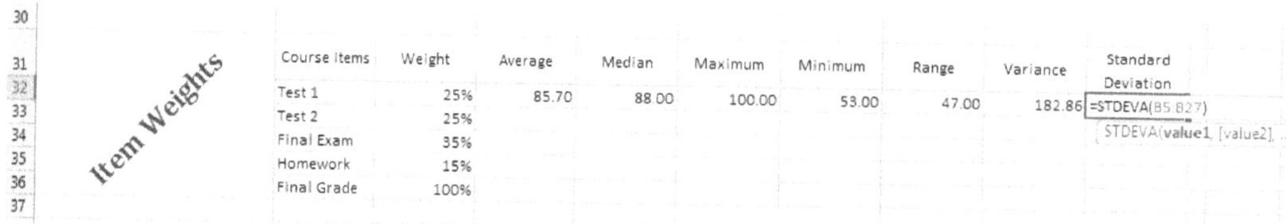

FIGURE 2-14B

FINDING THE FINAL GRADE FOR EACH STUDENT USING A WEIGHTED AVERAGE

Since the weights of each course item are different from one another, we cannot get a simple average using the **AVERAGE** function. Therefore, we must multiply each course item weight times each of the course item grades and then add those results. This is called a **Weighted Average**. For example, we must multiply C32 (which holds the weight of *Test 1*) times each *Test 1* score, C33 (which holds the weight of *Test 2*) times each Test 2, C34 (which holds the weight of the *Final Exam*) times each *Final Exam* score, and C35 (which holds the weight of *Homework averages*) times each *Homework Average*. Once we have done that, we must then add them all up. We will accomplish this by doing the following:

Step 1: Select Cell F5.

Step 2: Type = and then select Cell B5.

Step 3: Type * and then select Cell C32.

Step 4: Press the F4 key. When you do, it will apply the **Absolute Address** to cell C32, as we would want to do, because the *course item weight* must always be referenced in that cell regardless of how many rows to which we will copy this formula. Your formula will then read *=B5*$C$32*.

Step 5: Press the **plus sign** (+) on the **numeric keypad** of your computer.

Step 6: Select Cell C5.

Step 7: Type * and then select Cell C33.

Step 8: Press the F4 key. Your formula will then read *=B5*$C$32+C5*$C$33*.

Step 9: Press the **plus sign** (+).

Step 10: Then select Cell D5.

Step 11: Type * and then select Cell C34.

Step 12: Press the F4 key. It will then read *=B5*$C$32+C5*$C$33+D5*$C$34*.

Step 13: Press the **plus sign** (+).

Step 14: Then select Cell E5.

Step 15: Type * and then select Cell C35.

Step 16: Press the F4 key. It will then read *=B5*$C$32+C5*$C$33+D5*$C$34+E5*$C$35* (Figure 2-15).

Step 17: Press **Enter.** You will then see the value of 83.1 in cell F5. You may get slight variations of that number depending upon how much your spreadsheet rounds that number.

Step 18: Fill the function in cell F5 down to row 27 using the **Fill Handle**.

NOTE: You do not need to put parentheses around the *B12*$C$4* or any of the other grade items because Excel uses the **PEMDAS** system. **PEMDAS** stands for **Parentheses**, **Exponents**, **Multiplication**, **Division**, **Addition**, and **Subtraction**. Math needed within a parenthesis is performed first, then exponents are calculated, then the multiplication, then the division, then the addition, and then the subtraction.

It is the typical order of operations in any math calculation. That is why it is the order that the math is done in Excel. Therefore, since Excel does the multiplication first anyway, there is no need to add parentheses around each product.

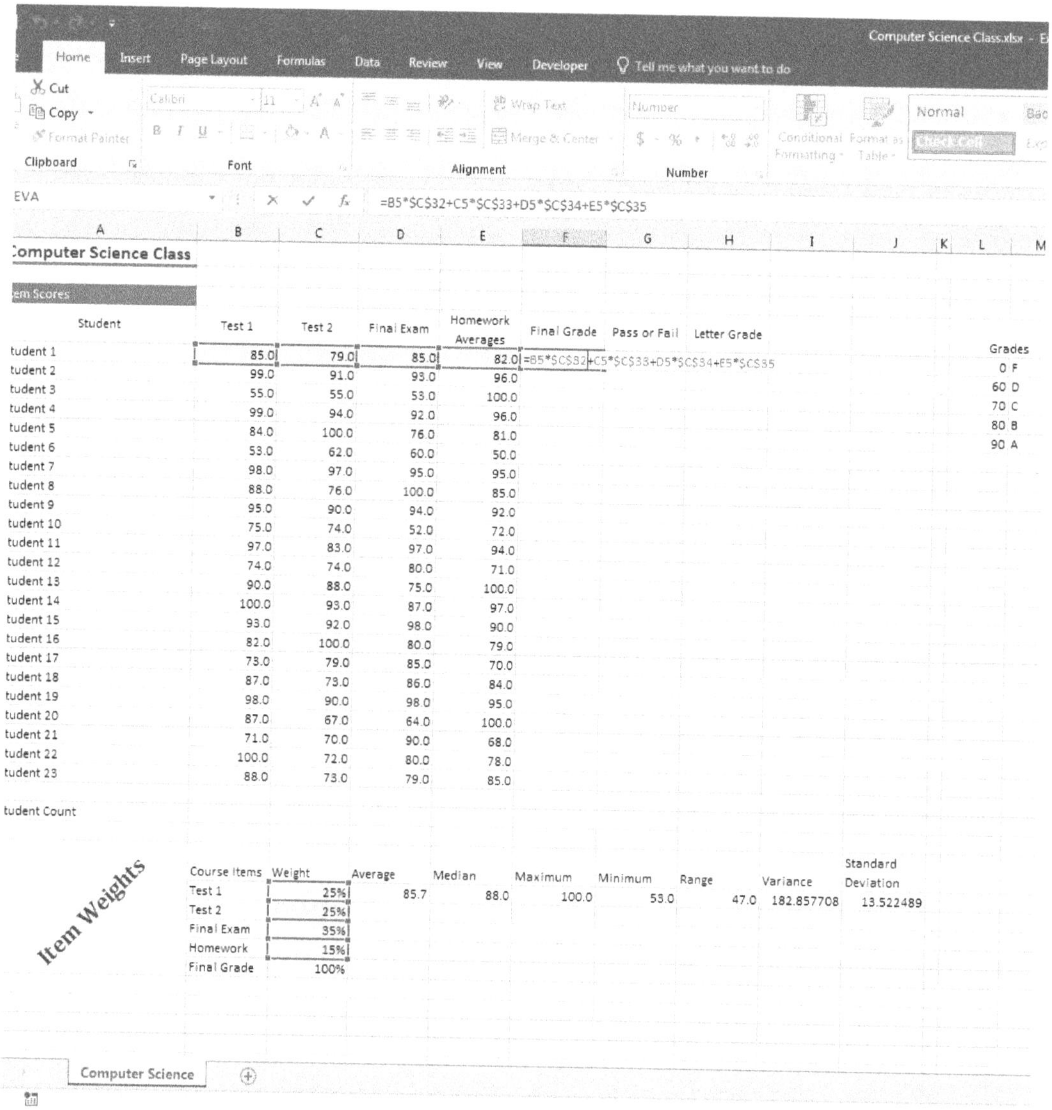

FIGURE 2-15

ANOTHER WAY TO FILL DATA DOWN

Sometimes using the mouse to pull the **Fill Handle** to copy formulas down to the cells below them may not be easy, especially if you have to fill the formula down to hundreds or thousands of rows. To fill down the formula you just created without using the fill handle, do the following:

Step 1: Select Cell F5 that has the formula you just created, as well as the cells below it, all the way down to Cell F27 (Figure 2-16). If you have to fill down many

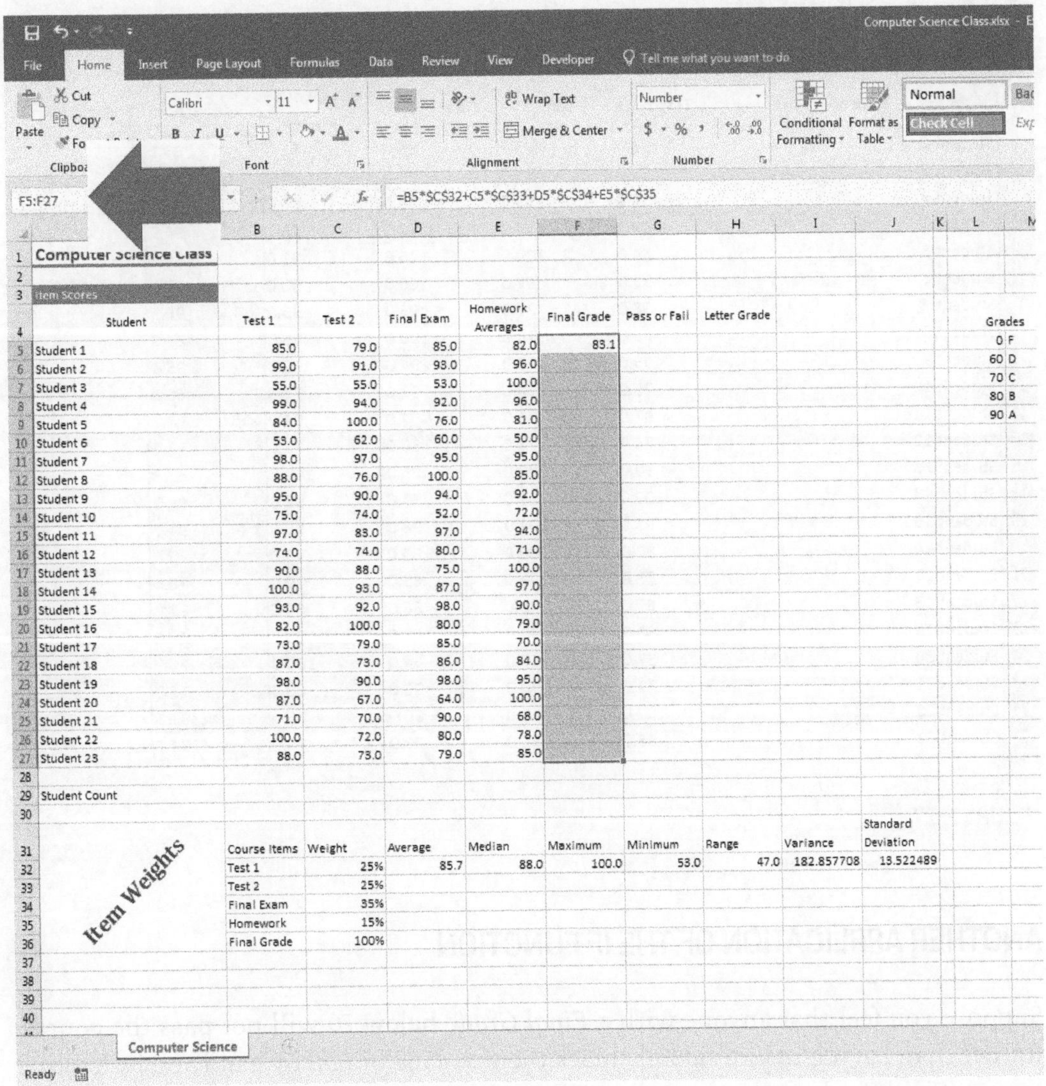

FIGURE 2-16

hundreds or thousands of cells, you can type in the range F5:F27 in the **Address Directory** box (Figure 2-16, Arrow) and Excel will select that range for you.

Step 2: Press **Ctrl D** (See Figure 2-17).

NOTE: The same method can be used to fill data to the right by selecting the formula and the cells to the right to which you want to fill and then pressing **Ctrl R**.

	A	B	C	D	E	F	G	
1	Computer Science Class							
2								
3	Item Scores							
4	Student	Test 1	Test 2	Final Exam	Homework Averages	Final Grade	Pass or Fail	Le
5	Student 1	85.0	79.0	85.0	82.0	83.1		
6	Student 2	99.0	91.0	93.0	96.0	94.5		
7	Student 3	55.0	55.0	53.0	100.0	61.1		
8	Student 4	99.0	94.0	92.0	96.0	94.9		
9	Student 5	84.0	100.0	76.0	81.0	84.8		
10	Student 6	53.0	62.0	60.0	50.0	57.3		
11	Student 7	98.0	97.0	95.0	95.0	96.3		
12	Student 8	88.0	76.0	100.0	85.0	88.8		
13	Student 9	95.0	90.0	94.0	92.0	93.0		
14	Student 10	75.0	74.0	52.0	72.0	66.3		
15	Student 11	97.0	83.0	97.0	94.0	93.1		
16	Student 12	74.0	74.0	80.0	71.0	75.7		
17	Student 13	90.0	88.0	75.0	100.0	85.8		
18	Student 14	100.0	93.0	87.0	97.0	93.3		
19	Student 15	93.0	92.0	98.0	90.0	94.1		
20	Student 16	82.0	100.0	80.0	79.0	85.4		
21	Student 17	73.0	79.0	85.0	70.0	78.3		
22	Student 18	87.0	73.0	86.0	84.0	82.7		
23	Student 19	98.0	90.0	98.0	95.0	95.6		
24	Student 20	87.0	67.0	64.0	100.0	75.9		
25	Student 21	71.0	70.0	90.0	68.0	77.0		
26	Student 22	100.0	72.0	80.0	78.0	82.7		
27	Student 23	88.0	73.0	79.0	85.0	80.7		
28								

FIGURE 2-17

ANOTHER APPLICATION OF THE IF FUNCTION

Suppose you feel that anyone with a *Final Grade* below *70* will not pass the course. Rather than having to individually mark them as *Pass* or *Fail* by hand and by looking at each *Final Grade*, you can use the **IF** function. In this example, there are very few students; therefore, using the IF function for this scenario probably isn't going to save much time. However, when an instructor has hundreds of students, or if you have a spreadsheet that is comparable to this one, it can be very helpful.

To show each student as *Pass* or *Fail*, do the following:

Step 1: Click on Cell G5.

Step 2: Type *=IF(F5<70,"Fail","Pass")*. See Figure 2-18. Do **NOT** fill the function down at this time. We will do it shortly. NOTE: The words *Fail* and *Pass* **MUST** be surrounded by quotation marks in the function because they are non-numeric.

Step 3: Press **Enter** and see your results.

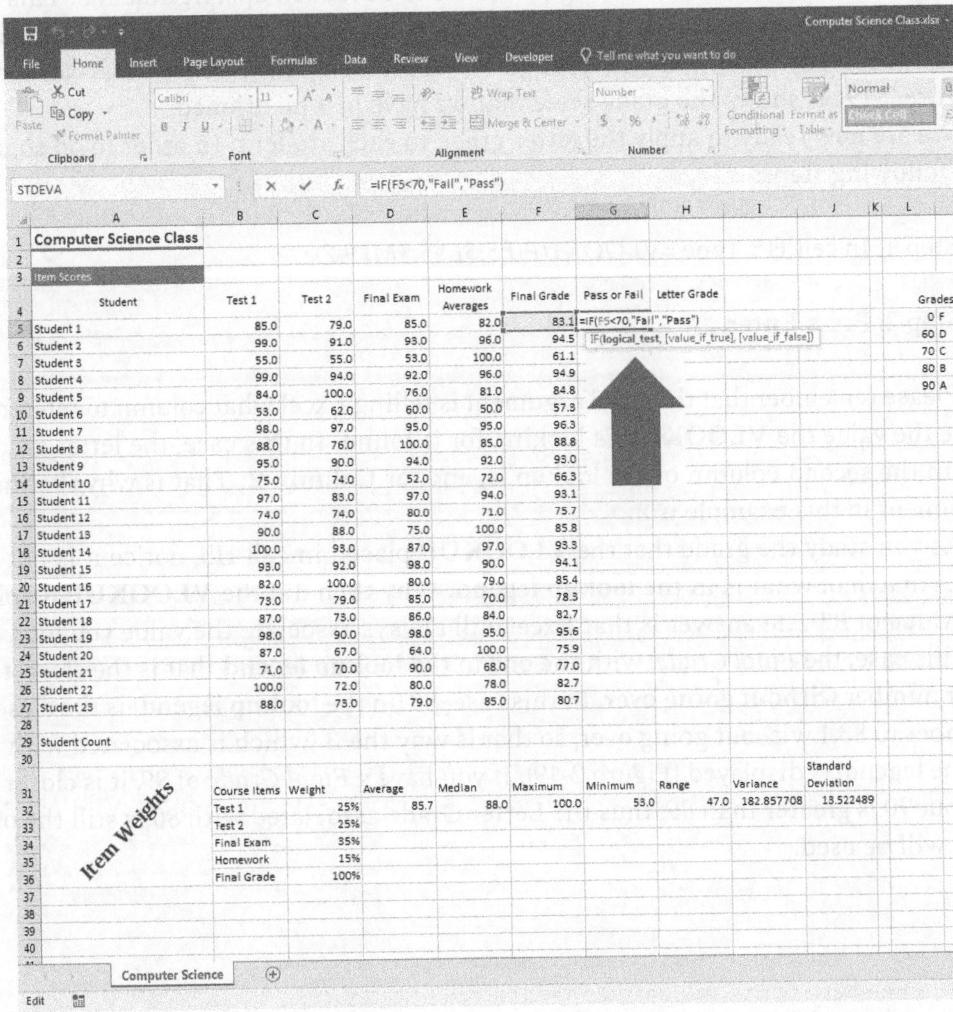

FIGURE 2-18

USING THE VLOOKUP FUNCTION

It is now time to enter the letter grades for each student. Suppose you consider a grade below *60* to be a failing grade, so you would thus assign an *F* to that student's grade. Also suppose that scores at or above *60* but less than *70* are given a *D*, scores at or above *70* but less than *80* are given a *C*, scores at or above *80* but less than *90* are given a *B*, and scores at or above *90* are awarded an *A*. Instead of entering them in manually, once again, allow Excel to do the work for you by using the **VLOOKUP** function. It has three arguments and works as follows:

=VLOOKUP(Argument 1: This is the value that is to be looked up; Argument 2: This tells **WHERE** the value in Argument 1 is to be looked up; Argument 3: This the column in a **lookup legend** that holds the value to be displayed). The **lookup legend** is what you already see in cells L5 through M9 of the *Computer Science Class* workbook (Figure 2-19). It shows the grade scale that we have just mentioned.

We are now ready to apply the letter grade to each student. In order to do so, use the following steps:

Step 1: In cell H5, type *=VLOOKUP(F5,$L$5:$M$9,2).*

Step 2: Press **Enter**.

Please remember that the third argument is telling Excel what column to reference once the value the **VLOOKUP** is looking for is found. In this case, the letter grades are in the second column of the lookup legend, or **Column 2**. That is why the third argument in this example is 2.

As you study the grade that the **VLOOKUP** placed in cell H5, notice it is not an exact match of what is in the **lookup legend**. Why then did the **VLOOKUP** display the value of *B*? The answer is that Excel will always associate the value you look up (in this case, the *Final Grade*) with the one in the **lookup legend** that is the closest to your number **without going over**. In this case, *80* (in the **lookup legend**) is the closest number to 83.1 without going over, so that is why the *B* (which is associated with *80* in the legend) is displayed (Figure 2-19). If you have a *Final Grade* of *89*, it is closer to *90*, but 90 is greater than *89*; thus the Letter Grade associated with *80* is still the one that will be used.

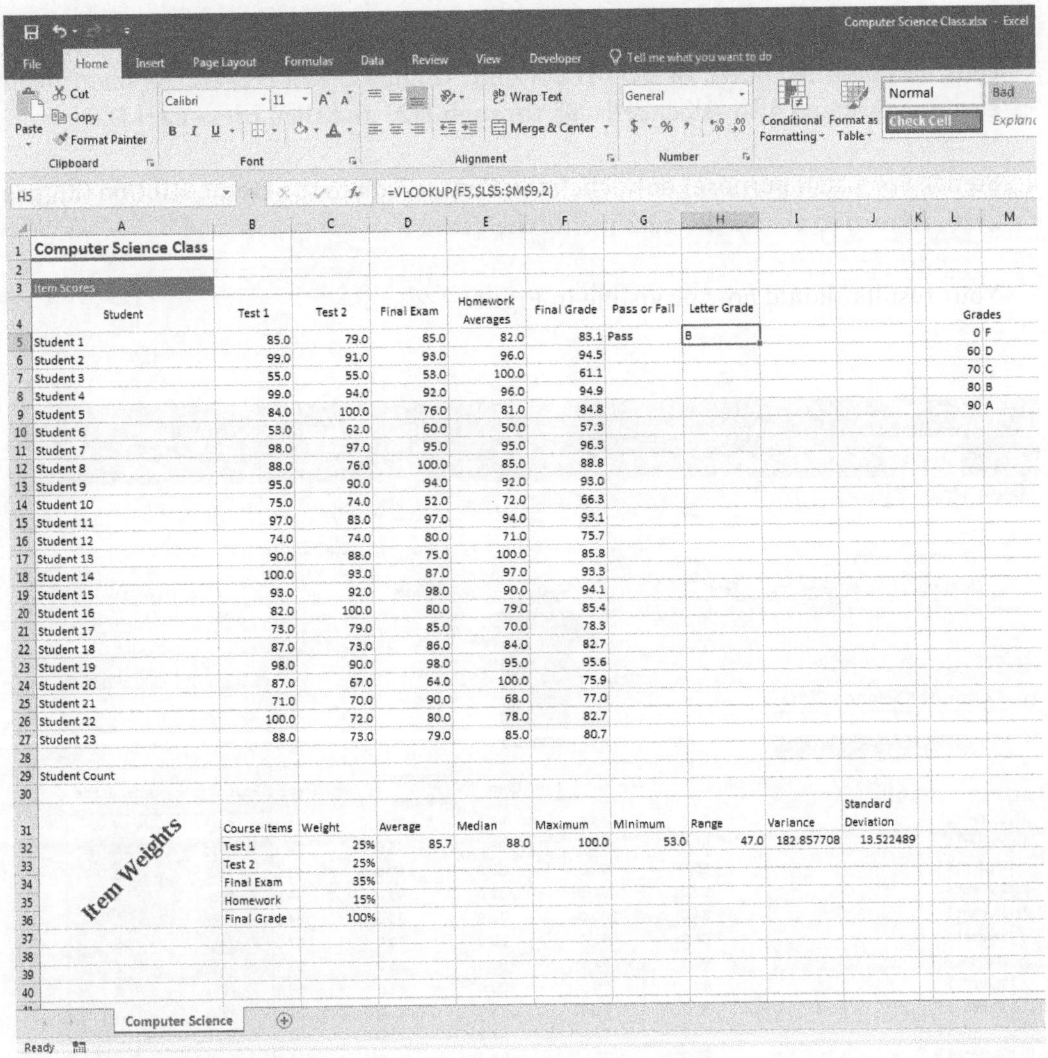

FIGURE 2-19

FILLING DOWN MULTIPLE FORMULAS

Now we will fill all of the formulas we have entered to their appropriate cells below them to show the statistics for each student. However, you don't have to fill down each of those formulas individually.

Let's fill down the *Pass or Fail* formula in cell G5 **AND** the *Letter Grade* formula in cell H5 to row 27 **at one time**.

To do this:

Step 1. Select cells G5 and H5.

Step 2. Pull the **Fill Handle** on the lower right of cell H5 down to row 27. **BOTH** cells will be filled down in **BOTH** columns (Figure 2-20). NOTE: As mentioned earlier, you could also select cells G5 through H27 and then press **Ctrl D**.

Step 3: For visual purposes only, click the **Center icon** in the **Home Ribbon** (Figure 2-20, Arrow) to center the data inside the cells.

Your results should now be visible in Figure 2-20.

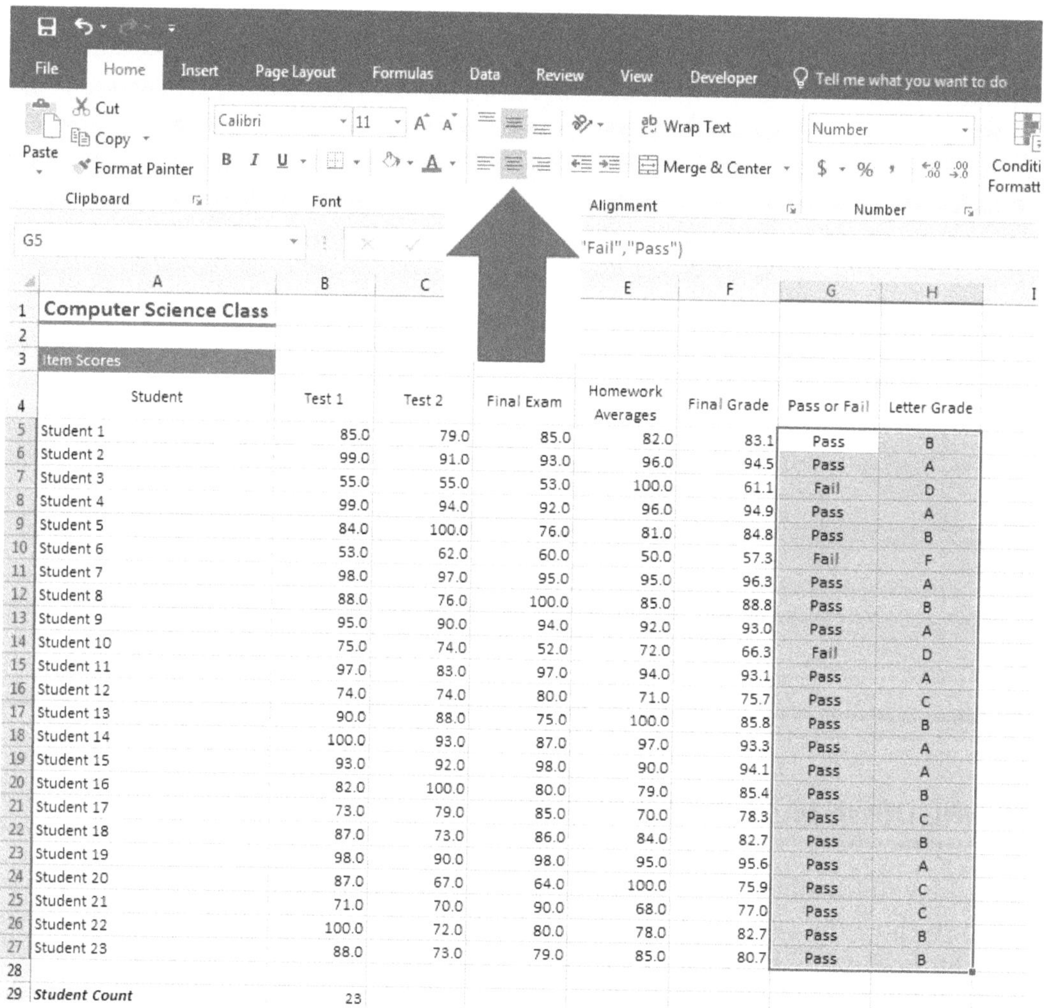

FIGURE 2-20

The reason that we have delayed the filling down of the **Average**, **Median**, **Maximum**, **Minimum**, **Range**, **Variance**, and **Standard Deviation** functions and formulas is because, for them, the process will be a little different. Please do the following steps:

Step 1: Select cells D32 through J32.

Step 2: Click on the **Fill Handle** to the lower right of cell J32. Recall that the **Fill Handle** is a **small black square** on the lower right of the cell, and when you have pointed to it, your mouse will become a small black cross.

Step 3: Click and drag the **Fill Handle** down to Row 36 (Figure 2-21).

Step 4: For appearance purposes only, click the **Add Decimal** icon in the **Home Ribbon** to make sure that all of the numbers have two decimal places. The results should be what you see in Figure 2-21, **but they need to be adjusted.**

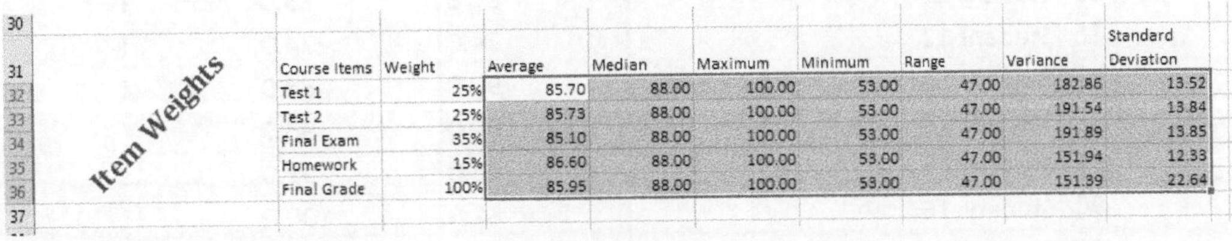

	Course Items	Weight	Average	Median	Maximum	Minimum	Range	Variance	Standard Deviation
Item Weights	Test 1	25%	85.70	88.00	100.00	53.00	47.00	182.86	13.52
	Test 2	25%	85.73	88.00	100.00	53.00	47.00	191.54	13.84
	Final Exam	35%	85.10	88.00	100.00	53.00	47.00	191.89	13.85
	Homework	15%	86.60	88.00	100.00	53.00	47.00	151.94	12.33
	Final Grade	100%	85.95	88.00	100.00	53.00	47.00	151.39	22.64

FIGURE 2-21

This is all fine and good, except that after filling these formulas down, the relative address has not given us the correct cell ranges for these calculations (except for the *Range* formula). Double-click cell D33. NOTE: Doing so puts you into what is called the **Edit Mode** of a cell, which will allow you to see what is called the **Rangefinder**. The **Rangefinder** will allow you to see the ranges you have referenced, and it will allow you to change the ranges of these functions quickly. You can also engage the same **Edit Mode** by pressing the **F2** function key at the top of your keyboard. It will clearly show that the relative address simply changed the cells' row references by one row, but the range in the **AVERAGE** function in cell D33, and in all of these functions, are **not** what is needed (Figure 2-22).

To move the ranges to the right place in order to get the proper cell references in these functions, use the following steps:

Step 1: Once again, make sure you are in the **Edit Mode** of cell D33, and thus make sure that the range is highlighted with the **Rangefinder** (Figure 2-22).

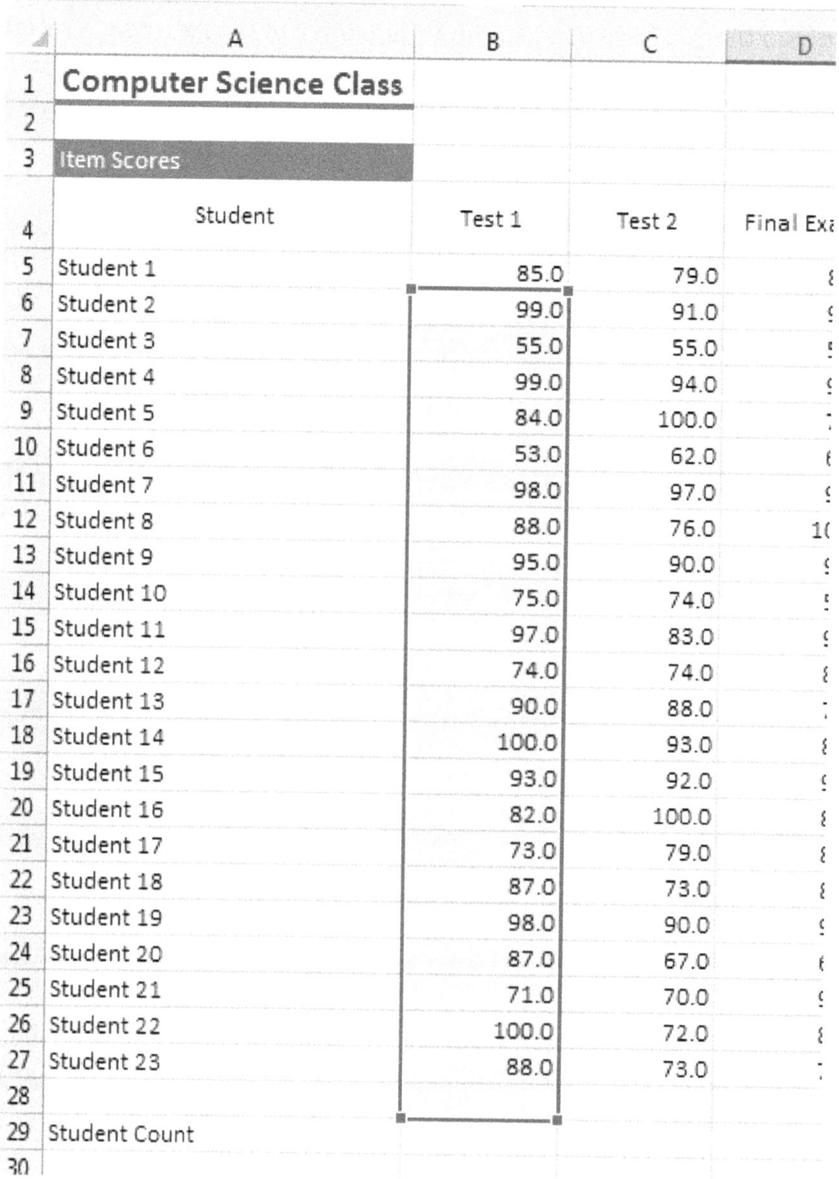

	A	B	C	D
1	**Computer Science Class**			
2				
3	Item Scores			
4	Student	Test 1	Test 2	Final Exa
5	Student 1	85.0	79.0	
6	Student 2	99.0	91.0	
7	Student 3	55.0	55.0	
8	Student 4	99.0	94.0	
9	Student 5	84.0	100.0	
10	Student 6	53.0	62.0	
11	Student 7	98.0	97.0	
12	Student 8	88.0	76.0	
13	Student 9	95.0	90.0	
14	Student 10	75.0	74.0	
15	Student 11	97.0	83.0	
16	Student 12	74.0	74.0	
17	Student 13	90.0	88.0	
18	Student 14	100.0	93.0	
19	Student 15	93.0	92.0	
20	Student 16	82.0	100.0	
21	Student 17	73.0	79.0	
22	Student 18	87.0	73.0	
23	Student 19	98.0	90.0	
24	Student 20	87.0	67.0	
25	Student 21	71.0	70.0	
26	Student 22	100.0	72.0	
27	Student 23	88.0	73.0	
28				
29	Student Count			
30				

FIGURE 2-22

Step 2: Click and drag the colored border of that range one (1) column to the right and one (1) row up until it is directly over the cell range of C5 through C27 (Figure 2-23). Please notice that when you do this, the cell range in the **AVERAGE** function in cell D33 will now have cells C5 through C27 in their range argument.

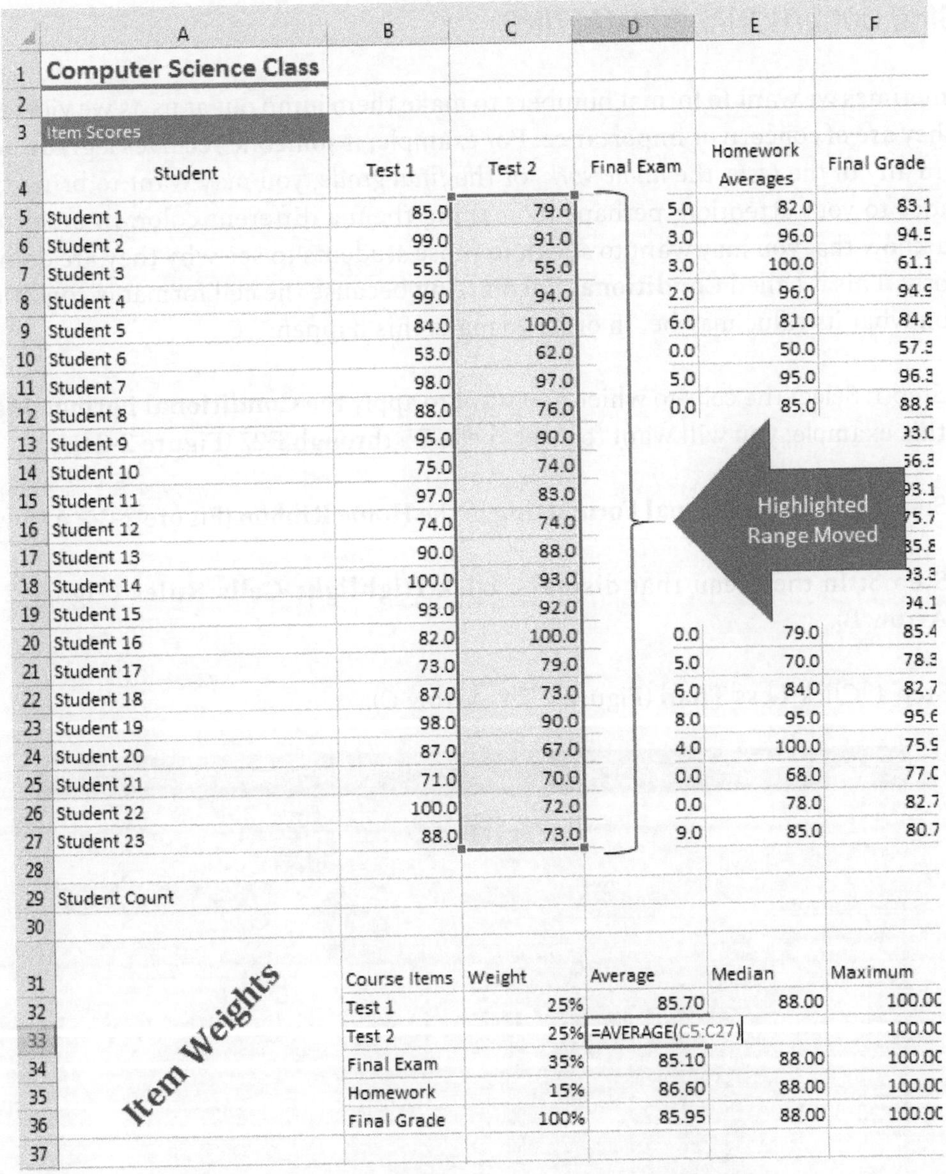

	A	B	C	D	E	F
1	**Computer Science Class**					
2						
3	Item Scores					
4	Student	Test 1	Test 2	Final Exam	Homework Averages	Final Grade
5	Student 1	85.0	79.0	5.0	82.0	83.1
6	Student 2	99.0	91.0	3.0	96.0	94.5
7	Student 3	55.0	55.0	3.0	100.0	61.1
8	Student 4	99.0	94.0	2.0	96.0	94.9
9	Student 5	84.0	100.0	6.0	81.0	84.8
10	Student 6	53.0	62.0	0.0	50.0	57.3
11	Student 7	98.0	97.0	5.0	95.0	96.3
12	Student 8	88.0	76.0	0.0	85.0	88.8
13	Student 9	95.0	90.0			93.0
14	Student 10	75.0	74.0			56.3
15	Student 11	97.0	83.0			93.1
16	Student 12	74.0	74.0			75.7
17	Student 13	90.0	88.0			85.8
18	Student 14	100.0	93.0			93.3
19	Student 15	93.0	92.0			94.1
20	Student 16	82.0	100.0	0.0	79.0	85.4
21	Student 17	73.0	79.0	5.0	70.0	78.3
22	Student 18	87.0	73.0	6.0	84.0	82.7
23	Student 19	98.0	90.0	8.0	95.0	95.6
24	Student 20	87.0	67.0	4.0	100.0	75.9
25	Student 21	71.0	70.0	0.0	68.0	77.0
26	Student 22	100.0	72.0	0.0	78.0	82.7
27	Student 23	88.0	73.0	9.0	85.0	80.7
28						
29	Student Count					
30						
31		Course Items	Weight	Average	Median	Maximum
32		Test 1	25%	85.70	88.00	100.00
33		Test 2	25%	=AVERAGE(C5:C27)		100.00
34		Final Exam	35%	85.10	88.00	100.00
35		Homework	15%	86.60	88.00	100.00
36		Final Grade	100%	85.95	88.00	100.00
37						

FIGURE 2-23

Repeat this process to find the **Average** for the *Final Exam* and then the *Homework Averages*. Then repeat the process to find the **Median**, **Maximum**, **Minimum**, **Variance**, and **Standard Deviation** for *Test 1*, *Test 2*, *Final Exam*, and *Home Averages*. It is not necessary to change the cell references for the **Range** calculation because a function range (or argument) was not used in that formula and the **Relative Address** will have given us the cells' row references needed.

USING CONDITIONAL FORMATTING

Sometimes we want to format numbers to make them jump out at us as we view them if they are of concern or importance. For example, if someone receives a grade below 70 in any of the *tests*, the *homework*, or the *final grade*, you may want to bring these grades to your attention, perhaps by making them a different color, in order to let you know that you may want to speak to those students to see why they are doing so poorly. This is called **Conditional Formatting** because the cell format is conditional upon what its value may be. In order to make this happen:

Step 1: Select the cells to which you want to apply the **Conditional Formatting**. In this example, you will want to select cells B5 through F27 (Figure 2-24).

Step 2: Click **Conditional Formatting** in the **Home Ribbon** (Figure 2-24, Arrow A).

Step 3: In the menu that displays, click **Highlight Cells Rules** (Figure 2-24, Arrow B).

Step 4: Click **Less Than** (Figure 2-24, Arrow C).

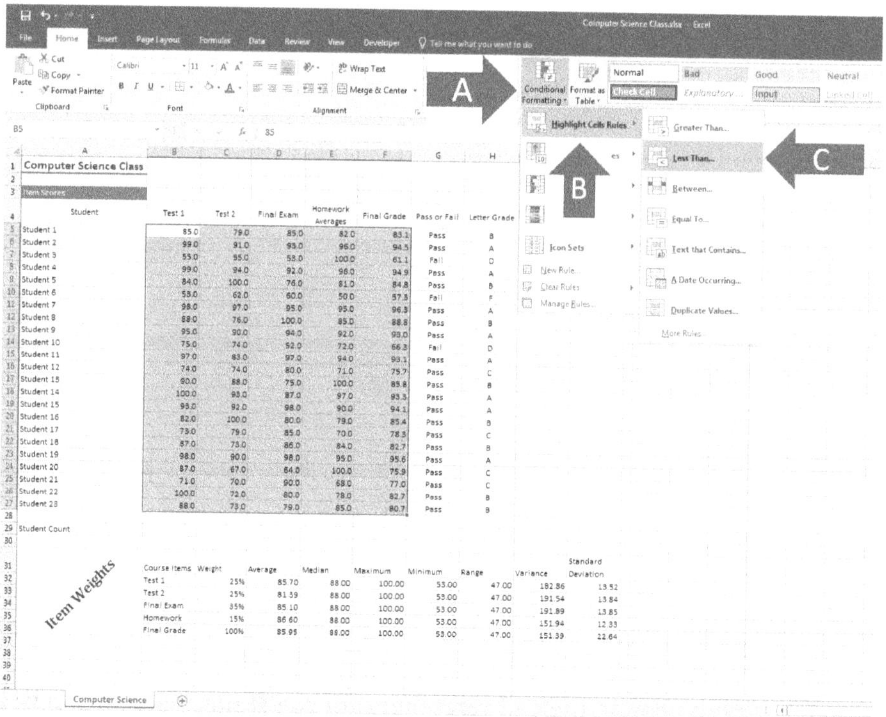

FIGURE 2-24

Step 5: A pop-up menu will then be displayed (Figure 2-25). When it does, type *70* in the box labeled **Format cells that are LESS THAN:** (Arrow A).

Step 6: Click the **with box** (Figure 2-25, Arrow B). It will show you that the **Light Red Fill with Dark Red Text** is the first choice of the preset color schemes (and it is the default setting), and it will be applied to the selected cells if you click **OK.** The cells will turn to that color scheme, as you can see in Figure 2-25.

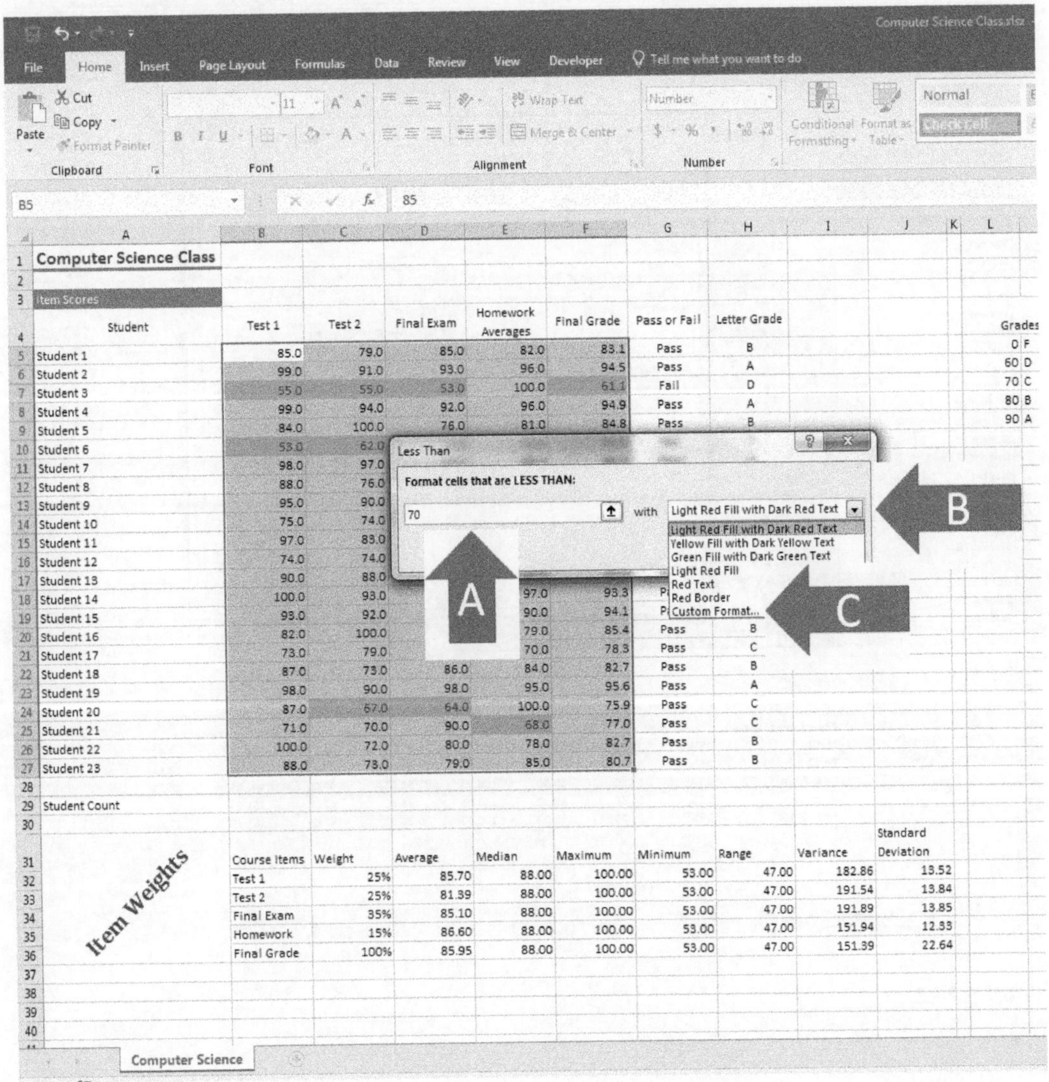

FIGURE 2-25

NOTE: You can use any color combination you want in this box, or you can customize the colors you want by clicking **Custom Format** (Figure 2-25, Arrow C). That would give you a pop-up menu (Figure 2-26) that will allow you to choose **ANY font color** and **fill** (or cell background) that you want that is not in the standard list.

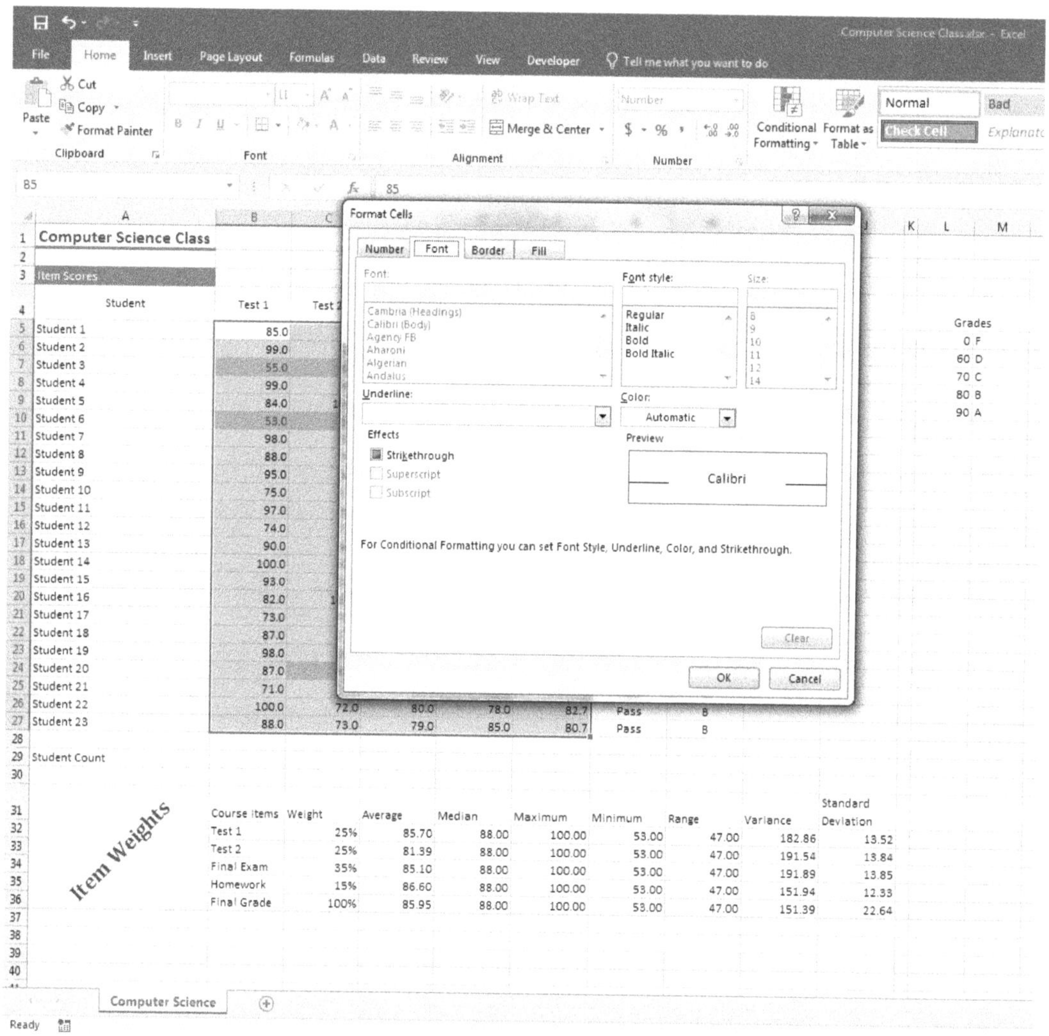

FIGURE 2-26

TO COUNT THE NUMBER OF ENTRIES

Counting the different number of students or any observations is very time consuming. If you want to know how many lines of data you have, or in this case if you want to know how many students there are in this class, **without** having to physically count them, you can do the following:

Step 1: Click on cell A29 and type the words *Student Count* (and make those words **bold** and ***italic***).

Step 2: In cell B29, type the following: =*COUNTA(A5:A27)*.

See the results in Figure 2-27.

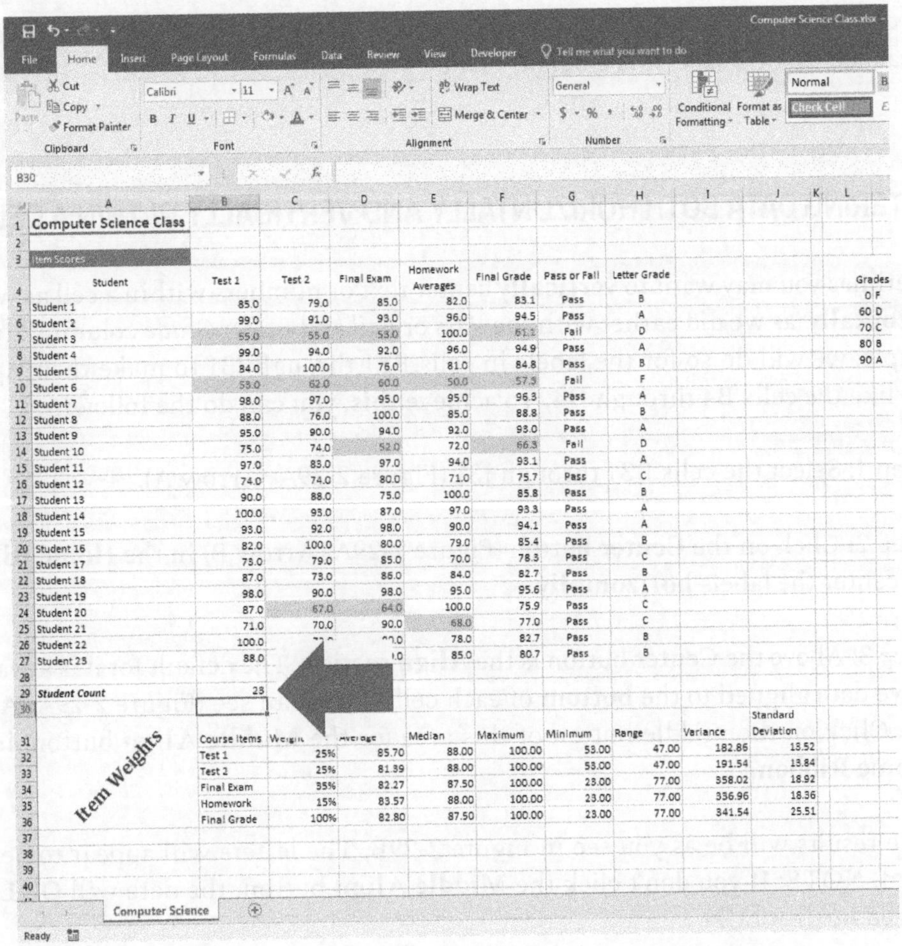

FIGURE 2-27

NOTE: If you want to count cells that do **NOT** have numeric values (or cells with text and alphanumeric data), you must put the **A** at the end of the **COUNT** function as **you** just did. If you had entered =*COUNT(A5:A27)*, you would get a zero because the **COUNT** function counts only **numeric values.**

TO CHANGE ROW HEIGHTS

In this example, we want to make row 30 taller to separate the Item Weights area from the other figures. To make the rows a different size, do the following:

Step 1: Place your mouse at the bottom of the cell (at the **row label separator**) (Figure 2-28, Arrow A).

Step 2: Click and drag the mouse downward until the **row height** you desire is reached (in this case, it is a height of *30*).

CENTERING DATA BOTH HORIZONTALLY AND VERTICALLY WITHIN A CELL

Sometimes you may want to **vertically** center text or numbers within a cell as well as **horizontally** (as we did earlier with the *Pass* or *Fail* and *Letter Grade* columns). In this example, we will do so for the labels in cells B31 through J31 to make them appear more like the cells B4 through H4. To achieve this, you can do the following:

Step 1: Select the cells B31 through J31 (Figure 2-29A, Arrow A).

Step 2: Click on the **Center** button (Figure 2-29A, Arrow B) in the **Home Ribbon** to center the labels **horizontally**.

Step 3: Above the **Center** button is the **Align** section. The default for it is always to have data aligned to the **bottom** of each cell as you can see (Figure 2-29A, Arrow C). Click on the middle button of this area (or the **Middle Align** button) in the **Home Ribbon.**

The results will be as you see in Figure 2-29B. The letters will appear to be suspended. NOTE: If you don't click the **Middle Align** button, the data will **ONLY** be centered **horizontally**.

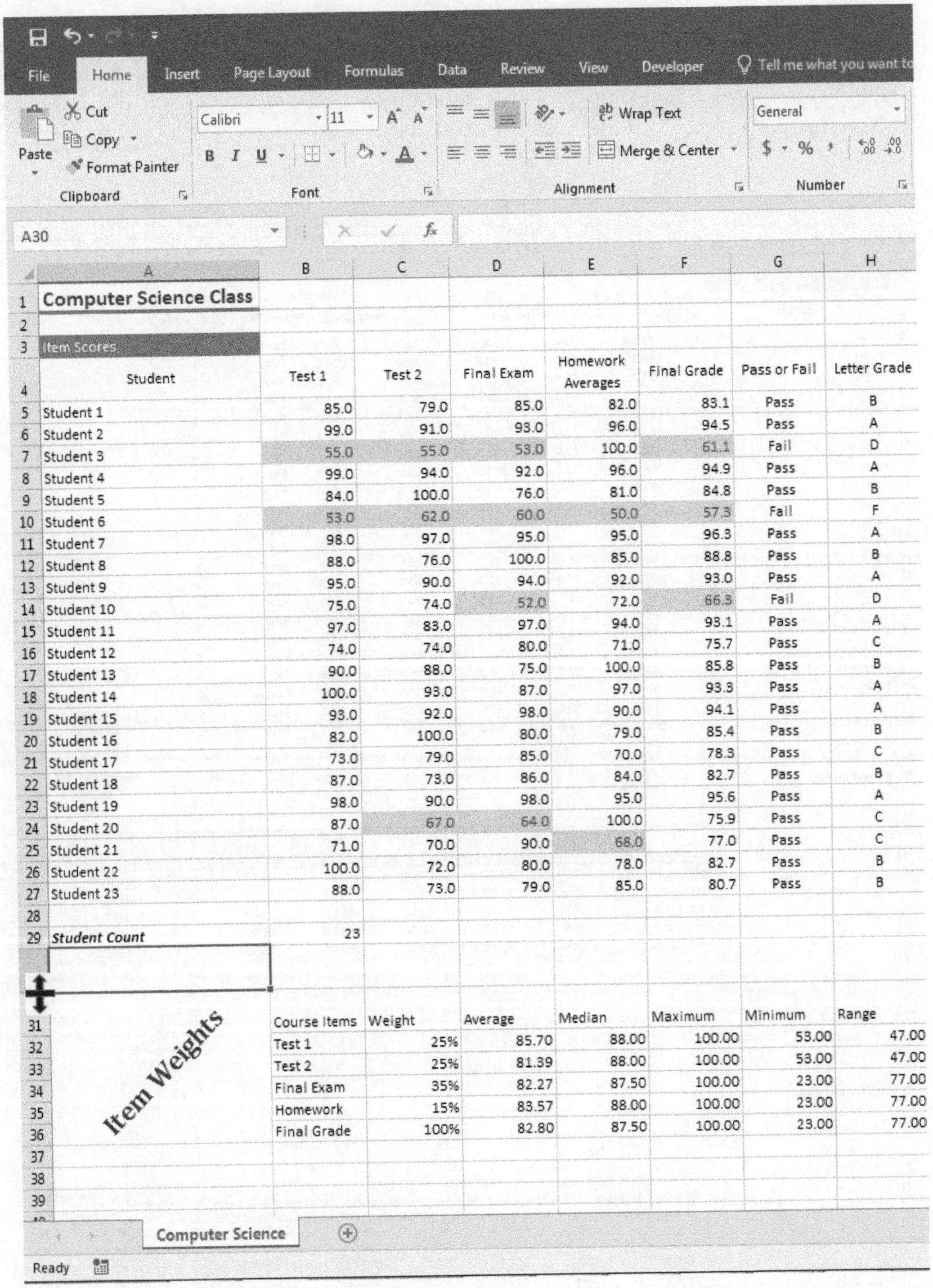

	A	B	C	D	E	F	G	H
1	**Computer Science Class**							
2								
3	Item Scores							
4	Student	Test 1	Test 2	Final Exam	Homework Averages	Final Grade	Pass or Fail	Letter Grade
5	Student 1	85.0	79.0	85.0	82.0	83.1	Pass	B
6	Student 2	99.0	91.0	93.0	96.0	94.5	Pass	A
7	Student 3	55.0	55.0	53.0	100.0	61.1	Fail	D
8	Student 4	99.0	94.0	92.0	96.0	94.9	Pass	A
9	Student 5	84.0	100.0	76.0	81.0	84.8	Pass	B
10	Student 6	53.0	62.0	60.0	50.0	57.3	Fail	F
11	Student 7	98.0	97.0	95.0	95.0	96.3	Pass	A
12	Student 8	88.0	76.0	100.0	85.0	88.8	Pass	B
13	Student 9	95.0	90.0	94.0	92.0	93.0	Pass	A
14	Student 10	75.0	74.0	52.0	72.0	66.3	Fail	D
15	Student 11	97.0	83.0	97.0	94.0	93.1	Pass	A
16	Student 12	74.0	74.0	80.0	71.0	75.7	Pass	C
17	Student 13	90.0	88.0	75.0	100.0	85.8	Pass	B
18	Student 14	100.0	93.0	87.0	97.0	93.3	Pass	A
19	Student 15	93.0	92.0	98.0	90.0	94.1	Pass	A
20	Student 16	82.0	100.0	80.0	79.0	85.4	Pass	B
21	Student 17	73.0	79.0	85.0	70.0	78.3	Pass	C
22	Student 18	87.0	73.0	86.0	84.0	82.7	Pass	B
23	Student 19	98.0	90.0	98.0	95.0	95.6	Pass	A
24	Student 20	87.0	67.0	64.0	100.0	75.9	Pass	C
25	Student 21	71.0	70.0	90.0	68.0	77.0	Pass	C
26	Student 22	100.0	72.0	80.0	78.0	82.7	Pass	B
27	Student 23	88.0	73.0	79.0	85.0	80.7	Pass	B
28								
29	***Student Count***	23						

	A	B	C	D	E	F	G	H
31	Item Weights	Course Items	Weight	Average	Median	Maximum	Minimum	Range
32		Test 1	25%	85.70	88.00	100.00	53.00	47.00
33		Test 2	25%	81.39	88.00	100.00	53.00	47.00
34		Final Exam	35%	82.27	87.50	100.00	23.00	77.00
35		Homework	15%	83.57	88.00	100.00	23.00	77.00
36		Final Grade	100%	82.80	87.50	100.00	23.00	77.00

FIGURE 2-28

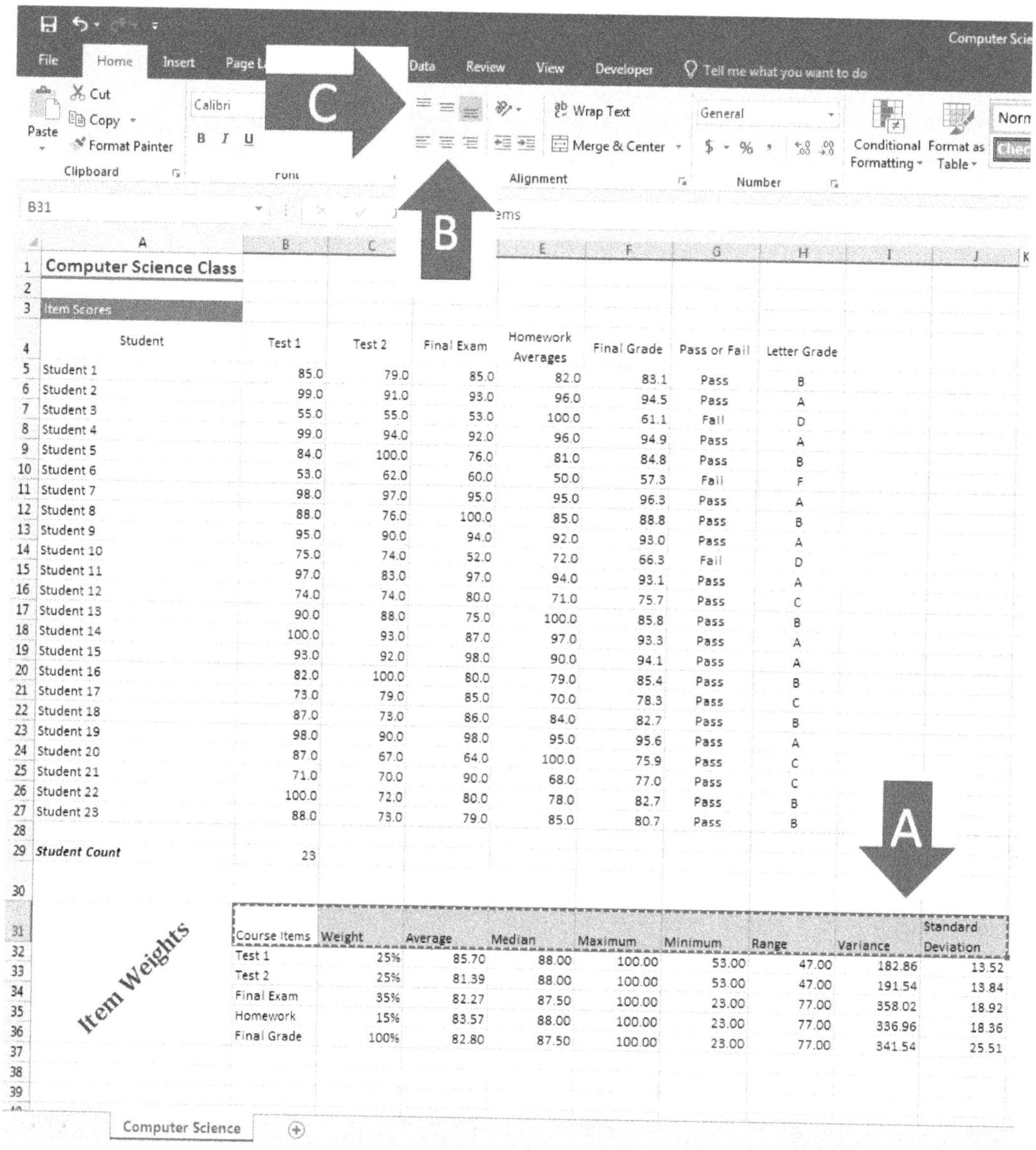

FIGURE 2-29A

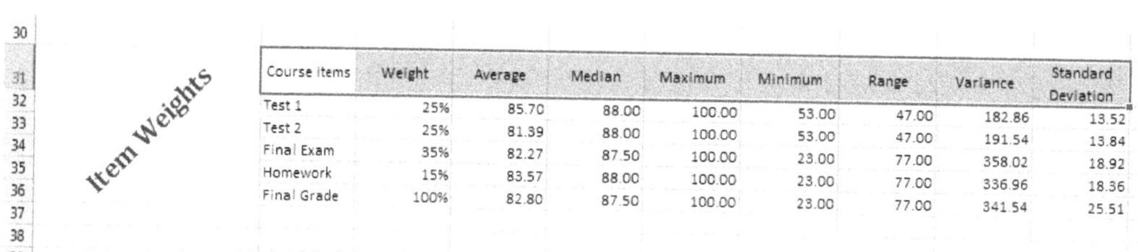

FIGURE 2-29B

ADDING A HEADER OR FOOTER

Headers and **Footers** can show things such as a **page number**, the **name** of the sheet you are printing, the **date** it was last updated, and more. NOTE: When you want to see your **Headers** or **Footers**, you must change the view of your spreadsheet to **Layout View** or **Print Preview.** All you have to do is decide whether you want the text in the top, bottom, right, left, or centered. There are a couple of ways to do so, but you can use one of those methods by doing the following steps:

Step 1: Click the **Insert Ribbon** (Figure 2-30, Arrow A).

Step 2: Click **Header & Footer** (Figure 2-30, Arrow B).

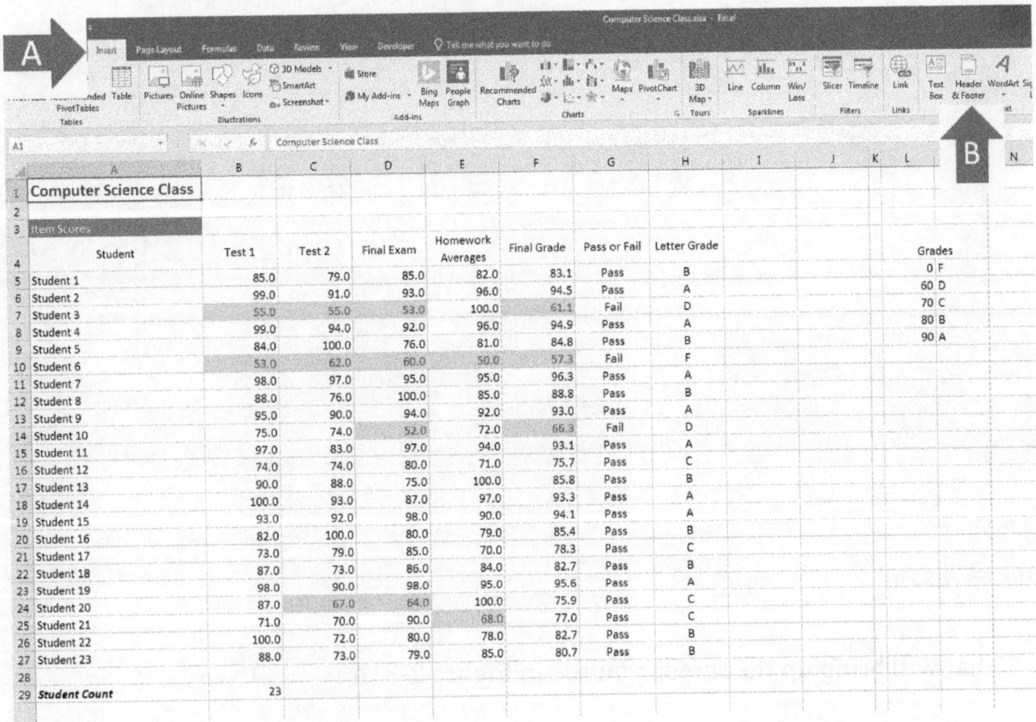

	Student	Test 1	Test 2	Final Exam	Homework Averages	Final Grade	Pass or Fail	Letter Grade
5	Student 1	85.0	79.0	85.0	82.0	83.1	Pass	B
6	Student 2	99.0	91.0	93.0	96.0	94.5	Pass	A
7	Student 3	55.0	55.0	53.0	100.0	61.1	Fail	D
8	Student 4	99.0	94.0	92.0	96.0	94.9	Pass	A
9	Student 5	84.0	100.0	76.0	81.0	84.8	Pass	B
10	Student 6	53.0	62.0	60.0	50.0	57.3	Fail	F
11	Student 7	98.0	97.0	95.0	95.0	96.3	Pass	A
12	Student 8	88.0	76.0	100.0	85.0	88.8	Pass	B
13	Student 9	95.0	90.0	94.0	92.0	93.0	Pass	A
14	Student 10	75.0	74.0	52.0	72.0	66.3	Fail	D
15	Student 11	97.0	83.0	97.0	94.0	93.1	Pass	A
16	Student 12	74.0	74.0	80.0	71.0	75.7	Pass	C
17	Student 13	90.0	88.0	75.0	100.0	85.8	Pass	B
18	Student 14	100.0	93.0	87.0	97.0	93.3	Pass	A
19	Student 15	93.0	92.0	98.0	90.0	94.1	Pass	A
20	Student 16	82.0	100.0	80.0	79.0	85.4	Pass	B
21	Student 17	73.0	79.0	85.0	70.0	78.3	Pass	C
22	Student 18	87.0	73.0	86.0	84.0	82.7	Pass	B
23	Student 19	98.0	90.0	98.0	95.0	95.6	Pass	A
24	Student 20	87.0	67.0	64.0	100.0	75.9	Pass	C
25	Student 21	71.0	70.0	90.0	68.0	77.0	Pass	C
26	Student 22	100.0	72.0	80.0	78.0	82.7	Pass	B
27	Student 23	88.0	73.0	79.0	85.0	80.7	Pass	B
28								
29	***Student Count***	23						

Grades	
0	F
60	D
70	C
80	B
90	A

FIGURE 2-30

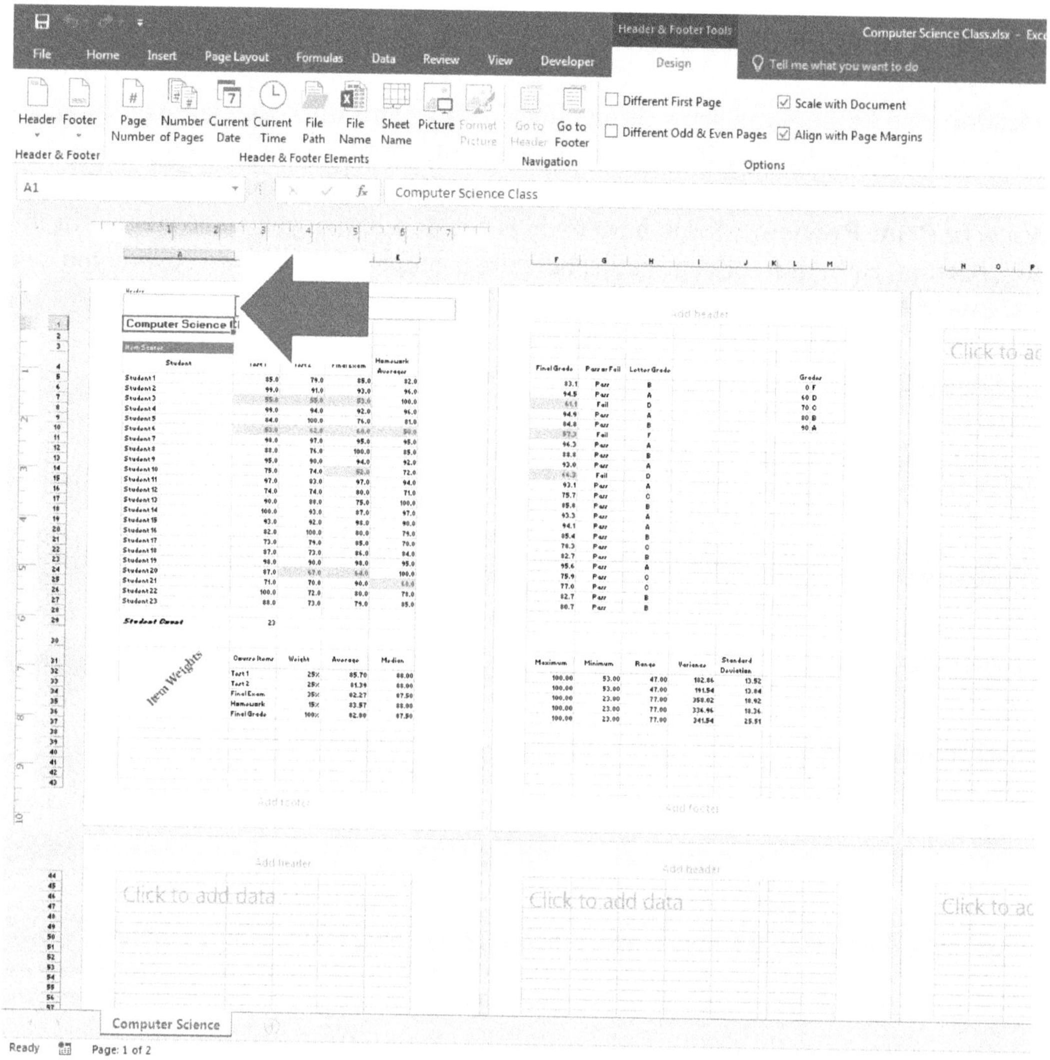

FIGURE 2-31A

That will bring up the screen you see in Figure 2-31A.

Step 3: You can click any one of the boxes (**left**, **right**, or **center**) and enter information for your header. You can type your name, anything about you, or the content of the spreadsheets. If you wish, you can click **Page Number**, **Number of Pages**, **Current Date**, **Current Time**, or any of the things shown in the **Header & Footer Tools** in the upper left of Figure 2-31A, and they will be displayed in the box you have selected.

Step 4: Press any key to get out of that box (such as the **Tab key**).

In this example, we will enter the date in the left of the **Header** (Figure 2-31A, Arrow).

Step 3: You can repeat the process by going to any one of the other boxes in the **center** or right. You can also click **Go to Footer** in that same area and do the same to the **Footer.**

Please notice all of the many things you can put into a **Header or Footer**, such as the **Current Time** that the spreadsheet was last printed, the **File Path**, the **Sheet Name**, and even pictures, such as your organization logo or whatever you wish. If you add the **Date** and **Time**, it will display the **date** and **time** that the spreadsheet was last printed.

To exit the **Header & Footer Tools,** simply click in any cell in the spreadsheet.

When you add a **header** or a **footer**, Excel will change to what is called the **Page Layout View**. This shows you what your printout would look like and the things that would show on printouts. To return to the **Normal View**, which will hide the **Header and Footer**, click the **Normal View** button in the lower right of your screen (Figure 2-31B, Arrow).

NOTE: Whenever you go to the **Page Layout View** or after you have been in the **Print Preview** (discussed later), Excel will put dotted lines around your spreadsheet. These don't hurt anything, and they don't mean that anything is wrong. They are only there to show you where the page breaks would be if you decided to print the spreadsheet at that time.

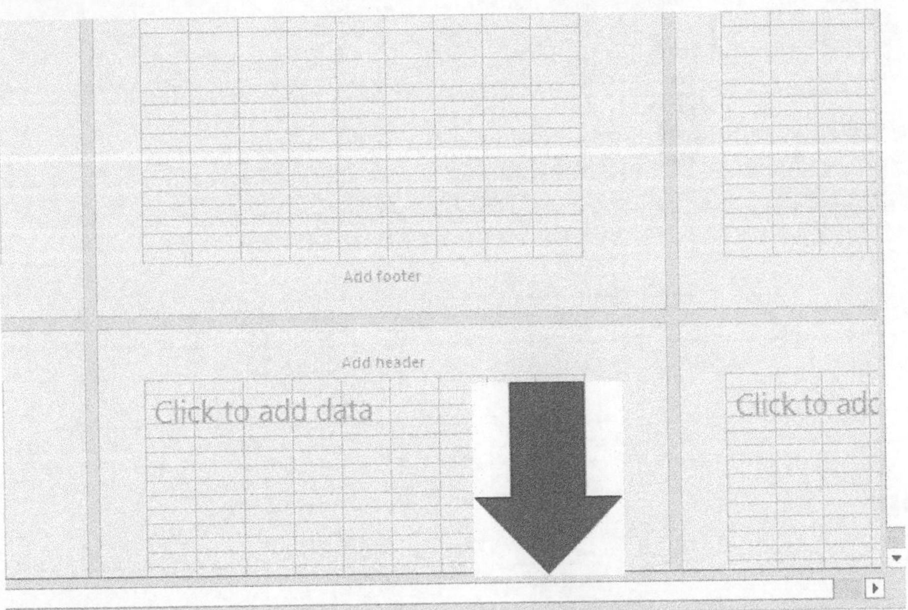

FIGURE 2-31B

THEMES

Sometimes users of Excel prefer to use a different format for all of the elements of their spreadsheet, including the **font** of the **row** and **column headings**. While this is not required in this assignment, the procedure for changing your spreadsheet **Theme** is something good for you to know. You can change it by doing the following steps:

Step 1: Click the **Page Layout Ribbon** (Figure 2-32, Arrow A). Notice the default **Theme** is **Office** (Figure 2-32, Arrow B), which is what most people use.

Step 2: Click the **Depth Theme** in the lower right of the **Theme** menu (Figure 2-32, Arrow C). You will notice that the spreadsheet numbers and labels will change. You will also see a difference in the **column headings** and **row headings**. If you would like a different one, repeat this process and choose another.

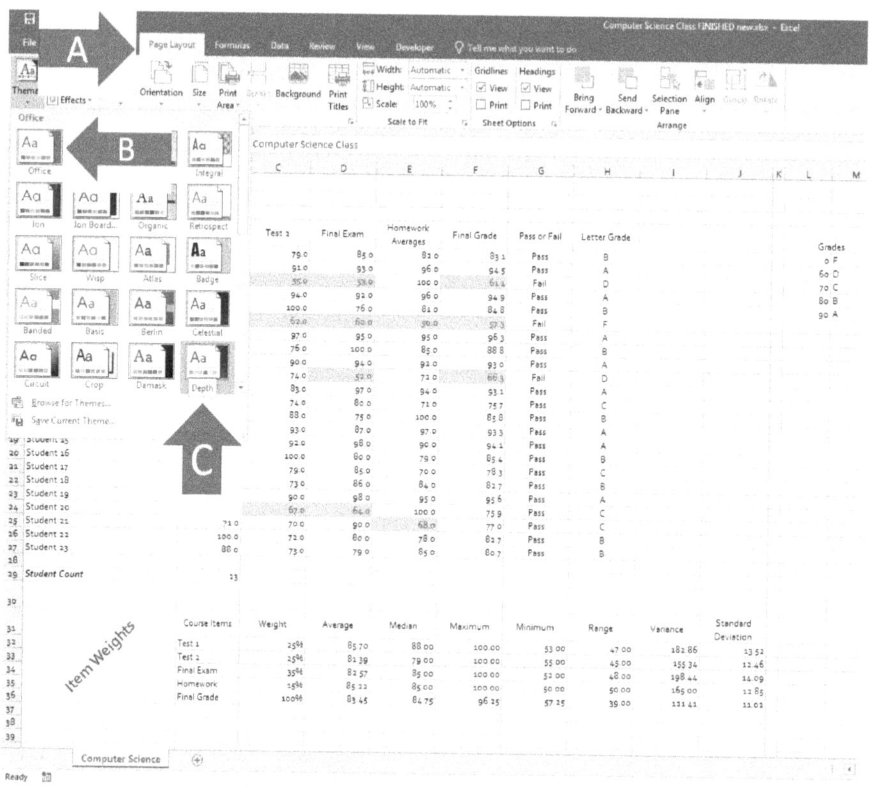

FIGURE 2-32

CHAPTER 2—ASSIGNMENT (SUMMARY OF TASKS)

1. Open the file in Excel called *Computer Science Class* that is associated with this book.
2. In cell F5, calculate the *Final Grade* score for *Student 1* using the weights given in cells C32 through C35. Make sure that it adds the products of each test score times their weights. In other words, make the cell F5 calculate the *Test 1 score times the Test 1 weight + Test 2 score times the Test 2 weight + Final Exam score times the Final Exam weight + Homework Average times the Homework weight.*

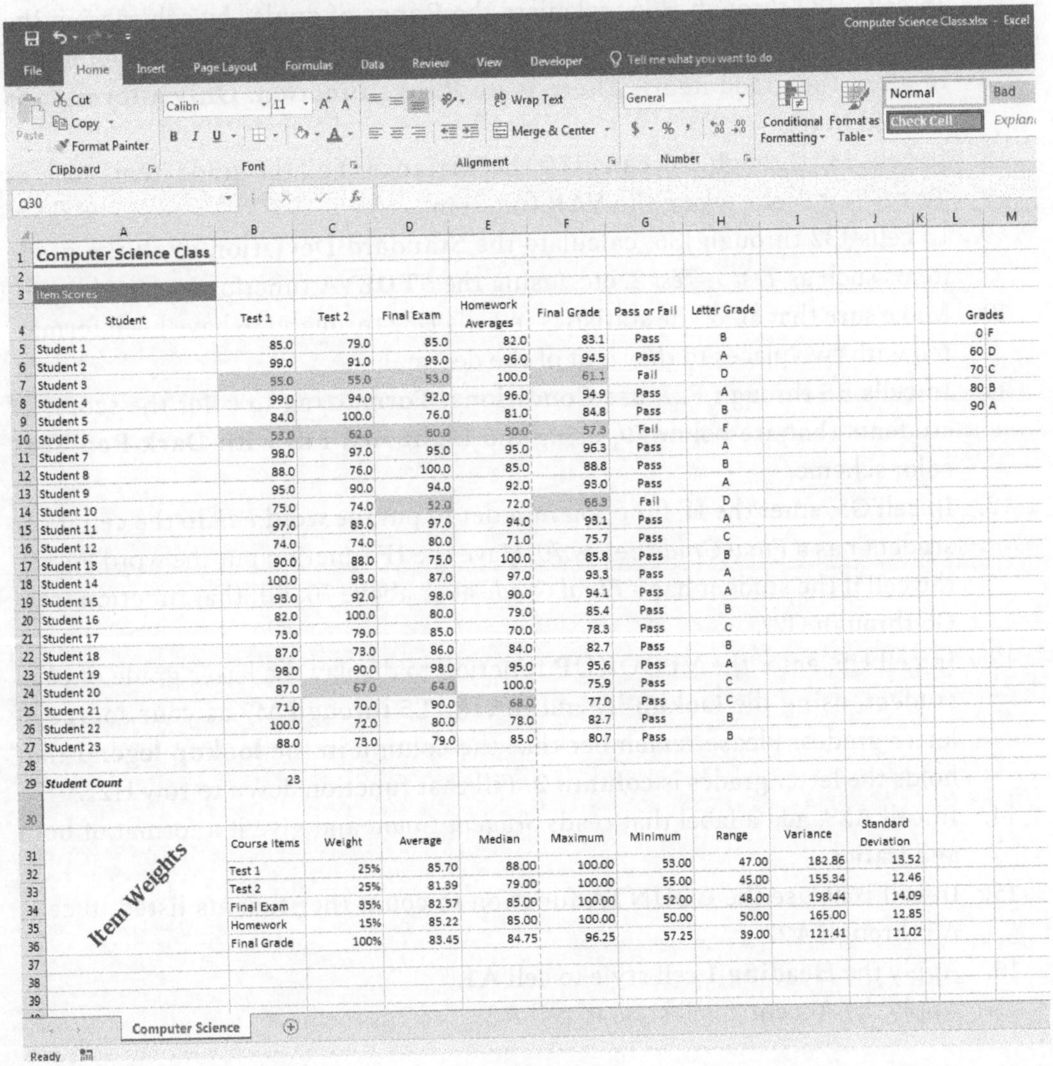

Computer Science Class

Item Scores

Student	Test 1	Test 2	Final Exam	Homework Averages	Final Grade	Pass or Fail	Letter Grade
Student 1	85.0	79.0	85.0	82.0	83.1	Pass	B
Student 2	99.0	91.0	93.0	96.0	94.5	Pass	A
Student 3	55.0	55.0	53.0	100.0	61.1	Fail	D
Student 4	99.0	94.0	92.0	96.0	94.9	Pass	A
Student 5	84.0	100.0	76.0	81.0	84.8	Pass	B
Student 6	53.0	62.0	60.0	50.0	57.3	Fail	F
Student 7	98.0	97.0	95.0	95.0	96.3	Pass	A
Student 8	88.0	76.0	100.0	85.0	88.8	Pass	B
Student 9	95.0	90.0	94.0	92.0	93.0	Pass	A
Student 10	75.0	74.0	52.0	72.0	66.3	Fail	D
Student 11	97.0	83.0	97.0	94.0	93.1	Pass	A
Student 12	74.0	74.0	80.0	71.0	75.7	Pass	C
Student 13	90.0	88.0	75.0	100.0	85.8	Pass	B
Student 14	100.0	93.0	87.0	97.0	93.3	Pass	A
Student 15	93.0	92.0	98.0	90.0	94.1	Pass	A
Student 16	82.0	100.0	80.0	79.0	85.4	Pass	B
Student 17	73.0	79.0	85.0	70.0	78.3	Pass	C
Student 18	87.0	73.0	86.0	84.0	82.7	Pass	B
Student 19	98.0	90.0	98.0	95.0	95.6	Pass	A
Student 20	87.0	67.0	64.0	100.0	75.9	Pass	C
Student 21	71.0	70.0	90.0	68.0	77.0	Pass	C
Student 22	100.0	72.0	80.0	78.0	82.7	Pass	B
Student 23	88.0	73.0	79.0	85.0	80.7	Pass	B
Student Count	23						

Grades	
0	F
60	D
70	C
80	B
90	A

Item Weights

Course Items	Weight	Average	Median	Maximum	Minimum	Range	Variance	Standard Deviation
Test 1	25%	85.70	88.00	100.00	53.00	47.00	182.86	13.52
Test 2	25%	81.39	79.00	100.00	55.00	45.00	155.34	12.46
Final Exam	35%	82.57	85.00	100.00	52.00	48.00	198.44	14.09
Homework	15%	85.22	85.00	100.00	50.00	50.00	165.00	12.85
Final Grade	100%	83.45	84.75	96.25	57.25	39.00	121.41	11.02

FIGURE 2-33

Be sure to use the **Absolute Address** when it is appropriate in order to be able to fill the formula down without errors. Fill the formula down to row 27.

3. In the cells D32 through D36, calculate the **Average** of all the grade items such as *Test 1*, *Test 2*, etc. (using the **AVERAGE** function).
4. In cells E32 through E36, calculate the **Median** of all the grade items such as *Test 1*, *Test 2*, etc. (using the **MEDIAN** function).
5. In cells F32 through F36, calculate the **Maximum** of all the grade items such as *Test 1*, *Test 2*, etc. (using the **MAX** function).
6. In cells G32 through G36, calculate the **Minimum** of all the grade items such as *Test 1*, *Test 2*, etc. (using the **MIN** function).
7. In cells H32 through H36, calculate the **Range** of grades by subtracting the minimum grades (in column G) from the maximum grades (in column F) of all of the grade items. **There is no Range function. Only a formula is needed here.**
8. In cells I32 through I36, calculate the **Variance** of all the grade items such as *Test 1*, *Test 2*, etc. (using the **VAR** function).
9. In cells J32 through J36, calculate the **Standard Deviation** of all the grade items such as *Test 1*, *Test 2*, etc. (using the **STDEVA** function).
10. Make sure that all of the statistics in cells D32 through J36 have been formatted with two places to the right of the decimal.
11. In cells B5 through F27, use **Conditional Formatting** to color the scores of students that are below *70*, using the **Light Red Fill with Dark Red Text** color scheme.
12. In cell G5, enter the **IF** function in order to put the word *Fail* in the cell if the student has a *Final Grade* below *70*. Have the **IF** function put the word *Pass* in the cell if the student has a *Final Grade* at or above 70. Fill that function from G5 through G27.
13. In cell H5, enter the **VLOOKUP** function to display the letter grade of each student, using the **lookup legend** in cells L5 through M9 as your source of letter grades. Please remember that the column in the **lookup legend** that holds the letter grades is **column 2**. Fill that function down to row H27.
14. In cell A29, add a label that reads *Student Count* and give it a format of **bold** and **italics**.
15. In cell B29, use the **COUNTA** function to count the students listed in cells A5 through A27.
16. Apply the **Heading 1** cell style to cell **A1**.
17. Apply the **Accent 1** cell style to cell **A3**.

18. Apply the **Title** cell style to cell **A31**. NOTE: Your computer may not have the same appearance or styles shown in this text. If so, choose the color scheme closest to what is requested.
19. **Merge and center** cells A31 through A36 and then rotate the text in that cell by using the **Rotate Counterclockwise** setting in the **Rotate** icon in the **Home Ribbon**.
20. Save the file and present it for grading.

CHAPTER 2—ALTERNATE ASSIGNMENT (SUMMARY OF TASKS)

Herzog Manufacturing uses the Moving Average method of forecasting sales for the upcoming month. It works by using a weighted average of sales from the last four months. The probability that the sales from four months ago will be a good indicator of sales for the upcoming month is rather low because many things can happen in four months. That's why the probability that sales from three months ago will be a good predictor is higher than that of four months ago. The same is true for sales from two months ago and from the most recent month. The task here is to calculate a forecast for next month based upon given probabilities.

1. Open the file in Excel called *Sales Forecast* that is associated with this book. Please notice that the probabilities for each month are given in cells C32 through C35.
2. In cell F5, calculate the *Next Month Forecast* for *Product 1* using the weights given in cells C32 through C35. Make sure that it adds the products of each test score times their weights. In other words, make the cell F5 calculate the <u>*Product 1 Sales 4 Months Ago in Thousands times the 4 Months Ago weight + Product 1 Sales 3 Months Ago in Thousands times the 3 Months Ago weight + Product 1 Sales 2 Months Ago in Thousands times the 2 Months Ago weight + Product 1 Sales Last Month in Thousands times the Last Month weight*</u>. Be sure to use the **Absolute Address** when appropriate in order to be able to fill the formula down. Fill the formula down to row 27.
3. Apply the **Accounting** format to cells F5 through F27.
4. In cells D32 through D36, calculate the **Average** of all the sales for each month such as *Sales 4 Months Ago in Thousands*, *Sales 3 Months Ago in Thousands*, etc. (using the **AVERAGE** function).
5. In cells E32 through E36, calculate the **Median** of all the sales for each month such as *Sales 4 Months Ago in Thousands*, *Sales 3 Months Ago in Thousands*, etc. (using the **MEDIAN** function).

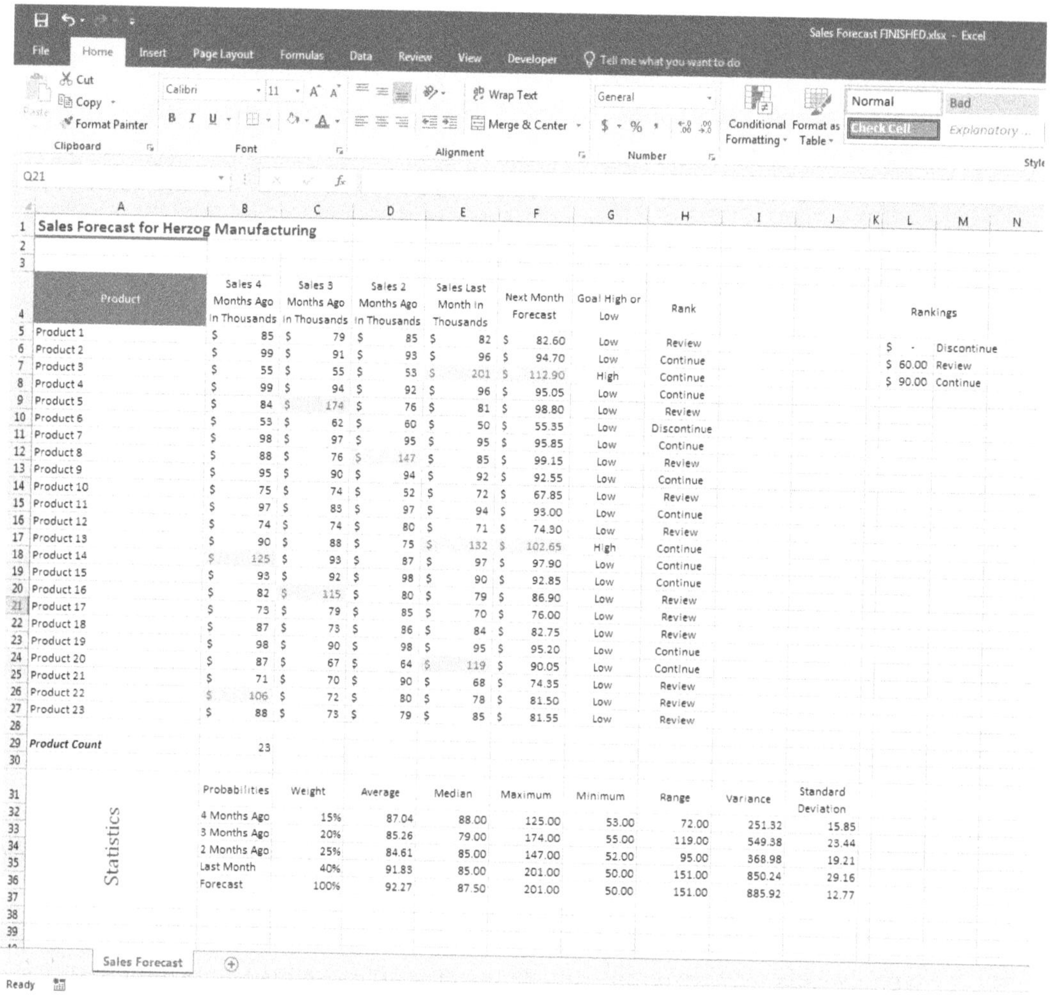

Sales Forecast for Herzog Manufacturing

Product	Sales 4 Months Ago in Thousands	Sales 3 Months Ago in Thousands	Sales 2 Months Ago in Thousands	Sales Last Month in Thousands	Next Month Forecast	Goal High or Low	Rank
Product 1	$ 85	$ 79	$ 85	$ 82	$ 82.60	Low	Review
Product 2	$ 99	$ 91	$ 93	$ 96	$ 94.70	Low	Continue
Product 3	$ 55	$ 55	$ 53	$ 201	$ 112.90	High	Continue
Product 4	$ 99	$ 94	$ 92	$ 96	$ 95.05	Low	Continue
Product 5	$ 84	$ 174	$ 76	$ 81	$ 98.80	Low	Review
Product 6	$ 53	$ 62	$ 60	$ 50	$ 55.35	Low	Discontinue
Product 7	$ 98	$ 97	$ 95	$ 95	$ 95.85	Low	Continue
Product 8	$ 88	$ 76	$ 147	$ 85	$ 99.15	Low	Review
Product 9	$ 95	$ 90	$ 94	$ 92	$ 92.55	Low	Continue
Product 10	$ 75	$ 74	$ 52	$ 72	$ 67.85	Low	Review
Product 11	$ 97	$ 83	$ 97	$ 94	$ 93.00	Low	Continue
Product 12	$ 74	$ 74	$ 80	$ 71	$ 74.30	Low	Review
Product 13	$ 90	$ 88	$ 75	$ 132	$ 102.65	High	Continue
Product 14	$ 125	$ 93	$ 87	$ 97	$ 97.90	Low	Continue
Product 15	$ 93	$ 92	$ 98	$ 90	$ 92.85	Low	Continue
Product 16	$ 82	$ 115	$ 80	$ 79	$ 86.90	Low	Review
Product 17	$ 73	$ 79	$ 85	$ 70	$ 76.00	Low	Review
Product 18	$ 87	$ 73	$ 86	$ 84	$ 82.75	Low	Review
Product 19	$ 98	$ 90	$ 98	$ 95	$ 95.20	Low	Continue
Product 20	$ 87	$ 67	$ 64	$ 119	$ 90.05	Low	Continue
Product 21	$ 71	$ 70	$ 90	$ 68	$ 74.35	Low	Review
Product 22	$ 106	$ 72	$ 80	$ 78	$ 81.50	Low	Review
Product 23	$ 88	$ 73	$ 79	$ 85	$ 81.55	Low	Review
Product Count	23						

Statistics

Probabilities	Weight	Average	Median	Maximum	Minimum	Range	Variance	Standard Deviation
4 Months Ago	15%	87.04	88.00	125.00	53.00	72.00	251.32	15.85
3 Months Ago	20%	85.26	79.00	174.00	55.00	119.00	549.38	23.44
2 Months Ago	25%	84.61	85.00	147.00	52.00	95.00	368.98	19.21
Last Month	40%	91.83	85.00	201.00	50.00	151.00	850.24	29.16
Forecast	100%	92.27	87.50	201.00	50.00	151.00	885.92	12.77

Rankings	
$ -	Discontinue
$ 60.00	Review
$ 90.00	Continue

FIGURE 2-34

6. In cells F32 through F36, calculate the **Maximum** of all the sales for each month such as *Sales 4 Months Ago in Thousands, Sales 3 Months Ago in Thousands*, etc. (using the **MAX** function).
7. In cells G32 through G36, calculate the **Minimum** of all the sales for each month such as *Sales 4 Months Ago in Thousands, Sales 3 Months Ago in Thousands*, etc. (using the **MIN** function).
8. In cells H32 through H36, calculate the **Range** of sales by subtracting the minimum sales in column G from the maximum sales of each of the months in column F. **There is no Range function. A formula is all that is needed here.**

9. In cells I32 through I36, calculate the **Variance** of all the sales for each month such as *Sales 4 Months Ago in Thousands, Sales 3 Months Ago in Thousands*, etc. (using the **VAR** function).
10. In cells J32 through J36, calculate the **Standard Deviation** of all the sales for each month such as *Sales 4 Months Ago in Thousands, Sales 3 Months Ago in Thousands*, etc. (using the **STDEVA** function).
11. Make sure that all of the statistics in cells D32 through J36 have been formatted with two places to the right of the decimal.
12. In cell G5, enter the **IF** function in order to put the word *High* in the cell if the *Forecast* is at or above *$100*. Have the **IF** function put the word *Low* in the cell if the *Forecast* is below *$100*. Fill that function from G5 through G27.
13. In cell H5, enter the **VLOOKUP** function to display the ranking of each product line based upon their forecasts, using the **lookup legend** in cells L5 through M9 as your source. Please remember that the column in the **lookup legend** that holds the rankings is column 2. The lookup legend assumes that products with forecasts below $60,000 will discontinued, at or above $60,000 and below $90,000 will be reviewed for discontinuation, and anything at or above $90,000 will be continued. Fill that function down to row 27.
14. In cells B5 through F27, use **Conditional Formatting** to show the sales that are above $100, using the **Light Green Fill and Dark Green Text.**
15. In cell A29, add the label that reads *Product Count*, with a format of **bold** and **italics**.
16. In cell B29, use the **COUNTA** function to count the different products listed in cells A5 through A27.
17. Apply the **Heading 1** cell style to cell **A1**.
18. Apply the **Accent 1** cell style to cell **A4**.
19. Apply the **Title** cell style to cell **A31**. NOTE: Your computer may not have the same appearance or styles shown in this text. If so, choose the color scheme closest to what is requested.
20. Merge and center cells A31 through A36, and then rotate the text in that cell by using the **Rotate Text Up** setting in the **Rotate** icon in the **Home Ribbon**.
21. Save the file and present it for grading.

CHAPTER THREE

Using the PMT (Monthly Payment) Function, the PV (Present Value) Function, Naming Cells, and Creating Borders

BY THE END of this chapter, it is our objective that the user will be able to:

1. Name cells.
2. Use the **PMT** function.
3. Use the **PV** function.
4. Create borders.

CHAPTER 3 TASK: USING EXCEL TO GENERATE LOAN MONTHLY PAYMENTS, AN AMORTIZATION SCHEDULE, AND PAYMENT POSSIBILITIES

At times you will want to take out a loan for a car, home, or some other major undertaking. In order to know if you can afford the monthly payments, Excel can calculate monthly payments, as well as other important data, to help you decide what kind of money you can afford to spend on such things. It will also help you to know what your payments would be at various rates of interest and whether you would be borrowing the money for ten, fifteen, twenty, twenty-five, or thirty years.

While we discuss these things, we will also discuss the concept of naming cells. It's often a good thing to be able to name cells rather than using their column letter and row number. This allows you to give them names that you can remember if you have to reference them in another part of a spreadsheet or in a completely different spreadsheet. A **named cell** also puts a bookmark in those cells in order to quickly hop from one place to another in a spreadsheet or workbook.

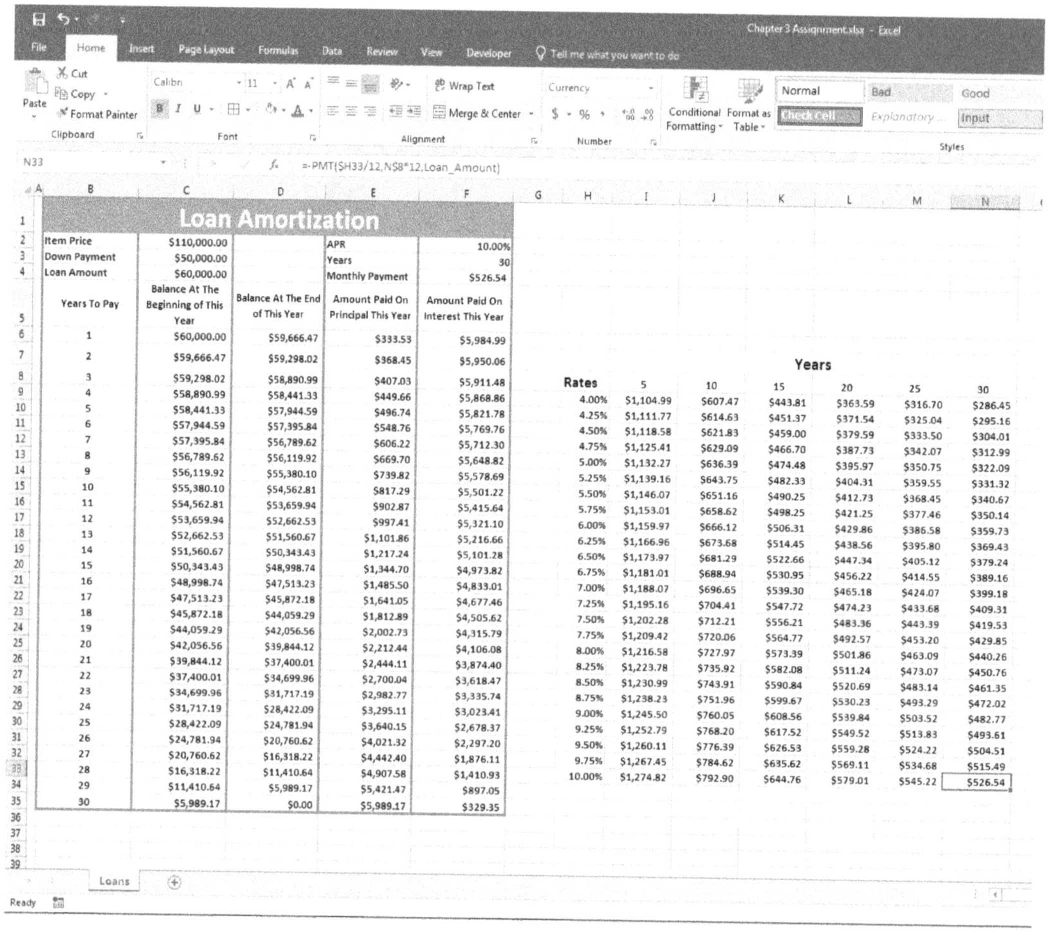

FIGURE 3-1A

BEGINNING AND PREPARING THE SPREADSHEET

Let's begin by starting a new spreadsheet and then adding a few things to prepare it for this task by using what has been learned so far in previous chapters.

1. Shrink column A to a width of approximately *1.00*. This will allow you to get a better view of the **spreadsheet borders** when they are applied later.
2. Make sure columns B1 through F1 each have a width of *17* characters.
3. Type the words *Loan Amortization* in cell B1, then **Merge and Center** those words across cells B1 through F1.
4. Apply the **Accent 2** cell style to the newly merged cell of B1. NOTE: Your computer colors may give you a different look than what you see here. If so, please use the closest cell style that you have available.

5. In cell B2, type the words *Item Price.*
6. In cell B3, type the words *Down Payment.*
7. In cell B4, type the words *Loan Amount.*
8. In cell E2, type *APR* (which stands for *Annual Percentage Rate*).
9. In cell E3, type the word *Years.*
10. In cell E4, type the words *Monthly Payment.*
11. Name the **Sheet Tab** *Loans.*

MAKE THE ENTIRE SPREADSHEET BOLD

To make the entire spreadsheet bold:

Step 1: Click on the **Select All Button** in the upper left of the spreadsheet where the row headings and the column headings meet (Figure 3-1B, Arrow). When you do, the entire spreadsheet will be dark. Any format you apply at that time will be applied to all data entered from this point forward.

Step 2: Click the **B** in the **Home Ribbon** (or the **Bold** icon).

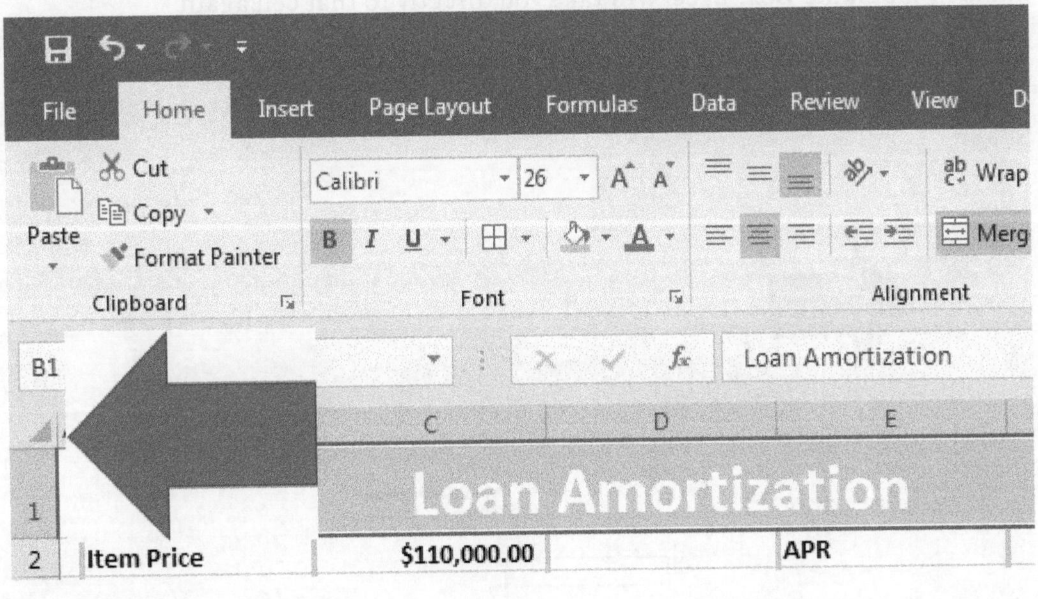

FIGURE 3-1B

NAMING CELLS ONE AT A TIME

If you give a cell a name that is easy to remember, you can also use it as a bookmark. Suppose, for example, you want to name the cell C2 as *Item Price*, since that is what will be entered there. Naming a cell is **NOT** the same as giving it a label. To name the cell:

Step 1: Click on cell C2.

Step 2: Click on the tiny arrow at the right of the **Name Box** (Figure 3-2, Arrow). It will have C2 already shown there. All you need to do is type over the C2 by entering *Item_Price*. NOTE: You will have to put an underscore between the two words as Excel does not permit you to leave spaces in a cell name. Excel also does not permit cell names to include symbols such as @, #, $, %, etc.).

Step 3: Press **Enter**.

When you press enter, you will see that the cell name will always be displayed in the **Name Box** when that cell is selected (Figure 3-3, Arrow).

If you move to another sheet and click in the **Name Box** again, you will see that the cell name remains there when you open the list in that box. By clicking on that name in the **Name Box**, Excel will take you directly to that cell again.

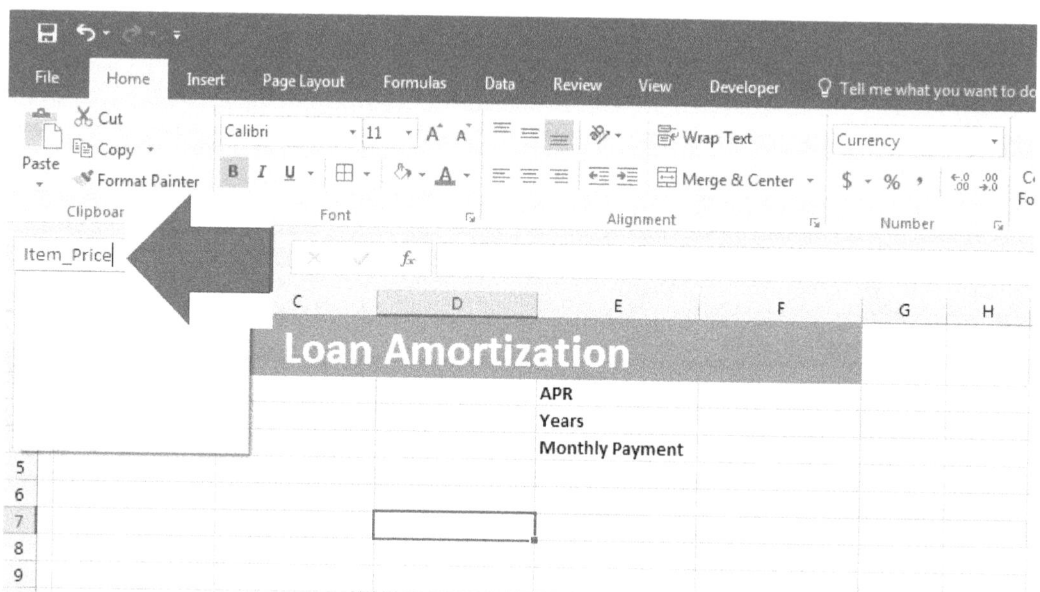

FIGURE 3-2

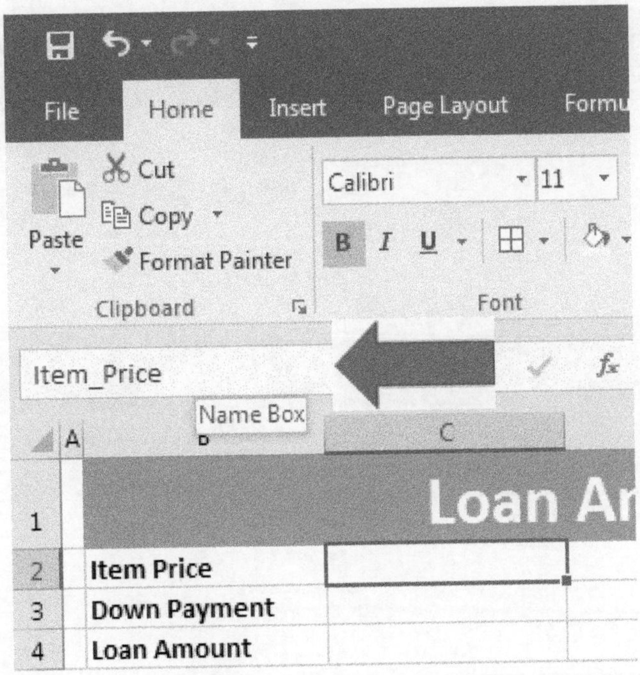

FIGURE 3-3

The problem is that it can be very time consuming to name each cell individually. If you want to give each of the cells a name using the labels that are to their left as their names, you can do so by doing the following:

Step 1: Select **ALL** of the cells in the range of **B2 through C4** (Figure 3-4). Do **NOT** select **only** the labels. Do **NOT** select **only** cells C2 through C4! NOTE: By selecting all of those cells, you will be naming cell C2 a second time.

Step 2: Select the **Formulas Ribbon** of Excel (Figure 3-5, Arrow A).

Step 3: Select **Create from Selection** (Figure 3-5, Arrow B).

Step 4: That will display a pop-up menu that reads **Create Name from Selection**. Make sure that only the **Left Column Option** is selected in that menu (Figure 3-5, Arrow C).

Step 5: Click **OK**.

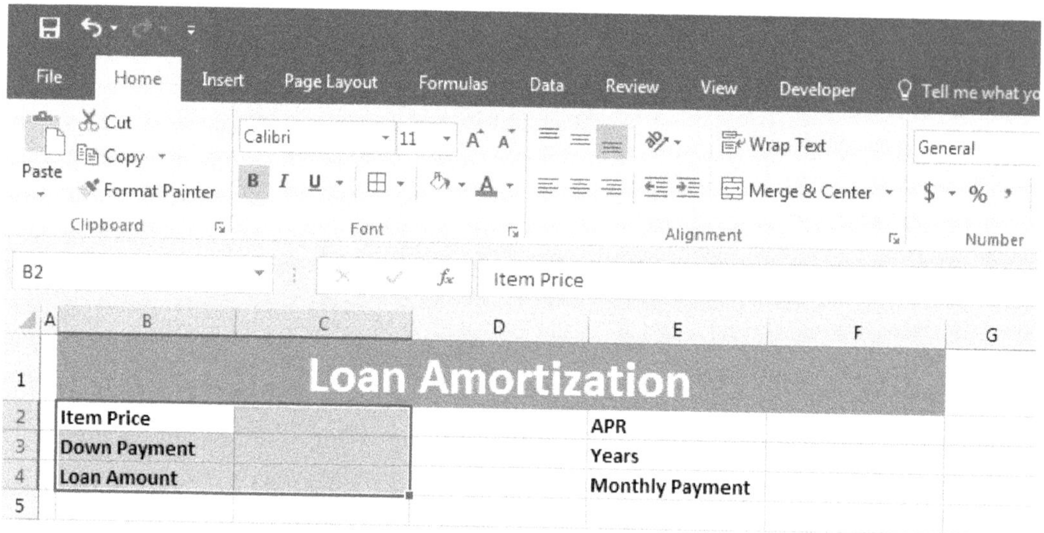

FIGURE 3-4

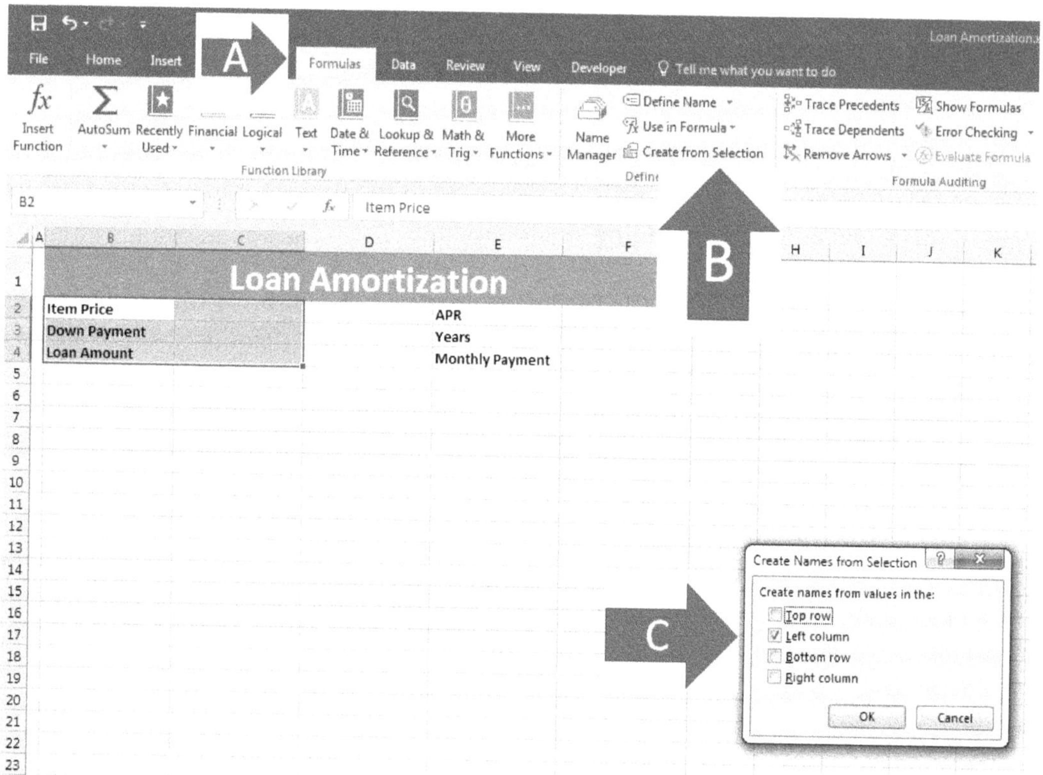

FIGURE 3-5

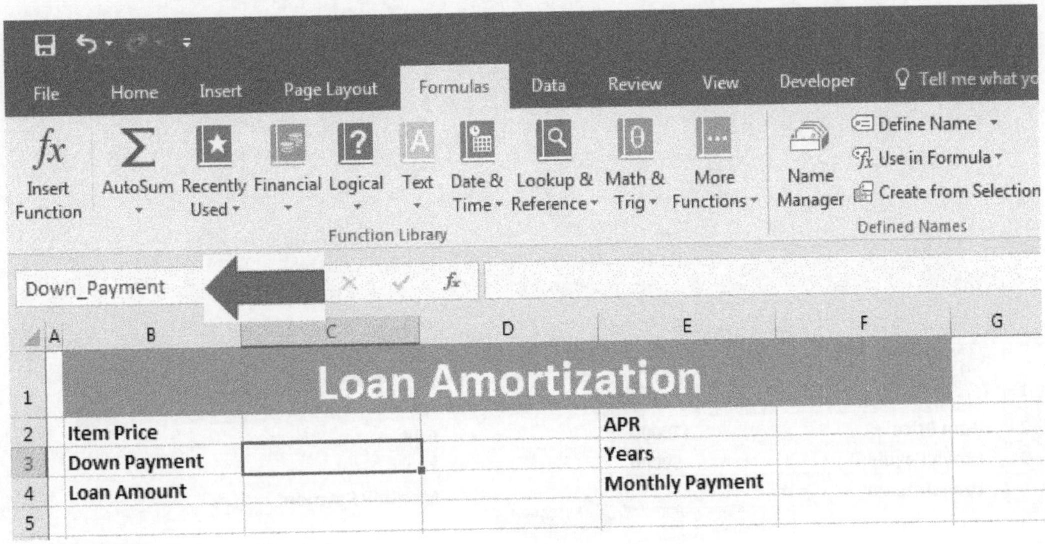

FIGURE 3-6

If you look to the left of the arrow in Figure 3-6, you will see that after selecting each of those cells, the name box will display their names. NOTE: The labels with two words have had their spaces automatically replaced by the underscores needed.

Repeat this process in order to name cell F2 *APR*, cell F3 *Years*, and cell F4 *Monthly_Payment*.

Now let's assume that we are buying a very small home for *$110,000*. Therefore, type *110000* (without the **dollar sign** [$]) into cell C2 as the *Item Price*. Also assume that we are going to put *$50,000* down on this loan. Therefore, type *50000* into cell C3 (without the **dollar sign** [$)]) as the *Down Payment*. We will format these cells with **dollar signs** ($) later.

In this example, with that *Item Price* and *Down Payment*, you would have a loan amount of *$60,000*; however, even though we **entered the item price and the down payment as raw data**, the *Loan Amount* in this example will be calculated.

Now, let's have the spreadsheet do the math for the loan amount with a formula but by using the **Named Cells** as cell references.

Step 1: In cell C4 type =.

Step 2: Begin typing the words *Item_Price*, but when you type the first few letters, a box will pop up showing you that you have named a cell that begins with those letters (in this case *Item_Price* as shown by the arrow in Figure 3-7).

Step 3: Double-click the *Item_Price* cell name, and it will place that into your formula. As mentioned earlier, whenever you type the first few letters of a prefix to

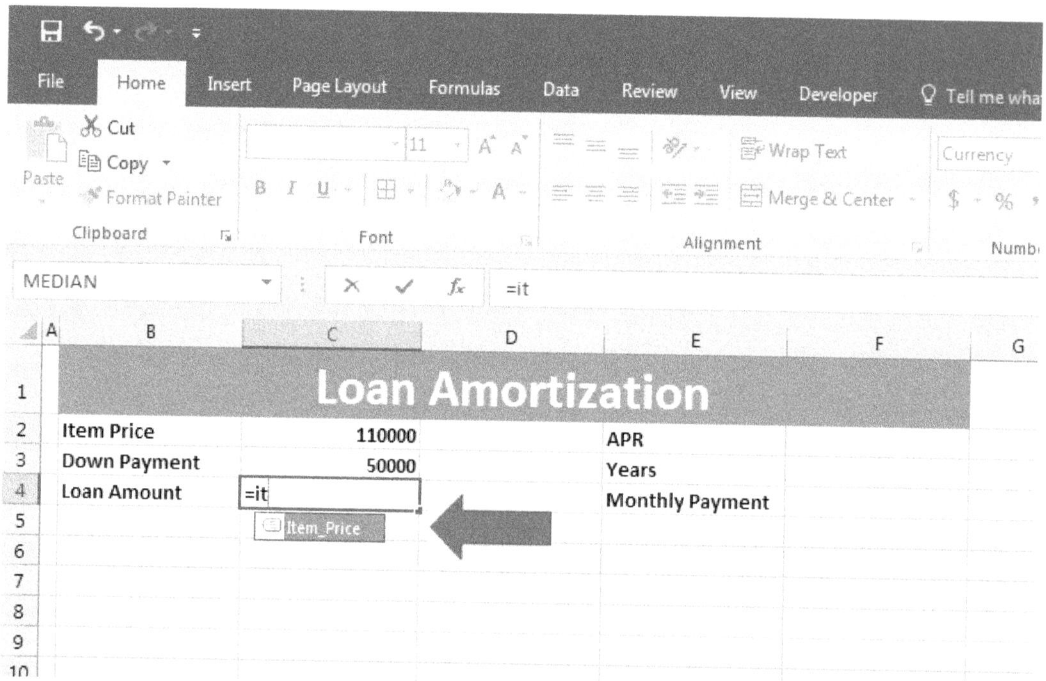

FIGURE 3-7

a function, the first few letters of the name of a named cell, or a named range of cells, the pop-up will list them and you can double-click them if you want to use them. Function prefixes are paired with the *fx* symbol, while the named cells are paired with what looks like a price tag.

Step 4: Press the **minus sign** (-) **key**.

Step 5: Begin typing the words *Down_Payment*, but when you type the first few letters, a box will pop up again showing you that you have named a cell that begins with those letters (Figure 3-8, Arrow). It will also show functions that begin with those letters.

Step 6: Double-click the *Down_Payment* cell name, and it will place it into the formula.

Step 7: Press **Enter**. The *Loan Amount* is now calculated (Figure 3-9, Arrow).

Make the format of cell C4 as **Currency**. You have now created a formula using named cells.

In cell F2, enter *10.00%*.

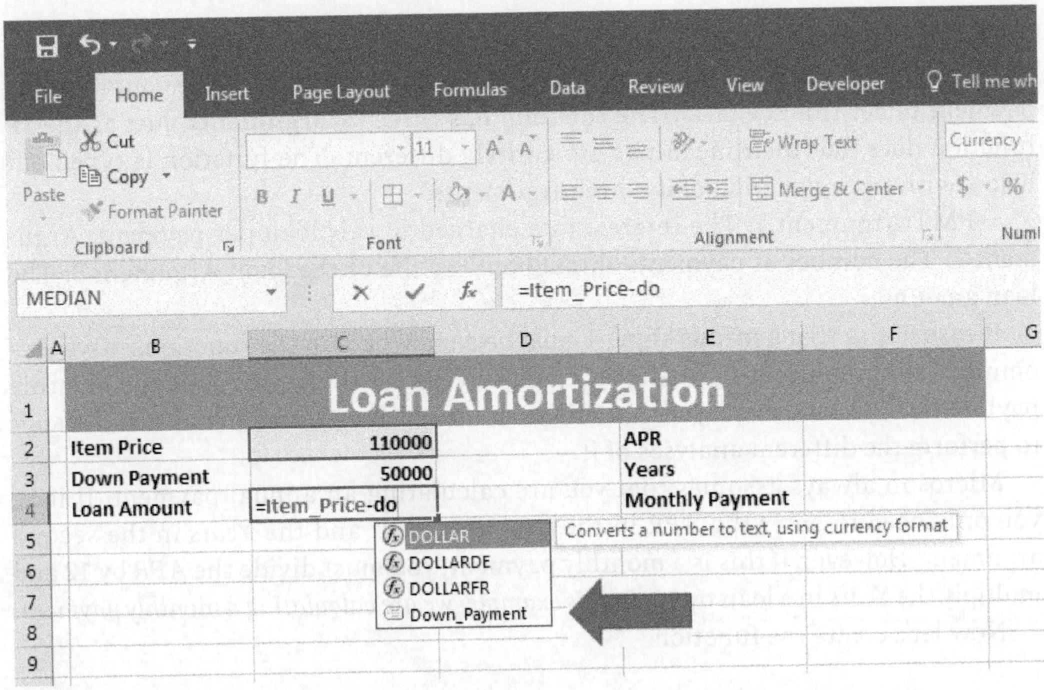

FIGURE 3-8

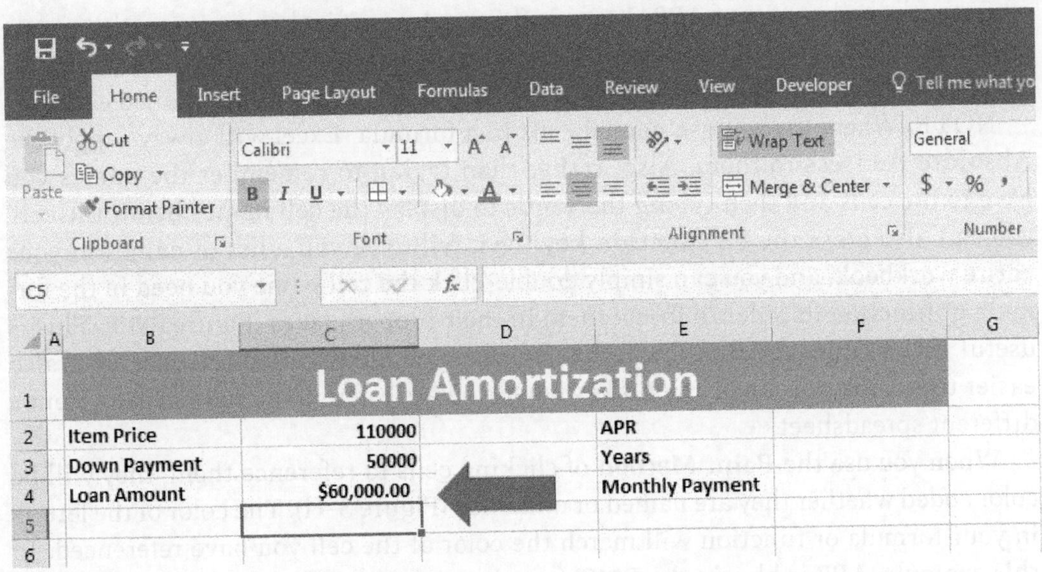

FIGURE 3-9

In cell F3, enter *30*.

Now let's use the **PMT** function. The **PMT** function will calculate your monthly payment under this scenario. The function has three (3) **arguments**, just as the **IF** function does, but their meanings are entirely different. The function is typed out below with an explanation of its three arguments:

=-PMT(Argument 1: The interest rate charged in this loan per payment; Argument 2: The number of payments throughout the life of the loan; Argument 3: The loan amount).

Please notice that a **minus sign** (-) must be somewhere in the function, so we recommend that you put it right after the = sign. Without the minus sign, the monthly payment will also be displayed as a **minus** or a **negative** (-), which makes it difficult to perform the different analyses of it.

Microsoft always assumes that you are calculating an annual payment. If it is, you only need to enter the *APR* in the first argument and the *Years* in the second argument. However, if this is a monthly payment, you must divide the *APR* by 12 and multiply the *Years* in a loan by 12. *In this example, we are calculating a monthly payment.*

Now let's create the function.

Step 1: In this example, click on cell F4.

Step 2: Enter =-PMT(APR/12,Years*12,Loan_Amount).

Enter the cell names of *APR*, *Years*, and *Loan Amount* into their respective places in the function (as we did earlier with the *Item_Price* and *Down_Payment* while calculating the *Loan_Amount*), and they will bring up the pop-ups mentioned earlier.

NOTE: Whenever you use named cells in a formula, Excel will always apply the **Absolute Address** to them. Also, rather than trying to remember the names you gave to the cells and then typing the words to display the cell names below the cell, you can also press the **F3 function key**. That will give you a list of names in your entire workbook, and you can simply double-click the cell name you need in the formula or function in order to insert them in their proper places (Figure 3-10). This is useful if you forget the first few letters of what you have named the cells. It is also easier to use the pop-up or the **F3 key** when you are referencing named cells from a different spreadsheet.

When you use the **Point Method** of clicking cells to reference them, they will be color coded whether they are named or unnamed (Figure 3-11). The color of the letters in your formula or function will match the color of the cell you have referenced. In this example, *APR* is blue in the **PMT** function as is the cell named *APR*. Using the **Point Method** will have the same effect as using the **F3** key or typing the first few letters and using the named cell displayed below the cell.

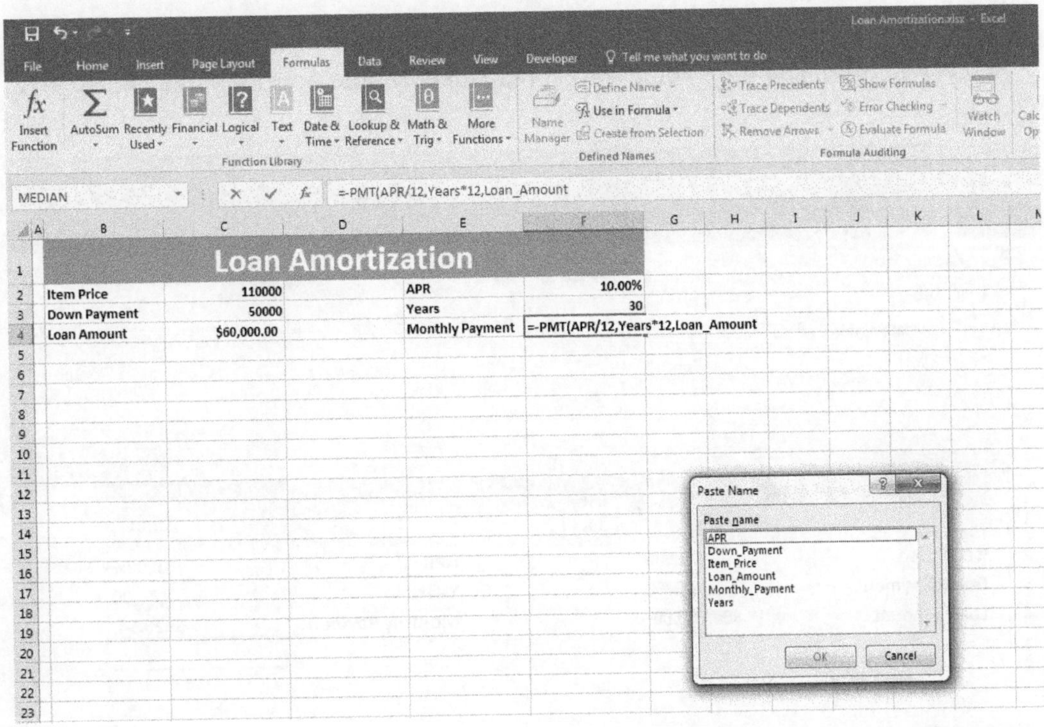

FIGURE 3-10

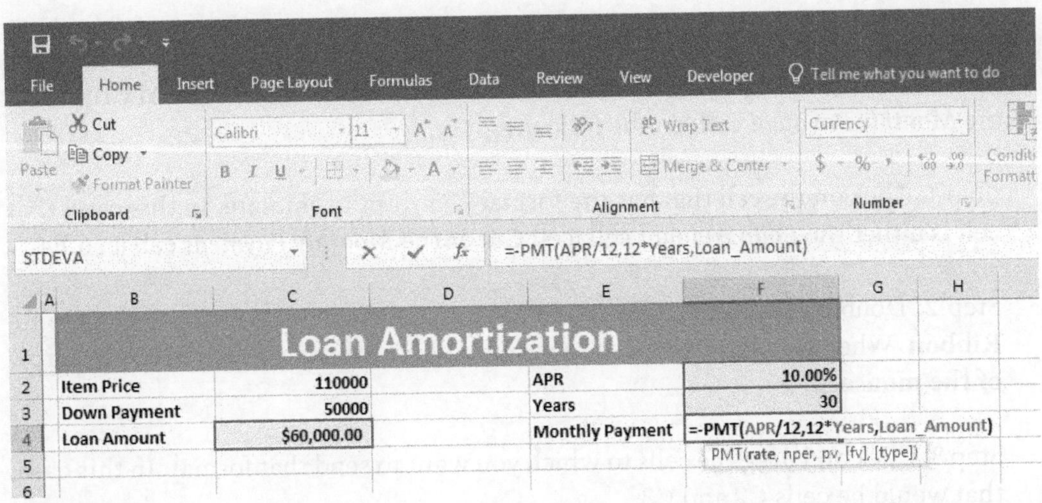

FIGURE 3-11

Step 3: Press **Enter.** (NOTE: If you forget to put the right parenthesis on a function, Excel will often add it for you.) Your spreadsheet should now look like Figure 3-12.

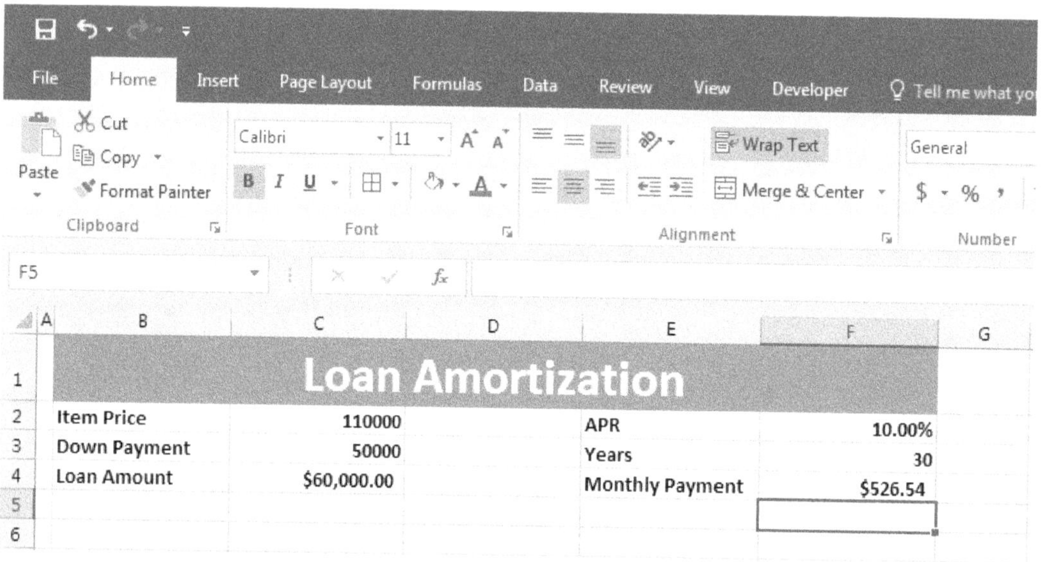

FIGURE 3-12

THE FORMAT PAINTER

If you would like the numbers in the *Item Price* and *Down Payment* cells to have the same currency format that is displayed in the *Loan Amount* cell or the **PMT** function in the *Monthly Payment* cell, do the following:

Step 1: Click on the cell that has the format you want to emulate. In this case, that cell is either the *Loan Amount* cell (cell C4) or the *Monthly Payment* cell (cell F4).

Step 2: Double-click on the **Format Painter** (Figure 3-13, Arrow) in the **Home Ribbon**. When you do, you will see a very small **paintbrush** attached to the **arrow of the mouse**.

Step 3: Click and drag the cells to which you want to send that format. In this case, that would be cells C2 and C3.

Step 4: Press **Esc**. That will stop the **Format Painter** from changing formats.

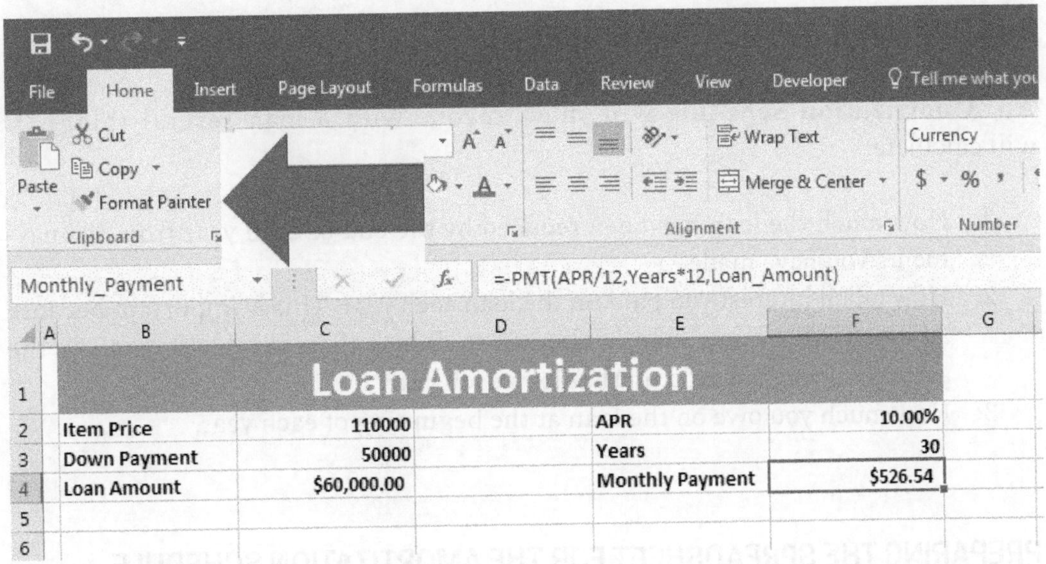

FIGURE 3-13

That will give you what you see in Figure 3-14. NOTE: If you click the **Format Painter** only once, it will allow you to paint a format to **only one cell** or cell range but not nonadjacent cells. On the other hand, double-clicking the **Format Painter** gives you unlimited places, adjacent and nonadjacent areas, where you can apply a format, and you only need to press **Esc** when you finish doing so.

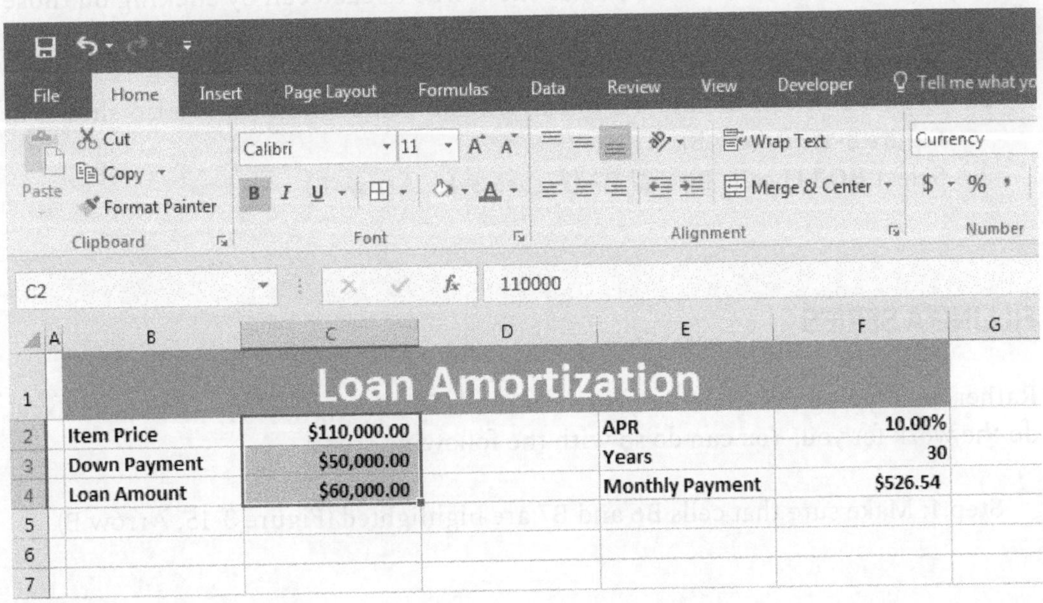

FIGURE 3-14

BUILDING AN AMORTIZATION SCHEDULE

An **Amortization Schedule** will show anyone with a loan several things. It will calculate:

1. How much the loan has been reduced by the end of each year from the payments you have made.
2. How much interest you paid on the loan each year. This is important because anyone with a home loan is allowed to reduce their taxable income by the amount of interest paid on that loan each year.
3. How much you owe on the loan at the beginning of each year.

PREPARING THE SPREADSHEET FOR THE AMORTIZATION SCHEDULE

Before we begin to enter formulas and functions needed for an **Amortization Schedule**, you will need to prepare the spreadsheet for this task by using what has been learned so far in previous chapters.

1. Type in the labels in cells B5 through F5 that are shown in Figure 3-15. You can either use the **Alt Enter** keystrokes to force the labels to wrap the text in those cells, or you can use the **Wrap Text** (Figure 3-15, Arrow A) from the **Home Ribbon**. Center the words in each cell by clicking on those cells, and then click the **Center Icon** in the **Home Ribbon** (also shown in Figure 3-15).
2. Type a 1 in cell B6 and a 2 in cell B7 and center those cells (also shown in Figure 3-15, Arrow B).
3. Select **BOTH** cells B6 and B7 (Figure 3-15, Arrow B).

FILLING A SERIES

Rather than typing in each of the years from cells B6 through B35, you can have Excel do the work for you. You can do so with the following steps:

Step 1: Make sure that cells B6 and B7 are highlighted (Figure 3-15, Arrow B).

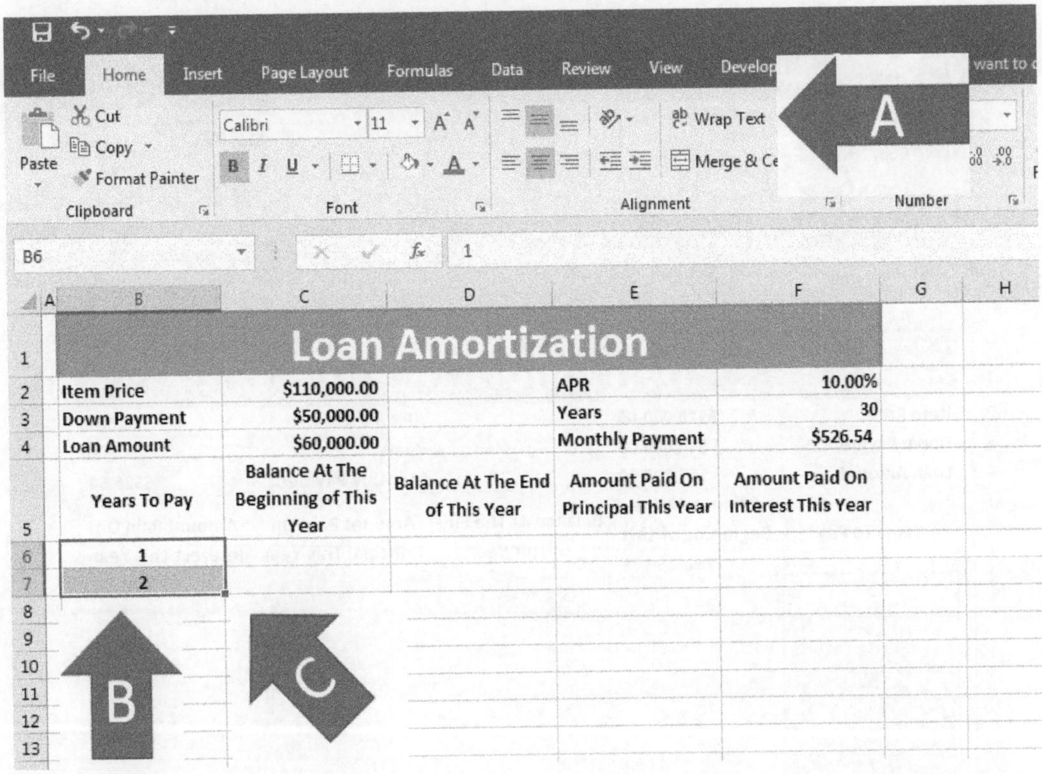

FIGURE 3-15

Step 2: Pull the **Fill Handle** (the very small box that you have used before) at the lower right of cell B7 (Figure 3-15, Arrow C) to create the series of numbers that represent the years of the loan. Pull the fill handle down to row 35 (Figure 3-16).

NOTE: When you fill a series with a **step value** (in this case, the **step value** is 1), make sure the first **TWO** cells are selected, as you have done here, before you pull the **Fill Handle**.

Let us now continue to build the *Amortization Schedule*.

The next thing to do is to enter =*Loan_Amount* in cell C6 (as shown in Figure 3-17). You will need to do this because the beginning balance of any loan will always be the loan amount.

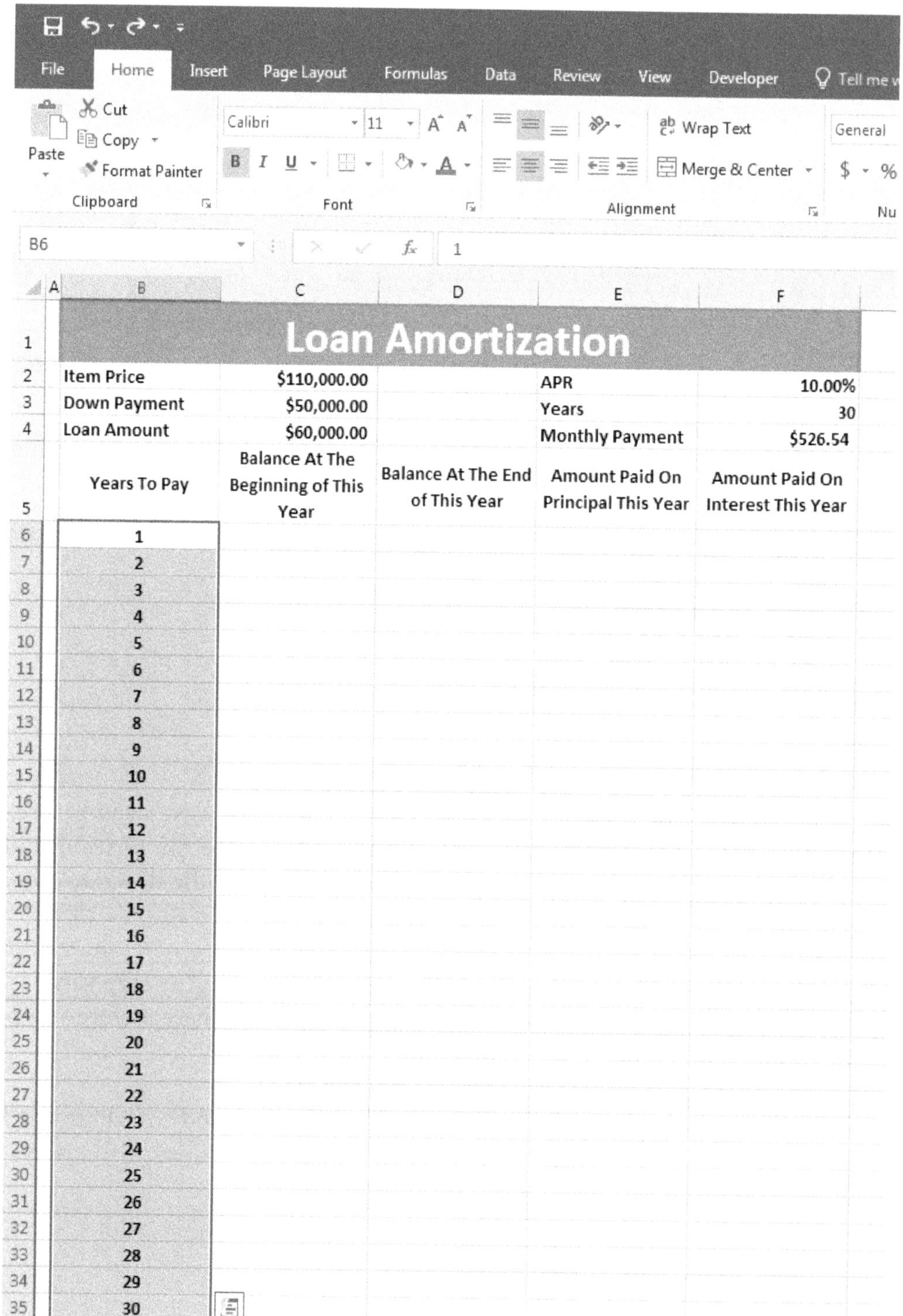

	B	C	D	E	F
1	Loan Amortization				
2	Item Price	$110,000.00		APR	10.00%
3	Down Payment	$50,000.00		Years	30
4	Loan Amount	$60,000.00		Monthly Payment	$526.54
5	Years To Pay	Balance At The Beginning of This Year	Balance At The End of This Year	Amount Paid On Principal This Year	Amount Paid On Interest This Year
6	1				
7	2				
8	3				
9	4				
10	5				
11	6				
12	7				
13	8				
14	9				
15	10				
16	11				
17	12				
18	13				
19	14				
20	15				
21	16				
22	17				
23	18				
24	19				
25	20				
26	21				
27	22				
28	23				
29	24				
30	25				
31	26				
32	27				
33	28				
34	29				
35	30				

FIGURE 3-16

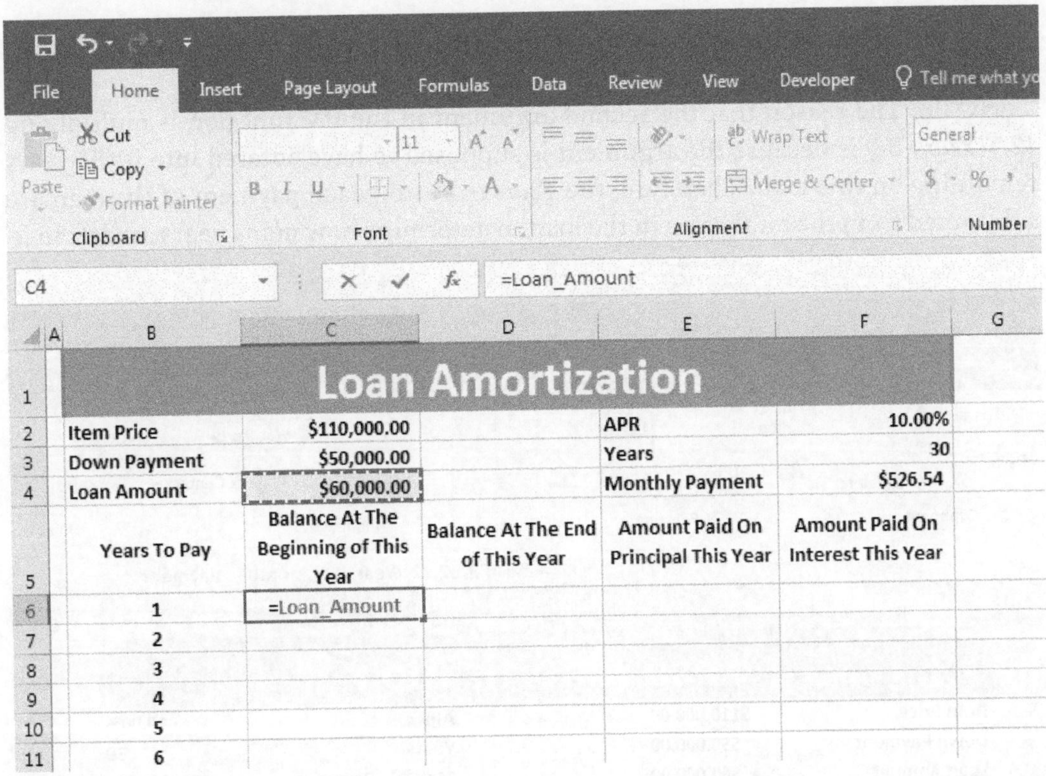

FIGURE 3-17

THE PV (OR PRESENT VALUE) FUNCTION

At this point, it is now important that you enter the **PV** function. The **PV**, or **Present Value** function, will tell you how much is owed on a loan at any given time depending upon:

1. How much the **monthly payment** is.
2. The **interest rate** charged to the borrower with each payment.
3. The **number of payments** remaining in the loan.

The **PV** function has three (3) arguments and is very similar to (but not the same as) the **PMT** function. The function is typed out below with an explanation of its three arguments:

=-PV(Argument 1: The interest rate charged in this loan per payment; Argument 2: The number of payments **REMAINING** in the life of this loan; Argument 3: The monthly payment of this loan).

At this time, type *=-PV(APR/12,12*(Years-B6),Monthly_Payment)* into D6 and then press **Enter**.

You will get the result you see in Figure 3-18. That is what would be owed at the end of the first year of the loan in this example.

NOTE: The reason that the second argument in the **PV** function is multiplying *12 * (Years-B6)* is because the argument is supposed to have entered into it the years remaining on this loan. Therefore, the *Years to Pay* on each payment of the loan are subtracted from the total years in the loan to determine how many years are left in it.

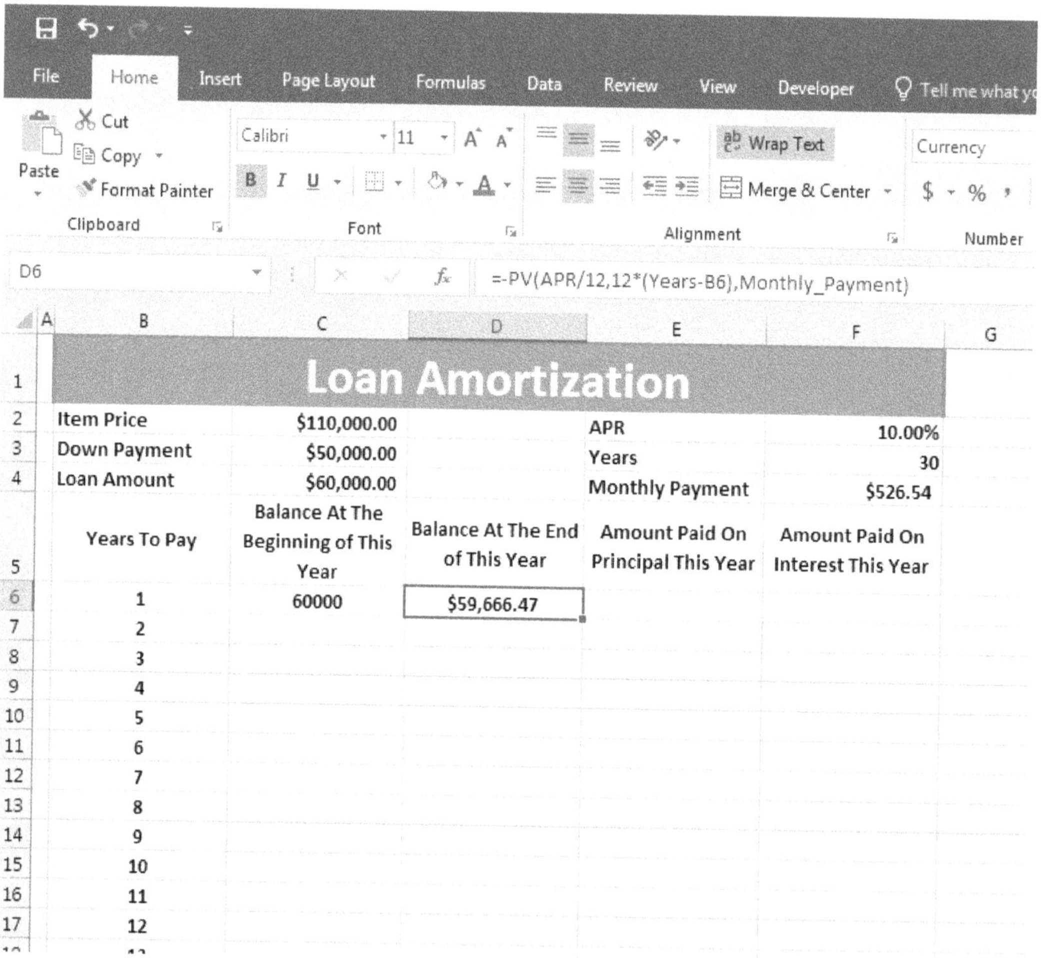

FIGURE 3-18

Next, in cell E6, type *C6-D6*. That will show how much has been paid on the principal in that year. The *Balance at the End of This Year* (in column D) subtracted from the *Balance at the Beginning of This Year* (in column C) will give you that number (Figure 3-19).

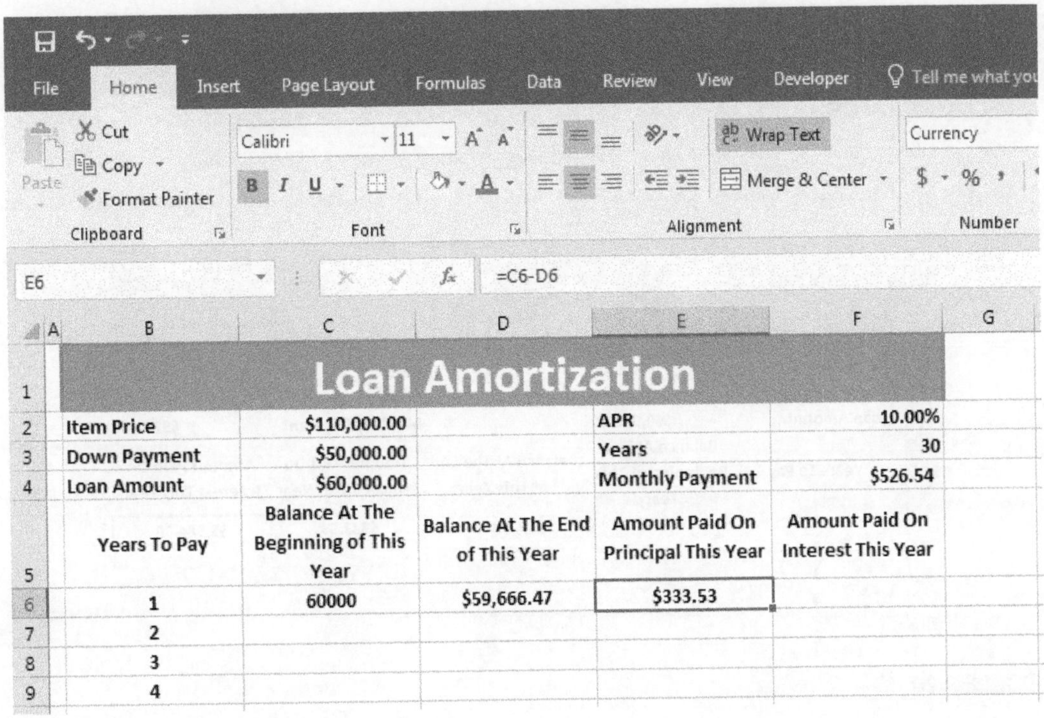

FIGURE 3-19

Every individual loan payment is comprised of only two things: the amount paid toward the **balance** (or **principal**) of the loan and the amount of **interest** paid in that payment. Therefore, the amount paid in **interest** in a given year will always be equal to all payments made, minus what is paid in **principal**. The amount paid in payments will be equal to *12*Monthly_Payment*. Therefore, in cell F6, type *=12*Monthly_Payment-E6* and you will get the result you see in Figure 3-20.

The beginning balance for the second year will always be the same as the ending balance of the previous year. Therefore, in cell C7, type *=D6*. The result will be what is seen in Figure 3-21.

Now pull the Fill Handle at the lower right of cell C7 all the way down to Row 35 (Figure 3-22). NOTE: When you fill down the formula, it will give you zeros. THAT'S OKAY because at this point there is no balance entered in the cells of column D yet. This will be fixed in the next step.

Now select cells D6 through F6, and then pull the fill handle at the lower right of cell F6. Fill those formulas all the way down to row 35. You will get the results seen in Figure 3-23. **You know you have done it right if the balance at the end of the last year goes to zero!**

Use the Format Painter to make cell C6 the same as D6.

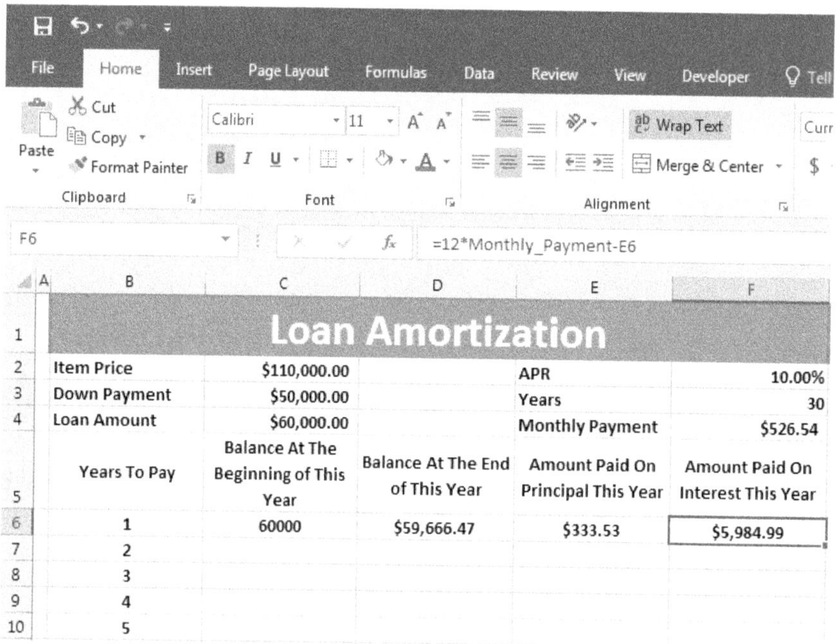

FIGURE 3-20

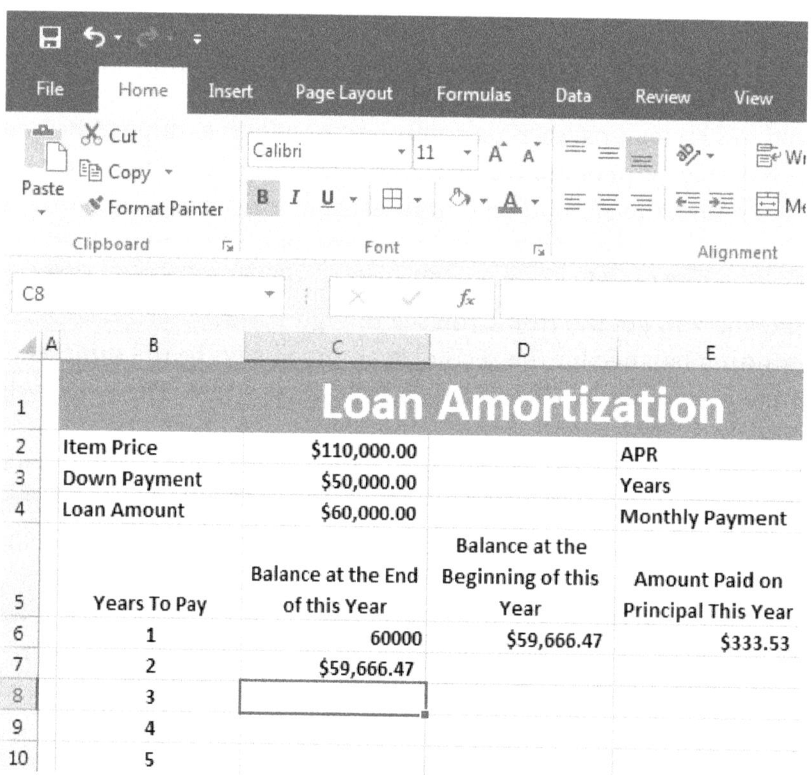

FIGURE 3-21

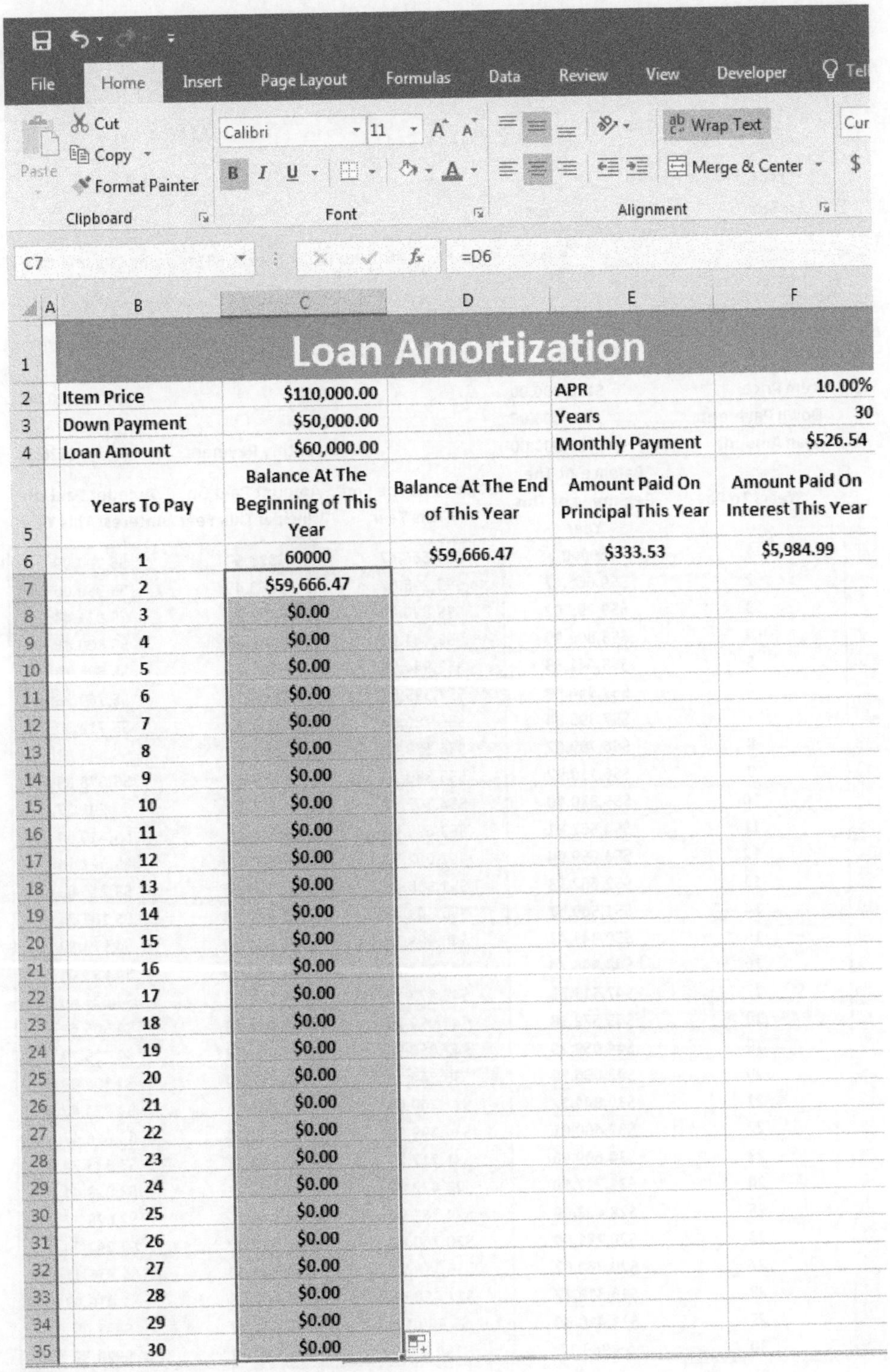

	A	B	C	D	E	F
1	Loan Amortization					
2		Item Price	$110,000.00		APR	10.00%
3		Down Payment	$50,000.00		Years	30
4		Loan Amount	$60,000.00		Monthly Payment	$526.54
5		Years To Pay	Balance At The Beginning of This Year	Balance At The End of This Year	Amount Paid On Principal This Year	Amount Paid On Interest This Year
6		1	60000	$59,666.47	$333.53	$5,984.99
7		2	$59,666.47			
8		3	$0.00			
9		4	$0.00			
10		5	$0.00			
11		6	$0.00			
12		7	$0.00			
13		8	$0.00			
14		9	$0.00			
15		10	$0.00			
16		11	$0.00			
17		12	$0.00			
18		13	$0.00			
19		14	$0.00			
20		15	$0.00			
21		16	$0.00			
22		17	$0.00			
23		18	$0.00			
24		19	$0.00			
25		20	$0.00			
26		21	$0.00			
27		22	$0.00			
28		23	$0.00			
29		24	$0.00			
30		25	$0.00			
31		26	$0.00			
32		27	$0.00			
33		28	$0.00			
34		29	$0.00			
35		30	$0.00			

FIGURE 3-22

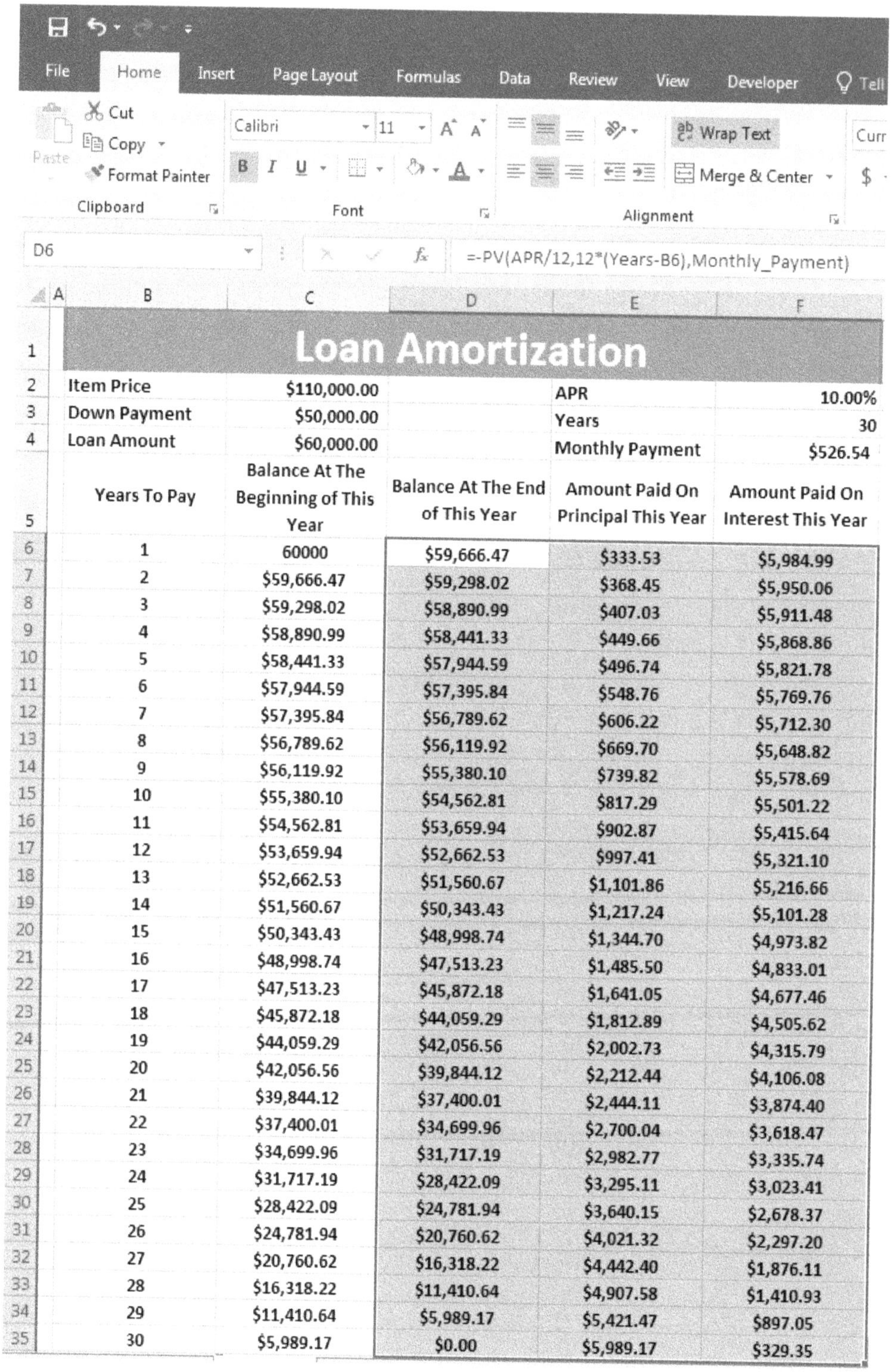

	A	B	C	D	E	F
1		Loan Amortization				
2		Item Price	$110,000.00		APR	10.00%
3		Down Payment	$50,000.00		Years	30
4		Loan Amount	$60,000.00		Monthly Payment	$526.54
5		Years To Pay	Balance At The Beginning of This Year	Balance At The End of This Year	Amount Paid On Principal This Year	Amount Paid On Interest This Year
6		1	60000	$59,666.47	$333.53	$5,984.99
7		2	$59,666.47	$59,298.02	$368.45	$5,950.06
8		3	$59,298.02	$58,890.99	$407.03	$5,911.48
9		4	$58,890.99	$58,441.33	$449.66	$5,868.86
10		5	$58,441.33	$57,944.59	$496.74	$5,821.78
11		6	$57,944.59	$57,395.84	$548.76	$5,769.76
12		7	$57,395.84	$56,789.62	$606.22	$5,712.30
13		8	$56,789.62	$56,119.92	$669.70	$5,648.82
14		9	$56,119.92	$55,380.10	$739.82	$5,578.69
15		10	$55,380.10	$54,562.81	$817.29	$5,501.22
16		11	$54,562.81	$53,659.94	$902.87	$5,415.64
17		12	$53,659.94	$52,662.53	$997.41	$5,321.10
18		13	$52,662.53	$51,560.67	$1,101.86	$5,216.66
19		14	$51,560.67	$50,343.43	$1,217.24	$5,101.28
20		15	$50,343.43	$48,998.74	$1,344.70	$4,973.82
21		16	$48,998.74	$47,513.23	$1,485.50	$4,833.01
22		17	$47,513.23	$45,872.18	$1,641.05	$4,677.46
23		18	$45,872.18	$44,059.29	$1,812.89	$4,505.62
24		19	$44,059.29	$42,056.56	$2,002.73	$4,315.79
25		20	$42,056.56	$39,844.12	$2,212.44	$4,106.08
26		21	$39,844.12	$37,400.01	$2,444.11	$3,874.40
27		22	$37,400.01	$34,699.96	$2,700.04	$3,618.47
28		23	$34,699.96	$31,717.19	$2,982.77	$3,335.74
29		24	$31,717.19	$28,422.09	$3,295.11	$3,023.41
30		25	$28,422.09	$24,781.94	$3,640.15	$2,678.37
31		26	$24,781.94	$20,760.62	$4,021.32	$2,297.20
32		27	$20,760.62	$16,318.22	$4,442.40	$1,876.11
33		28	$16,318.22	$11,410.64	$4,907.58	$1,410.93
34		29	$11,410.64	$5,989.17	$5,421.47	$897.05
35		30	$5,989.17	$0.00	$5,989.17	$329.35

FIGURE 3-23

BORDERS

Borders help make a spreadsheet easier to follow and comprehend. Borders can be placed around cells using many combinations. In this example, we want to put them down along the columns and around the perimeter of the entire amortization schedule.

In order to put colored borders around the cells, do the following:

Step 1: Select all of the cells in the amortization schedule (Figure 3-24).

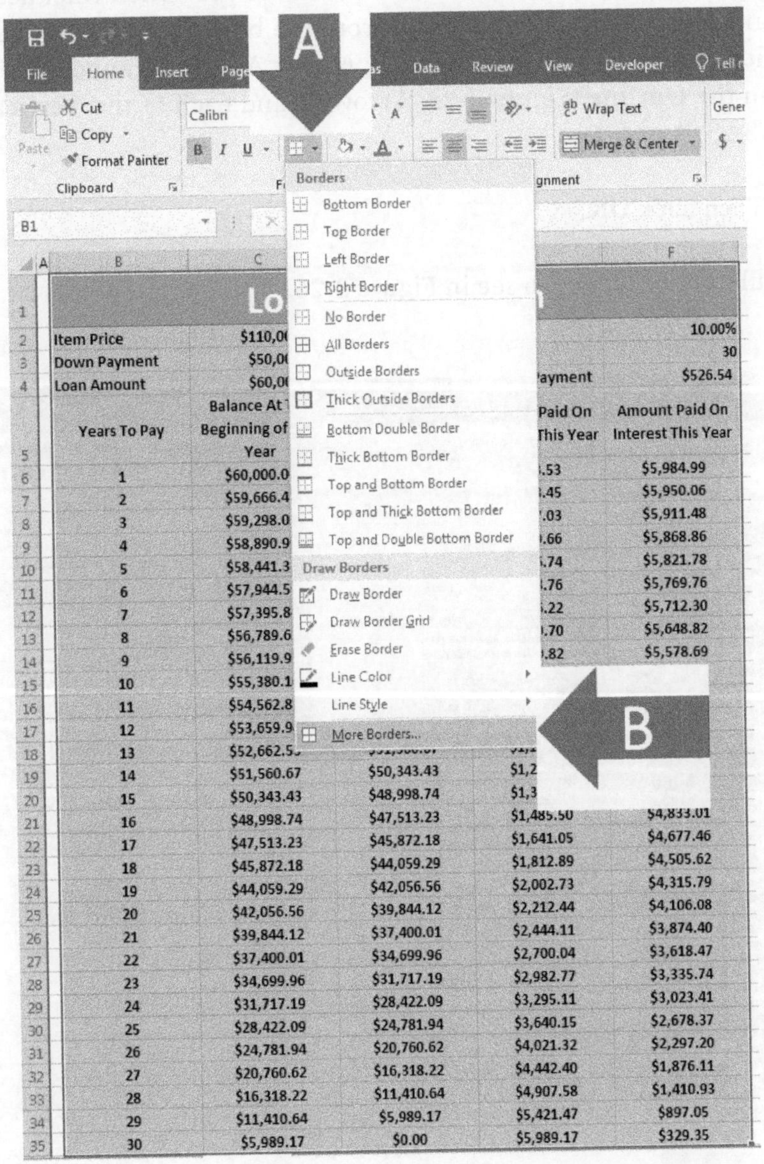

FIGURE 3-24

Step 2: Click the tiny arrow to the right of the **Border Box** in the **Home Ribbon** (Figure 3-24, Arrow A). Ordinarily, you can just choose any one of the **standard borders** in this list to create borders. In this example, however, the borders we want are not among the standard ones. To do something different, click **More Borders** (Figure 3-24, Arrow B).

Step 3: At that point, a **pop-up menu** will be displayed as you can see in Figure 3-25. When it does, you must now choose the **thickness**, the **color**, and then the **position** you want. In this case, choose the **Thick** thickness, which in this example is the thickness second from the bottom of the second column in the **Thickness Style box** (Figure 3-25, Arrow A), the **Orange** color (Arrow B), and then the **Outline** (Figure 3-25, Arrow C) and each of the **Vertical** positions (Figure 3-25, Arrow D).

Step 4: Then click **OK**.

The result will be what you see in Figure 3-26.

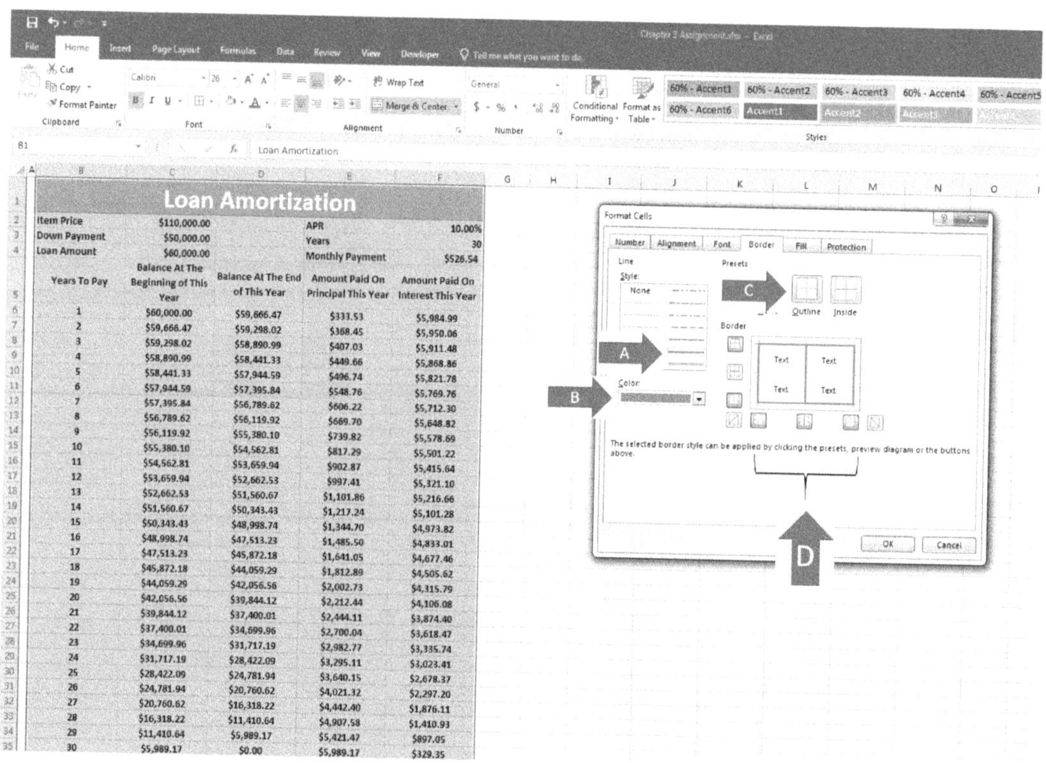

FIGURE 3-25

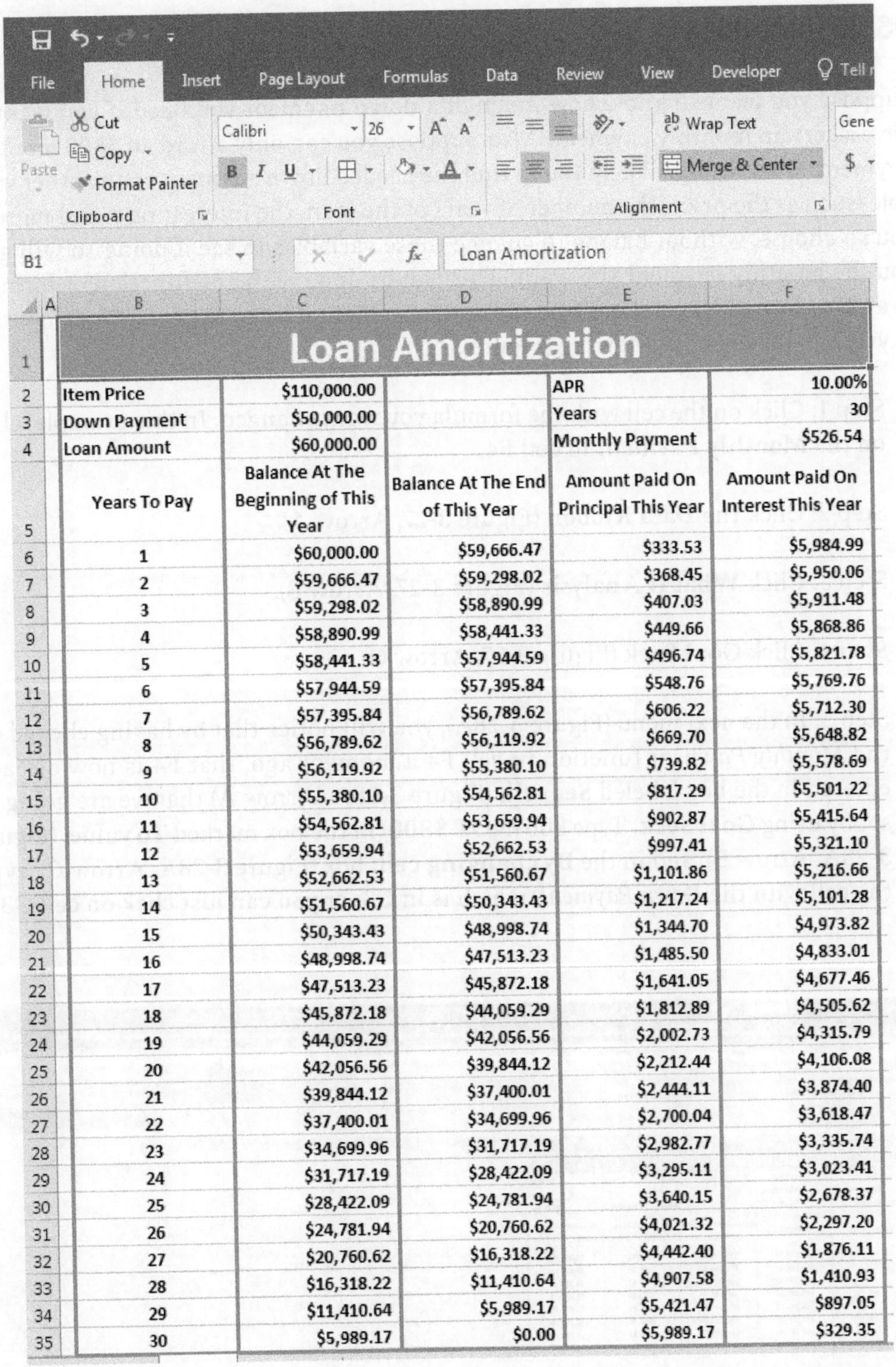

	B	C	D	E	F
1	Loan Amortization				
2	Item Price	$110,000.00		APR	10.00%
3	Down Payment	$50,000.00		Years	30
4	Loan Amount	$60,000.00		Monthly Payment	$526.54
5	Years To Pay	Balance At The Beginning of This Year	Balance At The End of This Year	Amount Paid On Principal This Year	Amount Paid On Interest This Year
6	1	$60,000.00	$59,666.47	$333.53	$5,984.99
7	2	$59,666.47	$59,298.02	$368.45	$5,950.06
8	3	$59,298.02	$58,890.99	$407.03	$5,911.48
9	4	$58,890.99	$58,441.33	$449.66	$5,868.86
10	5	$58,441.33	$57,944.59	$496.74	$5,821.78
11	6	$57,944.59	$57,395.84	$548.76	$5,769.76
12	7	$57,395.84	$56,789.62	$606.22	$5,712.30
13	8	$56,789.62	$56,119.92	$669.70	$5,648.82
14	9	$56,119.92	$55,380.10	$739.82	$5,578.69
15	10	$55,380.10	$54,562.81	$817.29	$5,501.22
16	11	$54,562.81	$53,659.94	$902.87	$5,415.64
17	12	$53,659.94	$52,662.53	$997.41	$5,321.10
18	13	$52,662.53	$51,560.67	$1,101.86	$5,216.66
19	14	$51,560.67	$50,343.43	$1,217.24	$5,101.28
20	15	$50,343.43	$48,998.74	$1,344.70	$4,973.82
21	16	$48,998.74	$47,513.23	$1,485.50	$4,833.01
22	17	$47,513.23	$45,872.18	$1,641.05	$4,677.46
23	18	$45,872.18	$44,059.29	$1,812.89	$4,505.62
24	19	$44,059.29	$42,056.56	$2,002.73	$4,315.79
25	20	$42,056.56	$39,844.12	$2,212.44	$4,106.08
26	21	$39,844.12	$37,400.01	$2,444.11	$3,874.40
27	22	$37,400.01	$34,699.96	$2,700.04	$3,618.47
28	23	$34,699.96	$31,717.19	$2,982.77	$3,335.74
29	24	$31,717.19	$28,422.09	$3,295.11	$3,023.41
30	25	$28,422.09	$24,781.94	$3,640.15	$2,678.37
31	26	$24,781.94	$20,760.62	$4,021.32	$2,297.20
32	27	$20,760.62	$16,318.22	$4,442.40	$1,876.11
33	28	$16,318.22	$11,410.64	$4,907.58	$1,410.93
34	29	$11,410.64	$5,989.17	$5,421.47	$897.05
35	30	$5,989.17	$0.00	$5,989.17	$329.35

FIGURE 3-26

USING GOAL SEEK

Suppose you want to know how much of a down payment you need if you want to have a certain monthly payment. Also, suppose you can only afford an $800 monthly payment. **Goal Seek** will help you to find the needed down payment or any other variable (such as the price, the number of years of the loan, the interest rate, and more) if you so choose, without having to change those variables to see if doing so will give you the monthly payment you want. This can be done with any function or formula in an Excel spreadsheet. Let's find the down payment needed to get an $800 monthly payment by doing the following (Figure 3-27):

Step 1: Click on the cell with the formula you want changed. In this example, click on the **Monthly Payment** in cell F4.

Step 2: Click the **Data Ribbon** (Figure 3-27, Arrow A).

Step 3: Click **What If Analysis** (Figure 3-27, Arrow B).

Step 4: Click **Goal Seek** (Figure 3-27, Arrow C).

Step 5: In the next menu (Figure 3-28A), you will notice that by having clicked on the *Monthly Payment* function in cell F4 a moment ago, that F4 is now already entered in the box labeled **Set cell:** (Figure 3-28A, Arrow A) that we are going to set by using **Goal Seek**. Type 800 (**NOT $800**) in the box marked **To value:** (Figure 3-28A, Arrow B), and in the **By changing cell:** box (Figure 3-28A, Arrow C), type the cell with the *Down Payment*, which is in C3, or you can just click on cell C3.

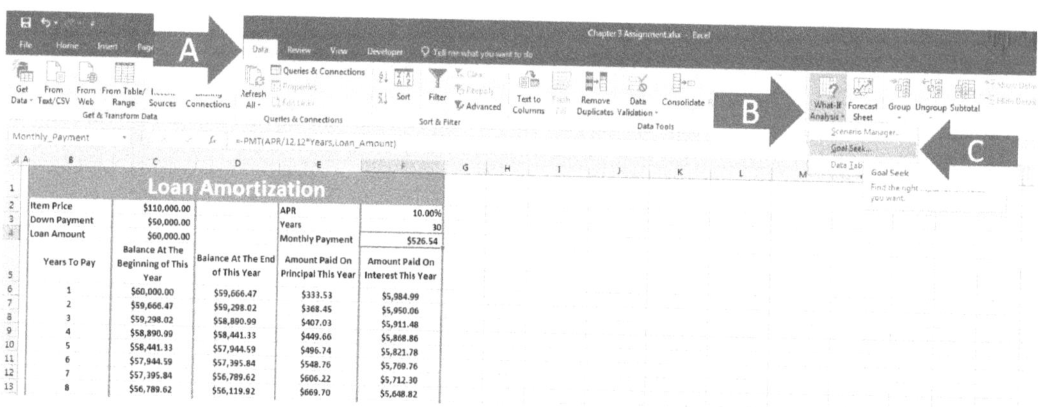

FIGURE 3-27

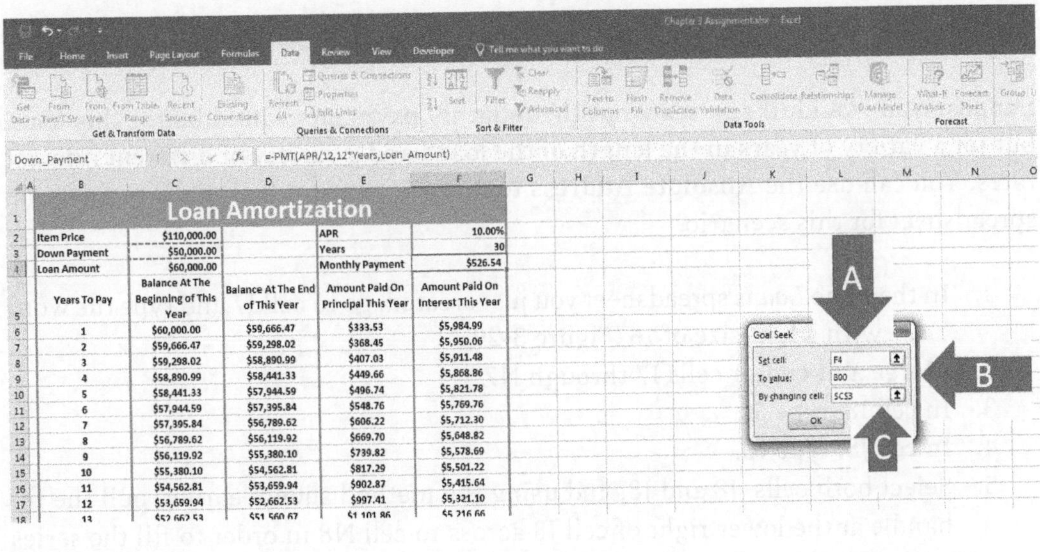

FIGURE 3-28A

Step 6: Click **OK**. **Goal Seek** will tell you that there are results (Figure 3-28B). Please notice that it also changed the down payment in cell C3, and by doing so, the monthly payment and all of the numbers in the **Amortization Schedule** have been adjusted accordingly.

The great thing about spreadsheets is that if you change any of the variables when considering a different loan, all of the numbers change instantly and automatically!

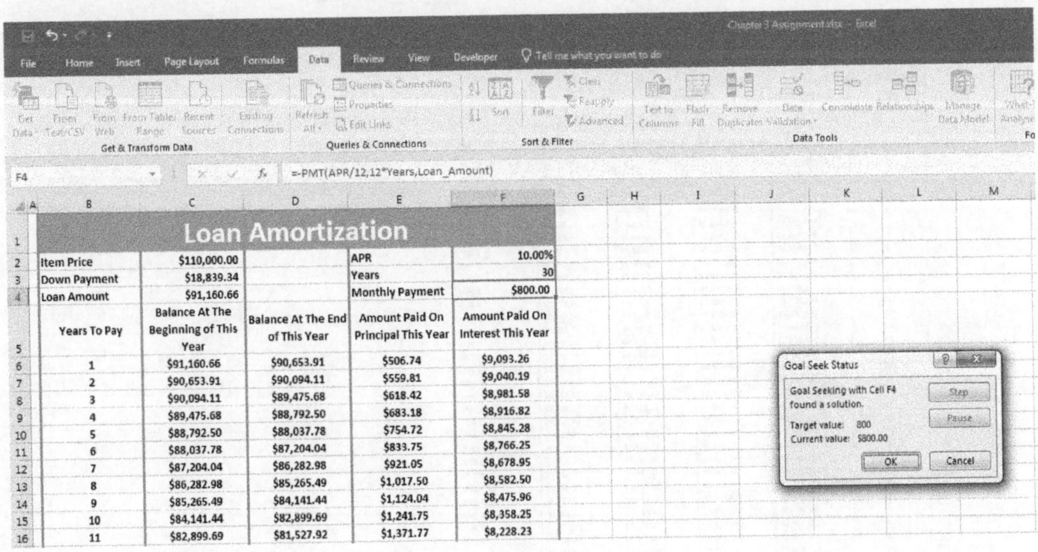

FIGURE 3-28B

ANOTHER APPLICATION OF LOANS AND ABSOLUTE ADDRESS

Suppose you would like to see what the monthly payments would be for this loan, but for a variety of years in the length of the loans and a variety of different interest rates. You can use the **Absolute Address** to make it happen. First, let's prepare the spreadsheet for this scenario.

1. In the same *Loans* spreadsheet you just created, go to cell I7 and type the word *Years* with a font size of *16* (Figure 3-29).
2. **Merge and center** cells I7 through N7.
3. In cell I8, type 5.
4. In cell J8, type *10*.
5. Select both cells, I8 and J8, and using the method shown earlier, pull the fill handle at the lower right of cell J8 across to cell N8 in order to fill the series so that it will now read, 5, 10, 15, 20, 25, and 30 (Figure 3-29).
6. In cell H8, enter the word *Rates* and center that word in the cell with a font of *14*.
7. In cell H9, enter *4.00%*.
8. In cell H10, enter *4.25%*.
9. Fill that series down to row 33, at which point the series will have reached a level of *10.00%*. Your spreadsheet should look like Figure 3-29 at this point.

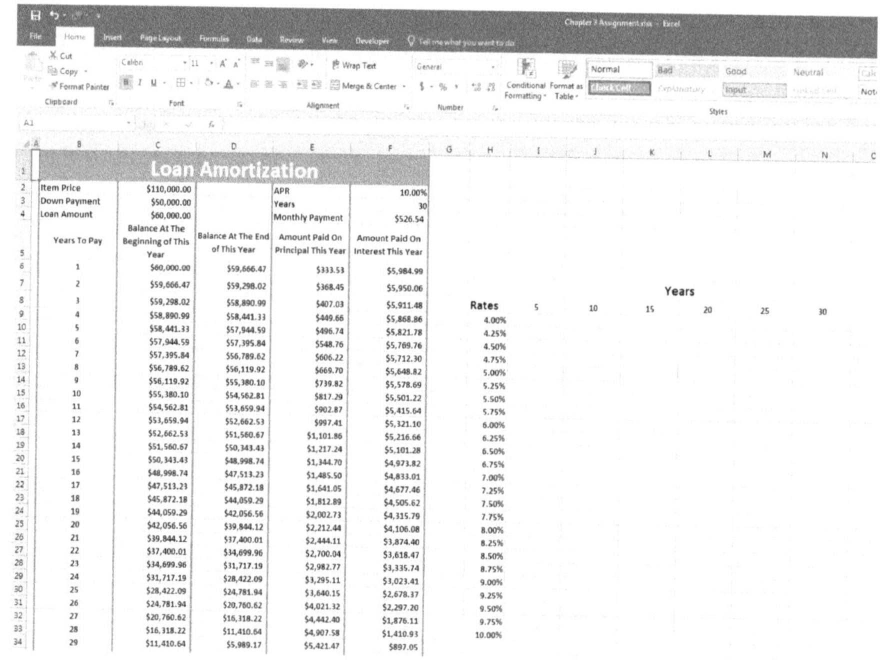

FIGURE 3-29

10. Place your mouse in cell I9 and type the following function that reads:
=-PMT($H9/12,I$8*12,Loan_Amount)

NOTE: When you click on cell H9 using the **Pointing Method**, press the **F4** key three (3) times, and Excel will put the **dollar sign** ($) to the left of the *H* in the H9 cell reference for you. When you click on cell I9 using the **Pointing Method**, press the **F4** key twice, and Excel will put the dollar ($) sign to the left of the *9* in the H9 cell reference for you. **Remember, whenever you want a column reference in a formula or function to remain unchanged, there must be a dollar ($) sign to its left. Whenever there is a dollar ($) sign to the left of the row reference, the row reference will also remain unchanged.** Pressing the **F4** key makes this happen, but you can also just type the dollar signs into the function arguments manually to place them at the left of the column reference or the row reference as needed.

Notice that by doing these changes to the **Absolute Address**, you are making sure that when you fill the function across, the reference to the **Rates** will not change. Also notice that by doing these changes to the **Absolute Address**, you are making sure that when you fill the function down, the reference to the **Years** will not change.

11. Fill the function to row 33.
12. Fill the functions in column I over to column N.
13. Widen columns H through N to a width of *12* characters. Your spreadsheet should now look like what you see in Figure 3-30.

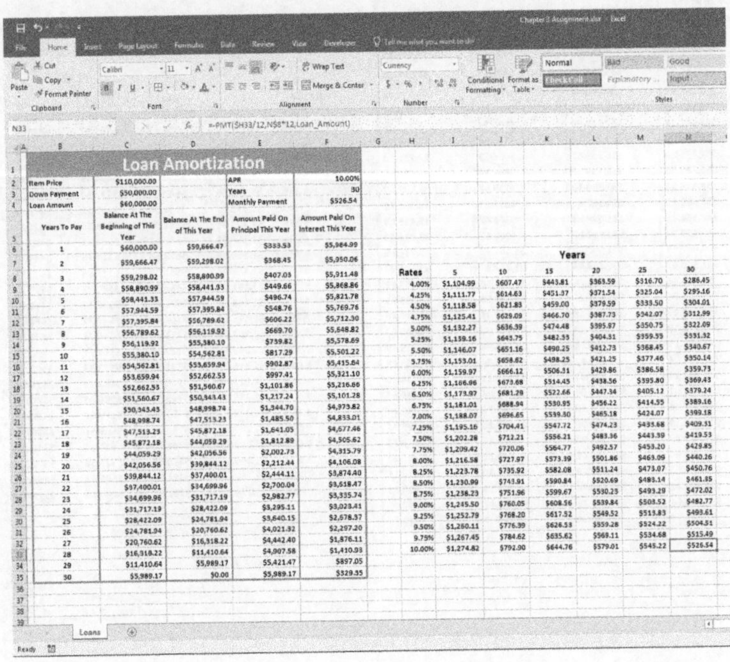

FIGURE 3-30

CHAPTER 3—ASSIGNMENT (SUMMARY OF TASKS)

Create the spreadsheet from this chapter using the labels as discussed and with the following specifications:

1. Shrink column A to a width of *1.00*.
2. Make sure columns B through F each have a width of *17* characters.
3. Type the words *Loan Amortization* in cell B1 and center it across cells B1 through F1.
4. In cell B2, type the words *Item Price.*
5. In cell B3, type the words *Down Payment.*
6. In cell B4, type the words *Loan Amount.*
7. In cell E2, type *APR.*
8. In cell E3, type the word *Years.*

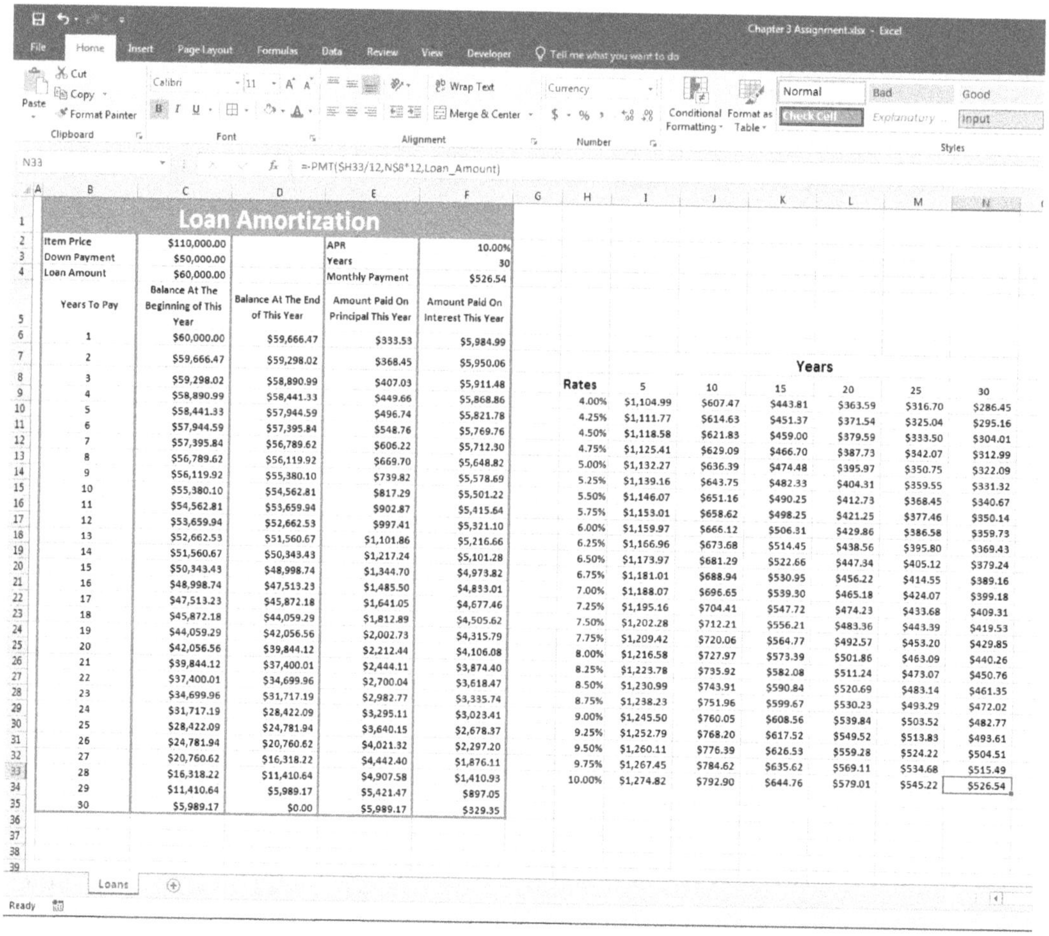

FIGURE 3-31

9. In cell E4, type the words *Monthly Payment.*
10. Name cells C2 through C4 and cells F2 through F4 whatever their labels are to their left (Using **Create from Selection** in the **Formulas ribbon**).
11. Put the labels in cells B5 through F5 as shown in Figure 3-31.
12. Generate a series of years in cells B6 through B35, also as shown in Figure 3-31.
13. In cell C2, enter *110000.*
14. In cell C3, enter *50000.*
15. In cell C4, create a formula that will calculate the *Loan_Amount* by subtracting the cell named *Down_Payment* from the cell named *Item_Price.* **Make sure you use the cell names needed in the formula.** For example, when you reference the price in cell C3, don't type in *C3,* but rather use the words *Down_Payment.*
16. In cell F2, type *10.00%.*
17. In cell F3, type *30.*
18. In cell F4, calculate the monthly payment by using the **PMT** function as follows:

    *=-PMT(APR/12,Years*12,Loan_Amount)*

Make sure you use the cell names needed in the function.

19. In cell C6, enter the formula that will give you the beginning balance of the loan for the first year by referencing the cell named *Loan_Amount.*
20. In cell D6, enter the function that will give the *Balance At the End of This Year* using the **PV** function as follows:

    =-PV(APR/12,12(Years-B6),Monthly_Payment)*
21. In cell E6, create a formula that will calculate the *Amount Paid on Principal This Year* by subtracting D6 from C6.
22. In cell F6, create a formula that will calculate the *Amount Paid on Interest This Year* as follows:

    *=12*Monthly_Payment-E6*
23. In Cell C7, enter the cell reference that has the ending balance of the previous year, which is D6.
24. Fill the formula in cell C7 down to row 35. **Do not be concerned that zeros will appear in all of them but the first cell. That will be corrected when the next step is completed.**
25. Fill cells D6, E6, and F6 down to row 35.
26. Make sure the format for all of the numbers is the same as those shown in Figure 3-31.
27. Apply **Accent 2** cell style to the **merged and centered** cell B1.
28. Make cell B1 a font size of *26.*
29. Add **vertical, orange borders** as shown in the picture in Figure 3-31.

30. Go to cell I7 and type the word *Years* with a font size of *16*.
31. **Merge and center** that word across cells I7 through N7.
32. In cell I8, type 5.
33. In cell J8, type *10*.
34. Fill the series you just entered into cells I8 and J8 across to cells N8.
35. In cell H8, enter the word *Rates* with a font of *14*.
36. In cell H9, enter *4.00%*.
37. In cell H10, enter *4.25%*.
38. Select both cells H9 and H10, and then fill that series down to row 33, at which point the series will have reached a level of 10.00% as you see in Figure 3-31.
39. Place your mouse in cell I9 and type the following function that reads:
    *=-PMT($H9/12,I$8*12,Loan_Amount)*
40. Fill the function to row 33, and then fill all of the functions in column I over to column N.
41. Widen columns H through N to a width of *10*. Your spreadsheet should now look like Figure 3-31.
42. Perform **Goal Seek**, using $800 as the preferred monthly payment. Do not put the dollar sign ($) in the **Goal Seek dialogue box. Save the results that you get by using Goal Seek!**
43. Submit the file and present it for grading.

CHAPTER 3—ALTERNATE ASSIGNMENT (SUMMARY OF TASKS)

Create a new spreadsheet with the following specifications:

1. Shrink column A to a width of *1.00*.
2. Make sure columns B through F each have a width of *17* characters.
3. Type the words *Loan Amortization* in cell B1, and center it across cells B1 through F1.
4. In cell B2, type the words *Item Price*.
5. In cell B3, type the words *Down Payment*.
6. In cell B4, type the words *Loan Amount*.
7. In cell E2, type *APR*.
8. In cell E3, type the word *Years*.
9. In cell E4, type the words *Monthly Payment*.
10. Name cells C2 through C4 and cells F2 through F4 whatever their labels are to the left (using **Create from Selection** in the **Formulas ribbon**).
11. Put the labels in cells B5 through F5 as shown in Figure 3-32.

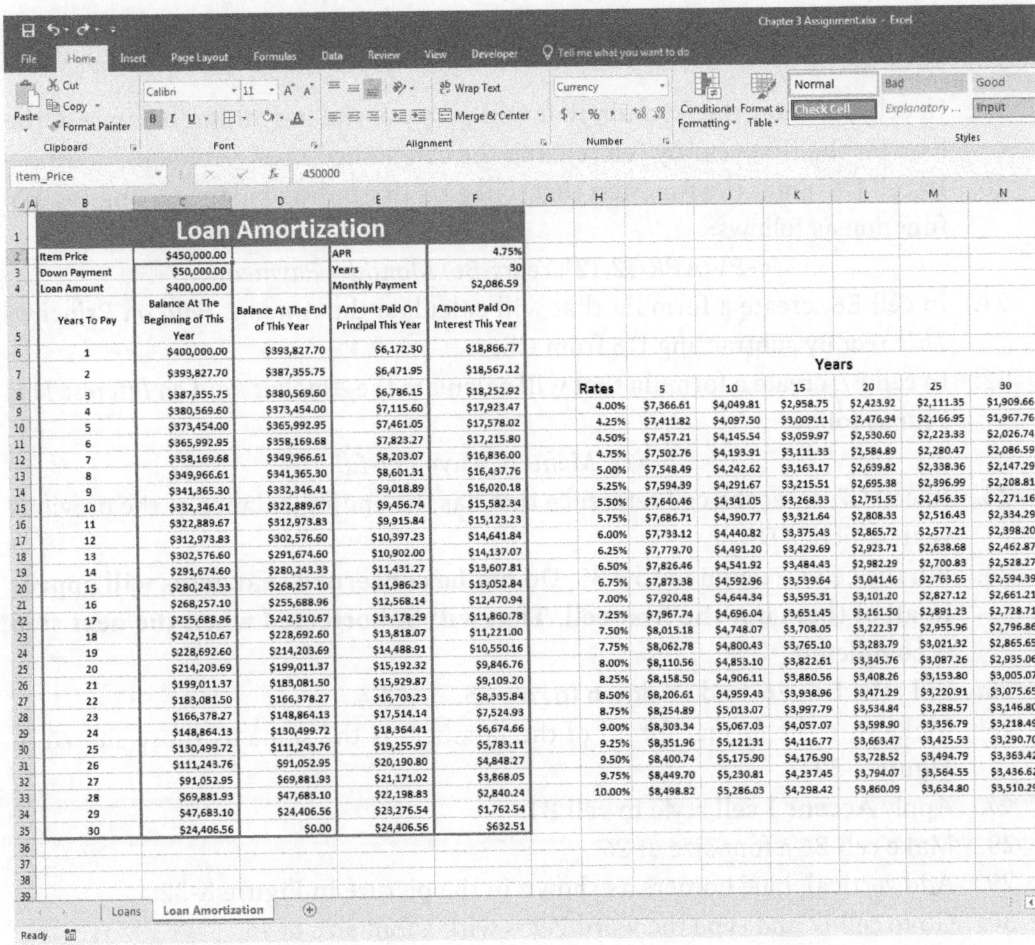

FIGURE 3-32

12. Generate a series of years in cells B6 through B35, also as shown in Figure 3-32.
13. In cell C2, enter *450000*.
14. In cell C3, enter *50000*.
15. In cell C4, create a formula that will calculate the *Loan Amount* by subtracting the cell named *Down_Payment* from the cell named *Item_Price*. **Make sure you use the cell names needed in the formula.** For example, when you reference the price in cell C3, don't type in *C3*, but rather use the words *Down_Payment*.
16. In cell F2, type *4.75%*.
17. In cell F3, type *30*.
18. In cell F4, calculate the monthly payment by using the **PMT function** as follows:

*=-PMT(APR/12,Years*12,Loan_Amount)*

Make sure you use the cell names needed in the function.

19. In cell C6, enter the formula that will give you the beginning balance of the loan for the first year by referencing the cell named *Loan_Amount.*
20. In cell D6, enter the function that will give the ending balance using the **PV** function as follows:

    =-PV(APR/12,12(Years-B6),Monthly_Payment)*
21. In cell E6, create a formula that will calculate the *Amount Paid on Principal This Year* by subtracting D6 from C6.
22. In cell F7, create a formula that will calculate the *Amount Paid on Interest This Year* as follows:

    *=12*Monthly_Payment-E6*
23. In cell C7, enter the cell reference that has the ending balance of the previous year, which is D6.
24. Fill the cell C7 down to row 35. **Do not be concerned that zeros will appear in all of them but the first cell. That will be corrected when the next step is completed.**
25. Fill cells D6, E6, and F6 down to row 35.
26. Make sure the format for all of the numbers is the same as those shown in Figure 3-32.
27. Apply **Accent 1** cell style to cell B1.
28. Make cell B1 a font size of *26*.
29. Add vertical, blue borders as shown in the picture in Figure 3-32.
30. Go to cell I7 and type the word *Years* with a font size of *16*.
31. Merge and center that word across cells I7 through N7.
32. In cell I8, type in *5*.
33. In cell J8, type in *10*.
34. Fill the series you just entered into cells I8 and J8 across to cells N8.
35. In cell H8, enter the word *Rates* with a font of *14*.
36. In cell H9, enter *4.00%*.
37. In cell H10, enter *4.25%*.
38. Select both cells H9 and H10, and then fill that series down to row 33, at which point the series will have reached a level of 10.00% as you see in Figure 3-32.
39. Place your mouse in cell I9, and type the following function that reads

    *=-PMT($H9/12,I$8*12,Loan_Amount)*
40. Fill the function to row 33, and then fill it over to column N.
41. Widen columns H through N to a width of *10*. Your spreadsheet should now look like Figure 3-32.

42. Perform **Goal Seek**, using $1,200 as the preferred monthly payment. Do not put the dollar ($) sign in the **Goal Seek dialogue box. Save the results that you get by using Goal Seek!**
43. Submit the file and present it for grading.

CHAPTER FOUR

Using the FV (Future Value) Function, Cell Indentation, Spreadsheet Protection, and Data Tables

BY THE END of this chapter, it is our objective that the user will be able to:

1. Create a **Data Table**.
2. Use the **FV** (**Future Value**) function.
3. Protect a spreadsheet.
4. Indent cells.

CHAPTER 4 TASK: CALCULATING AND ANALYZING INVESTMENT DATA FOR FUTURE NEEDS

Suppose you want to get an idea of how much money you will have in an investment if you put a certain amount of money into it each year. Suppose, for example, you are saving up for your children's college and that you expect to need $250,000 to put one of your children through college in eighteen years. Or maybe you have to replace the roof on your home or business in thirty years at a cost of $50,000. You can build a spreadsheet that will assist you in setting up these investments by using what is called the **Future Value** (**FV**) function.

Let's build a spreadsheet now, applying many of the concepts we have already learned in this book and then adding a few more.

1. Start a new spreadsheet.
2. In cell A1, type the words *Saving for College*, and make it bold with a font size of 22.
3. **Merge and center** those words across cells A1 through E1.
4. Make column A a width of *30*.
5. Make column B a width of *16*.
6. Make column C a width of *3*.

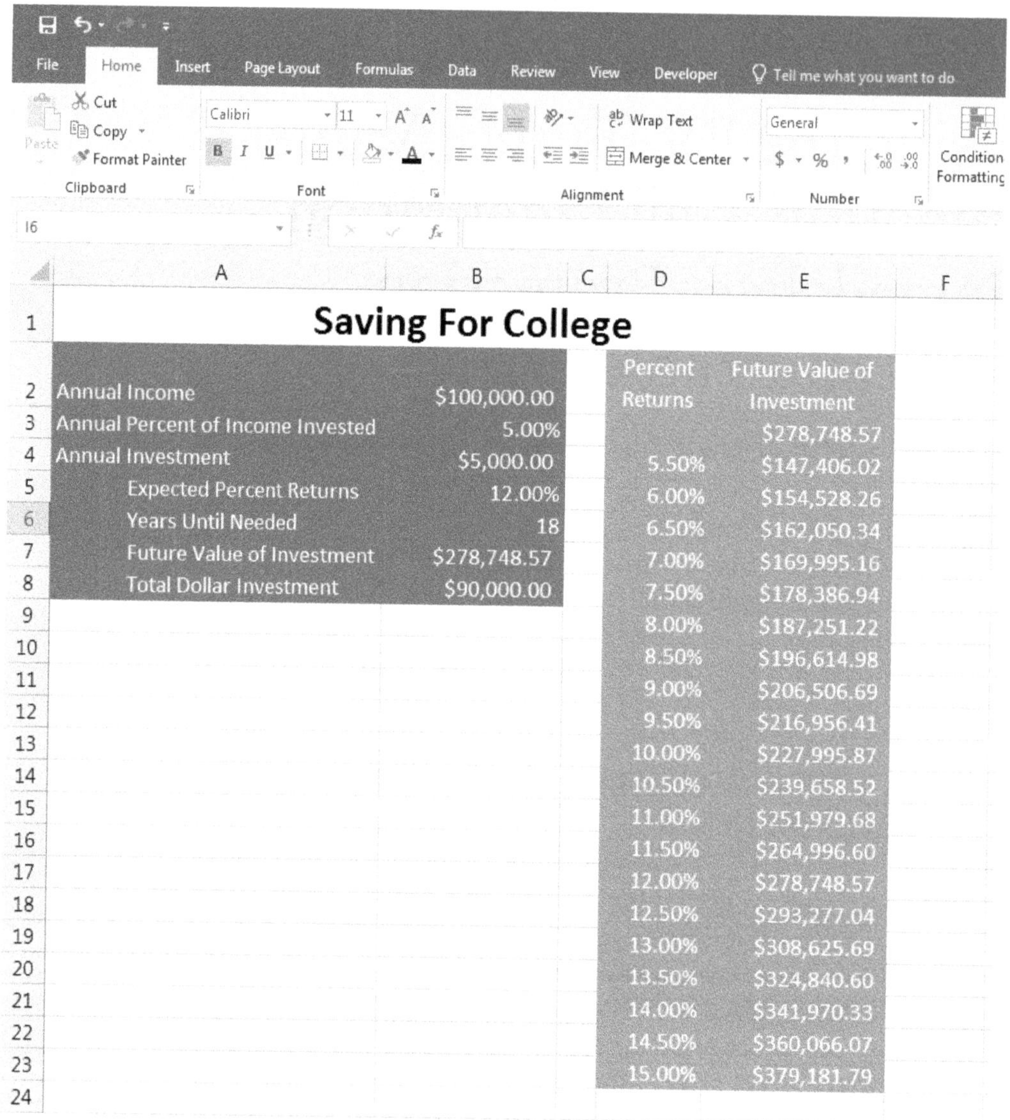

FIGURE 4-1

7. Make column D a width of *9*.
8. Make column E a width of *16*.
9. In cell A2, type the words *Annual Income*.
10. In cell A3, type the words *Annual Percent of Income Invested*.
11. In cell A4, type the words *Annual Investment*.
12. In cell A5, type the words *Expected Percent Returns*.
13. In cell A6, type the words *Years Until Needed*.
14. In cell A7, type the words *Future Value of Investment*.

15. In cell A8, type the words *Total Dollar Investment.*

TO INDENT DATA IN A CELL

There are times that you will likely want to **indent** the data in some spreadsheet cells in order to organize your spreadsheets better and to make them easier for users to follow. In this case, we are going to indent the text in cells A5 through A8 by five (5) characters by doing the following:

Step 1: Select the cell(s) in which you want indentation (cells A5 through A8).

Step 2: Click the **Increase Indent** icon in the **Home Ribbon** five (5) times (Figure 4-2, Arrow). NOTE: If you indent text too far, you can click the **Decrease Indent** icon (to the left of the **Increase Indent** icon) to remove some of the indentation.

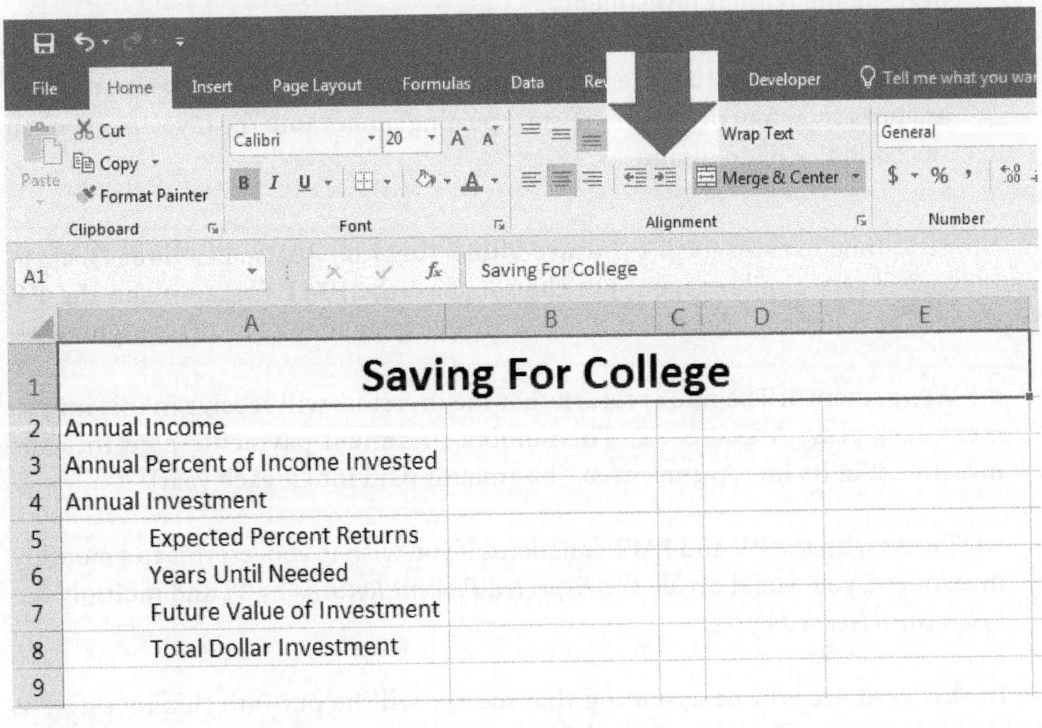

FIGURE 4-2

Let's continue to modify the spreadsheet in order by picking up where we left off.

16. Now make the names of cells the same as the labels at their left, using the **Create from Selection** command in the **Formulas Ribbon** as discussed in Chapter 3.
17. In cell B2, enter *100000* and apply the **Currency** dollars format to that cell (Figure 4-3). This is the income of the family who is saving for their children's college.
18. In cell B3, enter *5.00%* and make sure that it has **TWO** places to the right of the decimal. This *5.00%* represents the percent of their income they hope to invest each year for that college fund.
19. In cell B4, enter the formula:
    *=Annual_Income*Annual_Percent_of_Income_Invested*
    This will calculate how much they will be investing each year in dollars to the fund. Apply the **Currency** dollars format to that cell.
20. In cell B5, enter *12.00%* and make sure that it has **TWO** places to the right of the decimal. This is the average rate of return the family expects to receive over the life of their investment.
21. In cell B6, enter *18*, which represents the number of years they will have to wait until the child they are sending to college will be needing the funds. This assumes therefore that they will start putting money into the investment when the child is still an infant.

Now it is time to enter the **FV** (**Future Value**) function. It also has three (3) arguments and is very similar to, but not the same as, the **PMT** function and the **PV** function. The function is typed out below with an explanation of its three arguments:

=-FV(Argument 1: The rate of return that the investors will receive in this investment each year; Argument 2: The number of annual payments paid into the investment in its life; Argument 3: The amount paid into it each year).

NOTE: As with the **PV** and **PMT** functions, if you wish to convert this to a monthly investment, you would divide the *Expected Percent Returns* by 12 and multiply the *Years Until Needed* by 12.

In this case, we will be assuming that money will be put into this investment only once per year. Therefore, in cell B12, type in the **FV** (**Future Value**) function as follows:

=-FV(Expected_Percent_Returns,Years_Until_Needed,Annual_Investment)

The result of *$278,748.57* will be what you see in Figure 4-3.

NOTE: Just like the **PMT** and **PV** functions, Excel will automatically give the cell a **Currency** format.

Please note again that a **negative sign** (-) must be in the function. Otherwise, Excel will default to making the function negative. Thus, you need to add a negative to convert it to a positive number as you did for the **PMT** and **PV** functions.

You may want to know how much you will have **actually** invested into this venture in order for it to reach the value of *$278,748.57*. You can do so with the following steps:

Step 1: In cell B9, enter the formula:

*=Annual_Investment*Years_Until_Needed*

It will give you a value of $90,000.

Step 2: Using the **Format Painter**, paint the format of cell B7 to all of the other cells **with dollar values** if needed.

Your spreadsheet should now look like Figure 4-3.

At this time, apply the **Accent1 Cell Style** to cells A2 through B8.

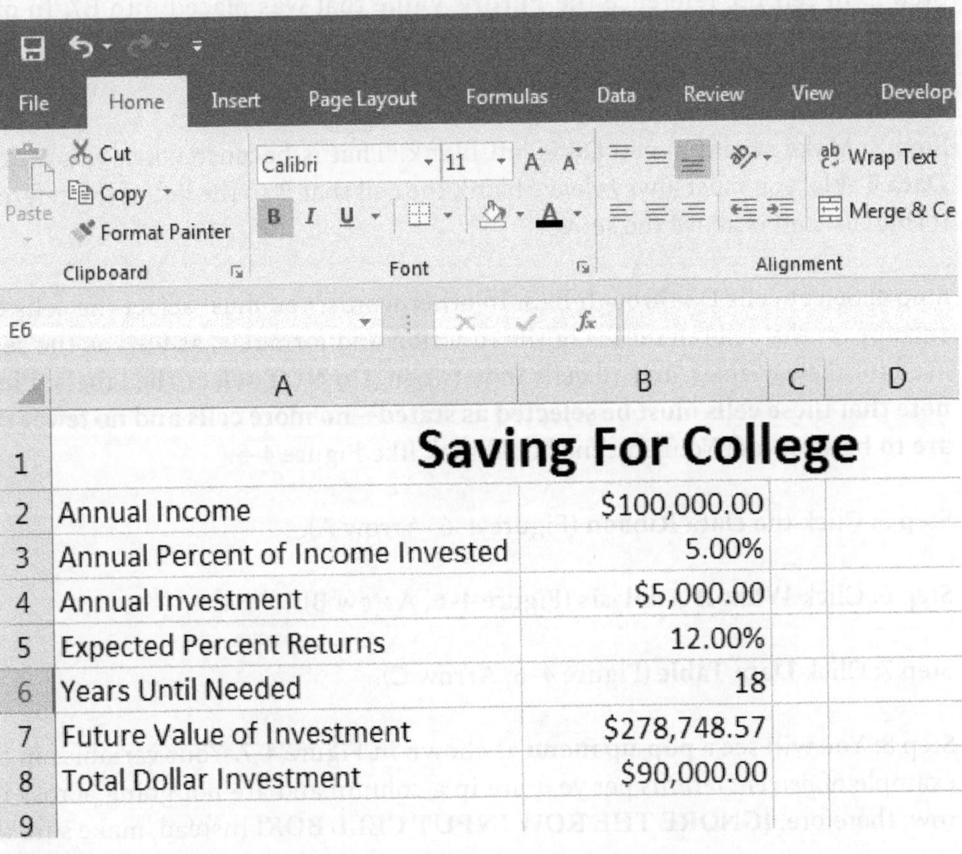

FIGURE 4-3

CREATING A DATA TABLE

A **Data Table** in Excel is a tool that is used to see different results or calculations with various scenarios. Suppose we want to see the **Future Value** of the college fund at different rates of return on the investments in which the fund is placed. One **slow** way to do so is to change the rate in cell B5 numerous times and then write down the results each time you change it. However, that would be very time consuming! Therefore, a **Data Table** would be more practical.

It is important to understand here that a **Data Table** is only reflecting calculations, functions, and formulas that have already been created somewhere else in the spreadsheet. In this case, it will reflect what has already been created in cells A2 through B8.

Creating a **Data Table** can be done by taking the following steps:

Step 1: Using methods already discussed in previous chapters, enter the labels in cells D2 and E2 that you see in Figure 4-4, while also entering the series of different *Percent Returns* in cells D4 through D23. NOTE: Make sure that the *Percent Returns* have a **percent format**.

Step 2: In cell E3, reference the **Future Value** that was placed into B7. In other words, the formula in cell E3 should read *=Future_Value_of_Investment* (Figure 4-4). Give cell E3 the **Currency** format.

Step 3: Make sure that cell D3 is left blank. That is because when you create a **Data Table**, you must always leave blank the cell that is to the left of the reference formula(s) and is above the series.

Step 4: Select cells D3 through E23. In other words, you must select the cells that encompass the cell references of the function and formulas, as well as the series used for the variables, and all cells in between. **Do NOT select the labels. Please note that these cells must be selected as stated—no more cells and no fewer cells are to be selected!** Your screen should look like Figure 4-5.

Step 5: Click the **Data Ribbon** (Figure 4-6, Arrow A).

Step 6: Click **What If Analysis** (Figure 4-6, Arrow B).

Step 7: Click **Data Table** (Figure 4-6, Arrow C).

Step 8: You will see a **pop-up menu** as shown in Figure 4-7. Your variables in this example of *percent returns* per year are in a column and are not going across in a row; therefore, **IGNORE THE ROW INPUT CELL BOX!** Instead, make sure that

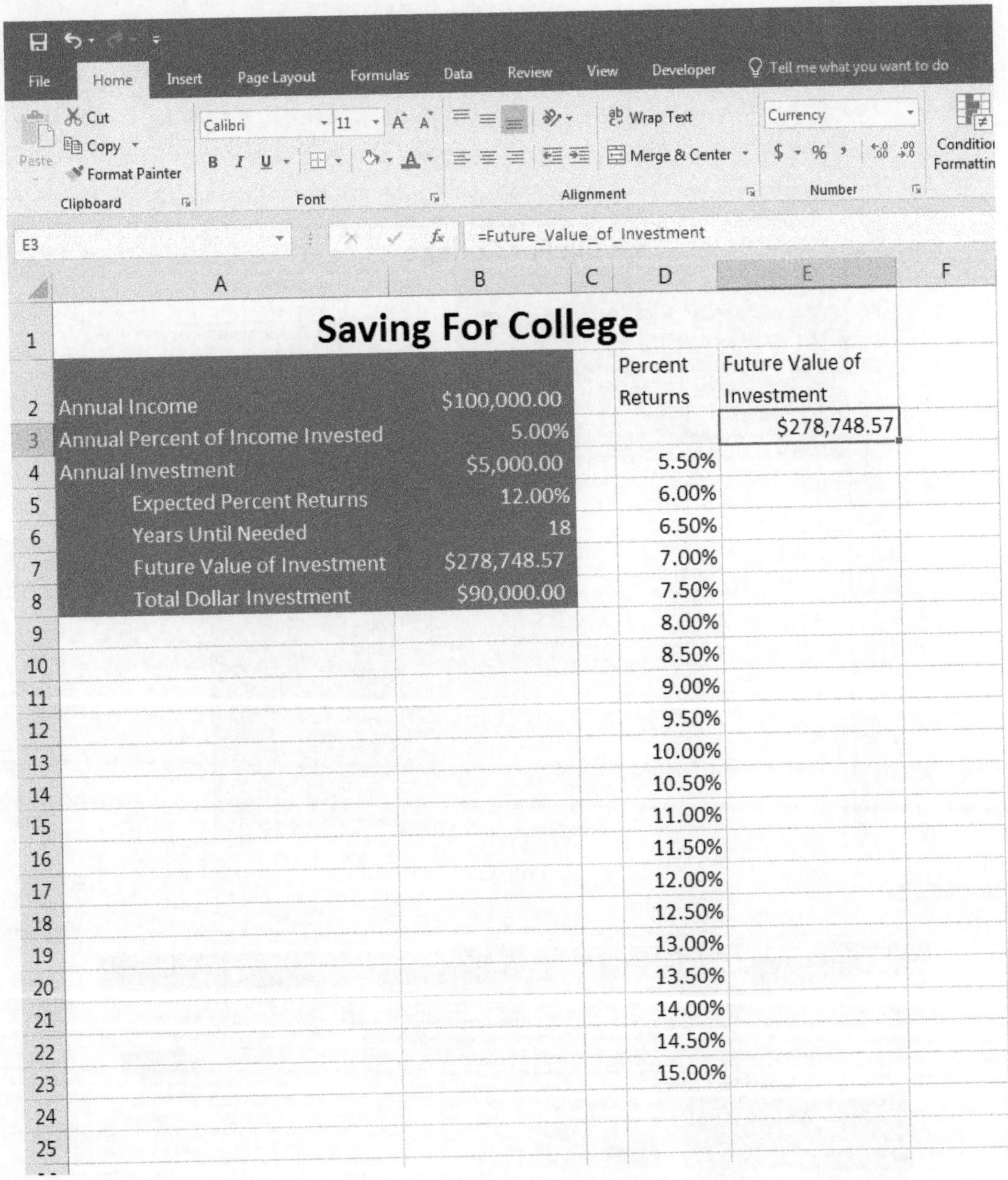

FIGURE 4-4

in the **Column Input Cell box** you tell Excel to have each of the percent returns in column D sent to the *Expected Percent Returns* cell (cell B5) in order to see what the future values will be at each of the different rates of return (Figure 4-7, Arrow).

Step 9: Click **OK**. Your screen should look like what you see in Figure 4-8.

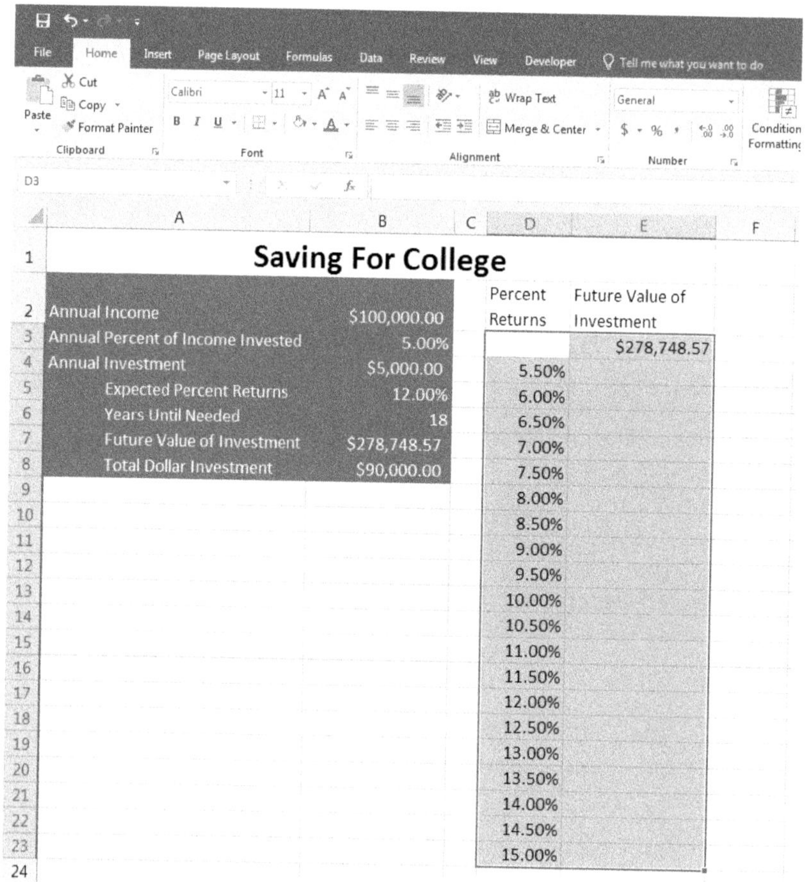

FIGURE 4-5

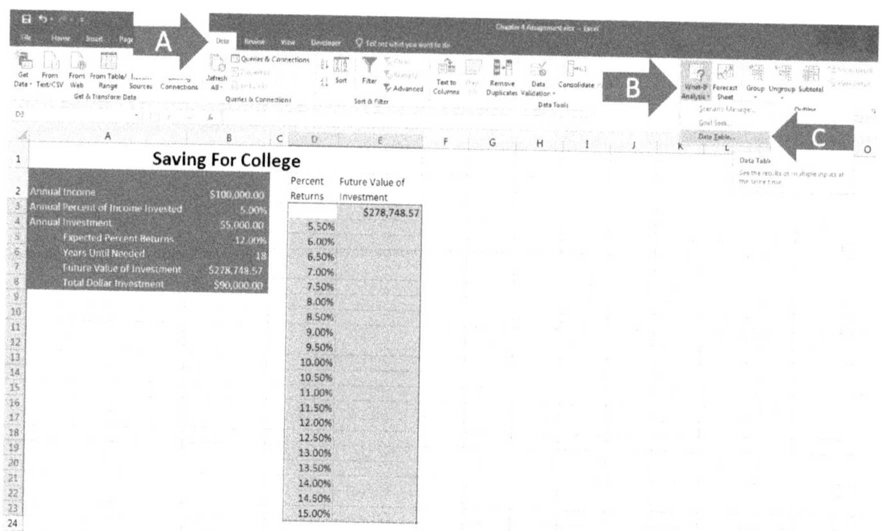

FIGURE 4-6

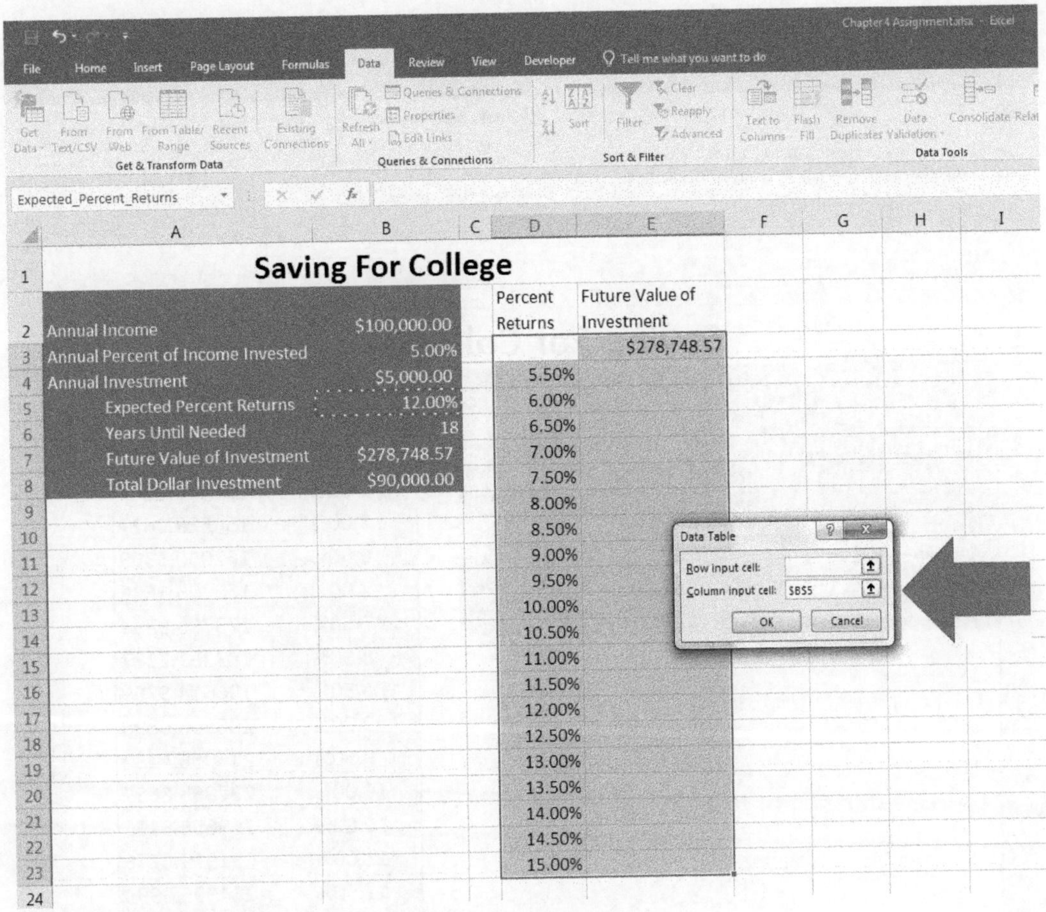

FIGURE 4-7

NOTE: If you make a mistake and must redo the **Data Table**, you cannot erase it without first selecting **all of it** and then pressing **delete. You CANNOT delete a PORTION of a Data Table.**

Now make the spreadsheet look like Figure 4-9 (so it will be a little easier to read) by doing the following:

1. Paint the currency value in cell B7 to the cells E4 through E23.
2. Center the labels inside each of cells D2 and E2.
3. Apply the **Accent6 Cell Style** to cells D2 through E23.

Your spreadsheet should look like what you see in Figure 4-9.

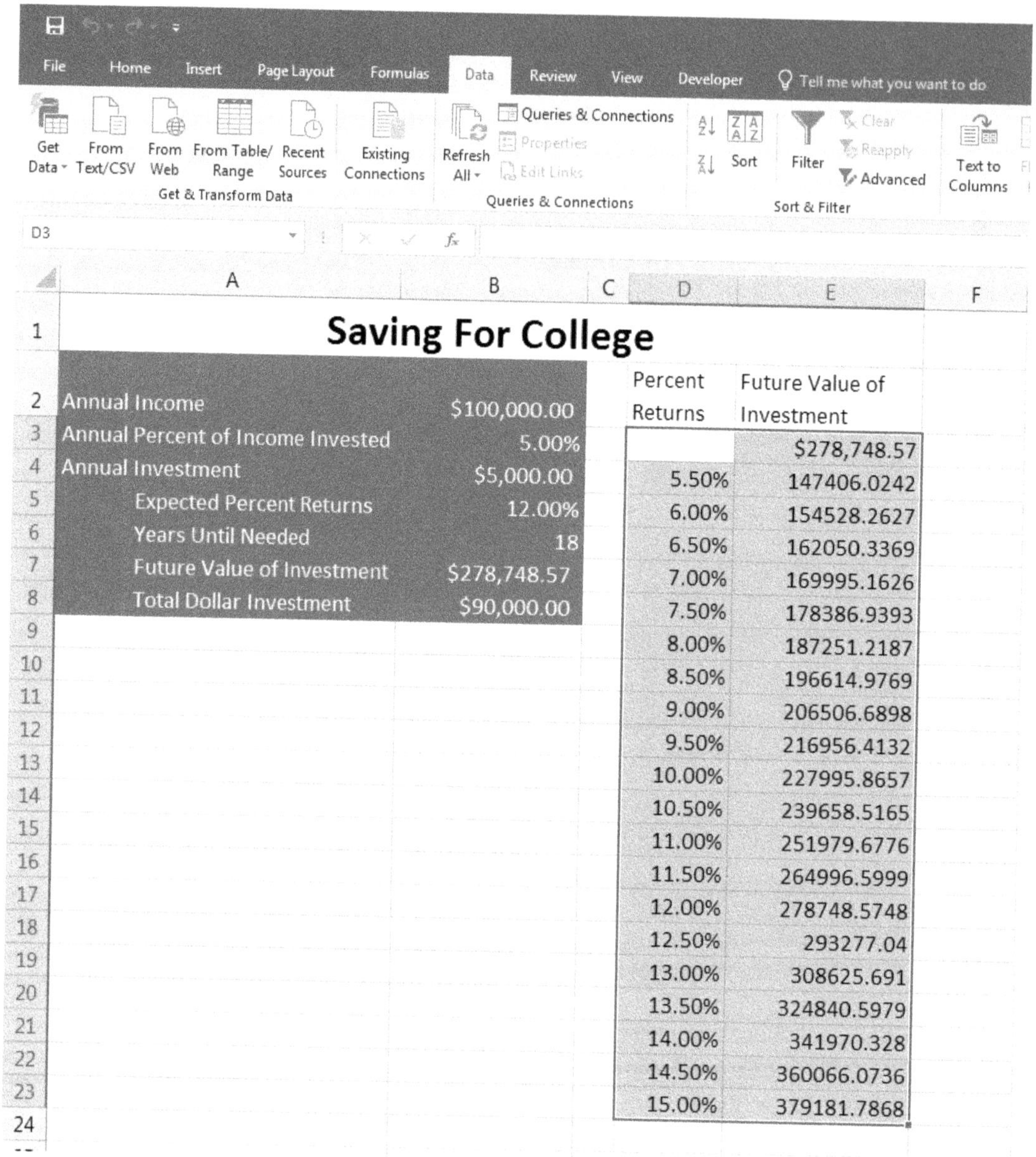

File
Home
Insert
Page Layout
Formulas
Data
Review
View
Developer
Tell me what you want to do
Get Data
From Text/CSV
From Web
From Table/ Range
Recent Sources
Existing Connections
Get & Transform Data
Refresh All
Queries & Connections
Properties
Edit Links
Queries & Connections
Sort
Filter
Clear
Reapply
Advanced
Sort & Filter
Text to Columns
D3
Saving For College
Annual Income $100,000.00
Annual Percent of Income Invested 5.00%
Annual Investment $5,000.00
Expected Percent Returns 12.00%
Years Until Needed 18
Future Value of Investment $278,748.57
Total Dollar Investment $90,000.00
Percent Returns
Future Value of Investment
$278,748.57
5.50% 147406.0242
6.00% 154528.2627
6.50% 162050.3369
7.00% 169995.1626
7.50% 178386.9393
8.00% 187251.2187
8.50% 196614.9769
9.00% 206506.6898
9.50% 216956.4132
10.00% 227995.8657
10.50% 239658.5165
11.00% 251979.6776
11.50% 264996.5999
12.00% 278748.5748
12.50% 293277.04
13.00% 308625.691
13.50% 324840.5979
14.00% 341970.328
14.50% 360066.0736
15.00% 379181.7868

FIGURE 4-8

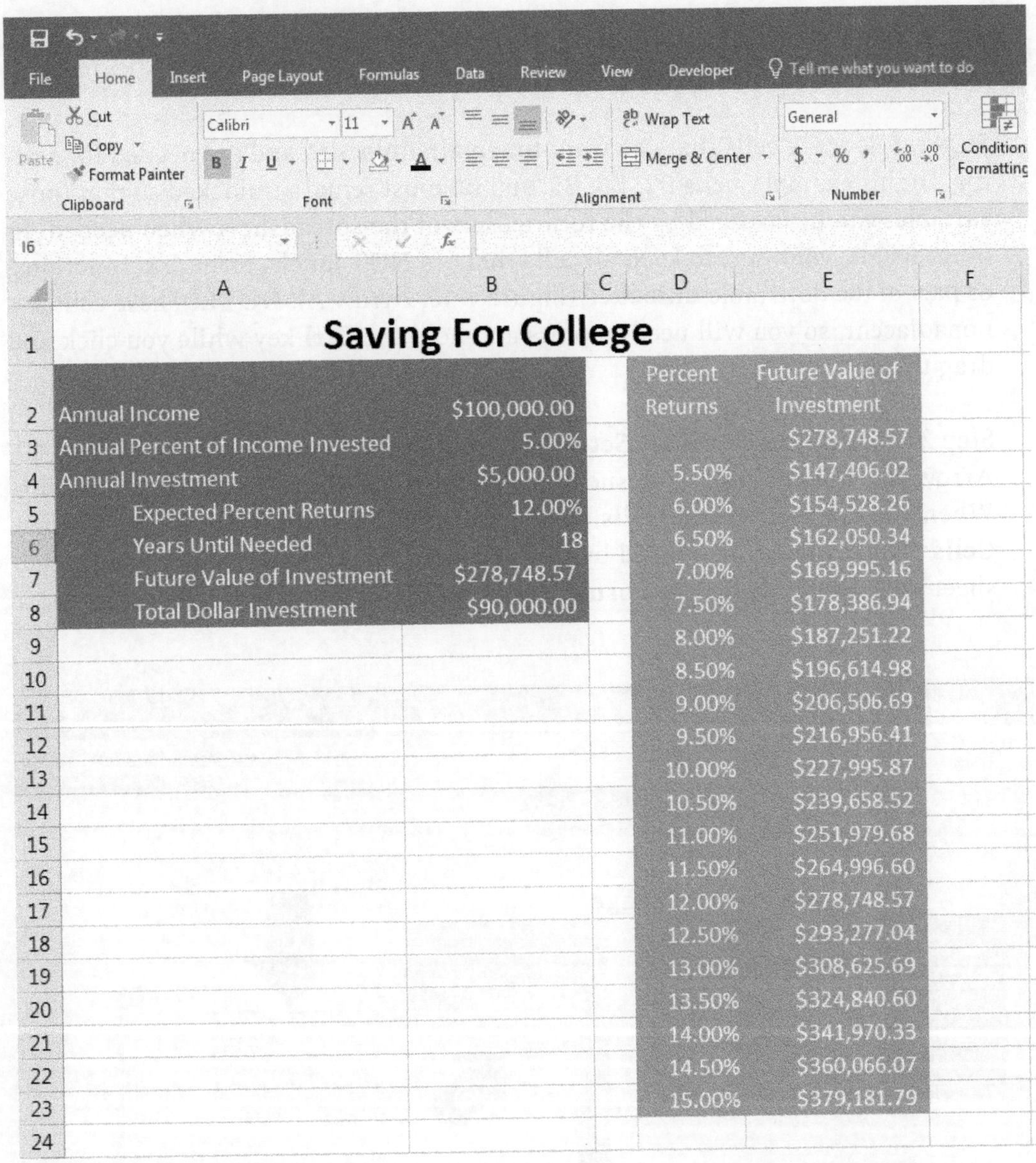

FIGURE 4-9

PROTECTING THE SPREADSHEETS

Sometimes when people use spreadsheets, they accidentally type over some of the important labels, functions, and formulas that are in place. To avoid this, the sheets can be locked. However, if we lock the **entire** sheet, we cannot enter the variables. In this case, the variables are the *Annual Income, the Annual Percent of Income Invested*, the *Expected Percent Returns*, and the *Percent Returns*. Therefore, before we lock the sheet, we must tell Excel to keep those cells **unlocked** for later data entry. Here are the steps to **keep** cells **unlocked** when the sheet is locked **later**:

Step 1: Select the cells that you want to remain unlocked and are needed for editing. In this example, cells B2, B3, B5, and B6 must remain unlocked so that those variables can be changed by the future users of the spreadsheet when needed. In other words, you want to keep all cells that are **NOT** labels, formulas, functions, or part of the data table **unlocked** (Figure 4-10, Arrow A). NOTE: These cells are nonadjacent, so you will need to press and hold the **Ctrl** key while you click and drag them.

Step 2: Click on the tiny **Font Settings** button in the **Home Ribbon** (Figure 4-10, Arrow B). It is not labeled as such, but please click where Arrow B is pointing. When you do, the **Format Cells** pop-up menu will appear. NOTE: The **Format Cells** menu can also be opened by right-clicking the selected cells in the spreadsheet and then choosing the **Format Cells** option.

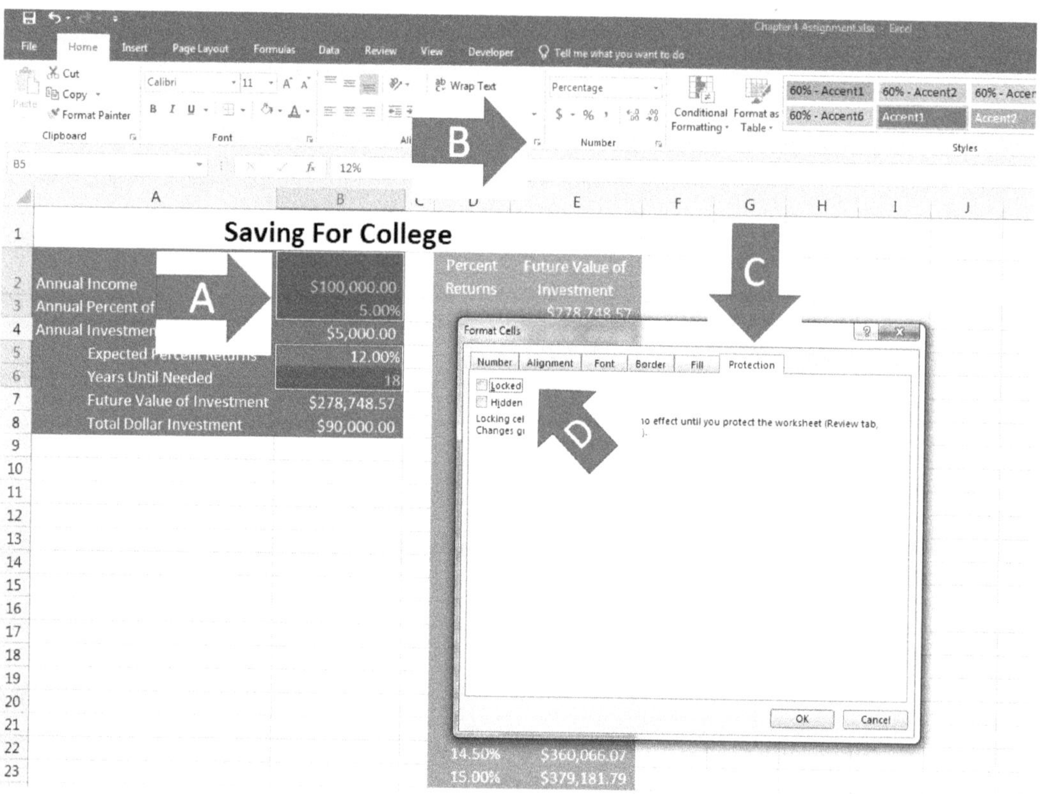

FIGURE 4-10

Step 3: Click the tab marked **Protection** (Figure 4-10, Arrow C).

Step 4: Uncheck the **Locked** box (Figure 4-10, Arrow D). Notice that all cells as a default are set to lock when the spreadsheet is locked. By unchecking the **Locked** box in the **Protection** menu, the selected sheets will not lock when you lock the sheet later.

Step 5: Click **OK**.

Now it's time to lock the rest of the sheet. To do so:

Step 1: Click **Review** in the **Home Ribbon** (Figure 4-11, Arrow A).

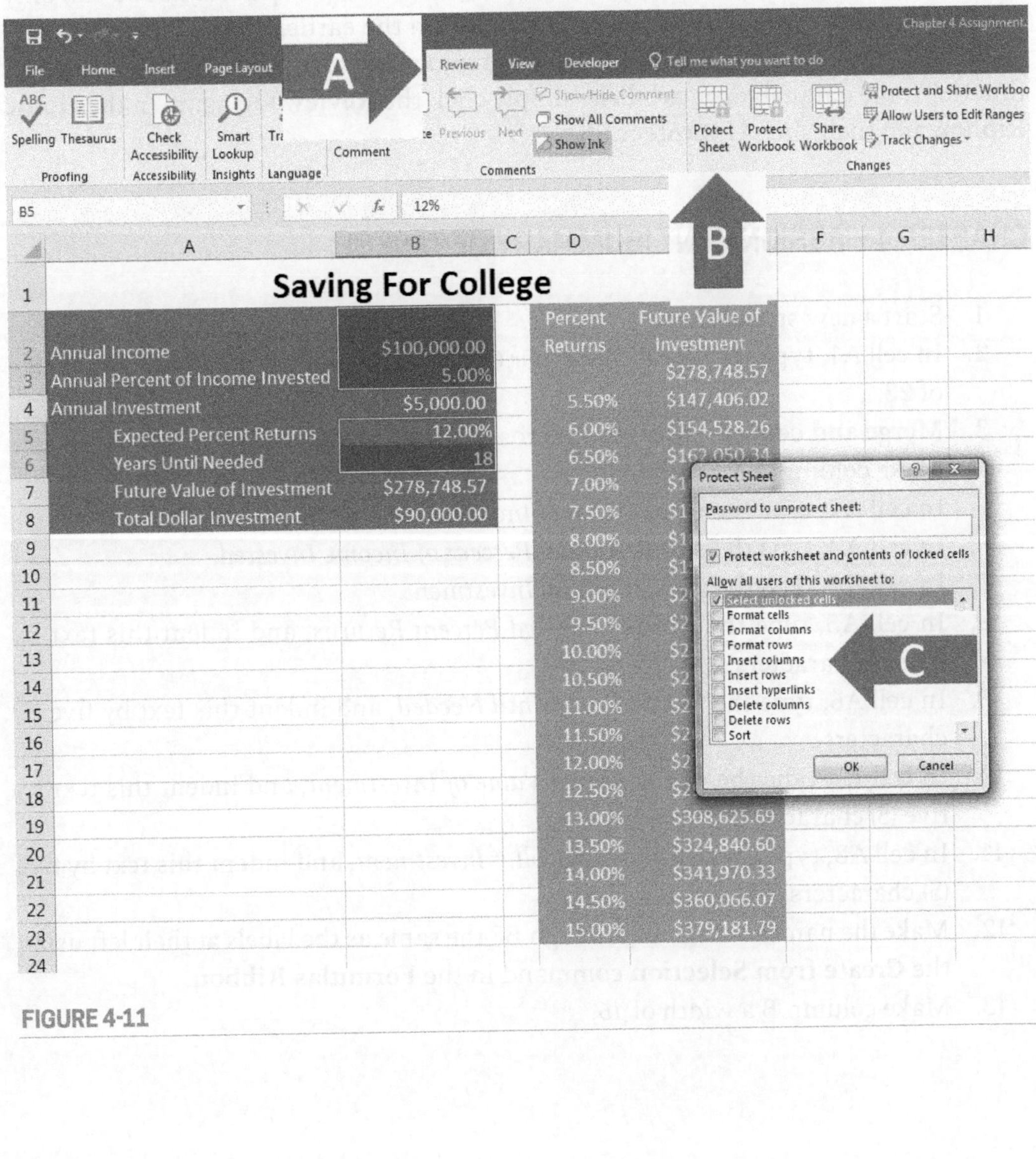

FIGURE 4-11

Step 2: Click **Protect Sheet** (Figure 4-11, Arrow B). That will bring you the **Protect Sheet pop-up menu**. NOTE: As you can see in this menu, you can also lock the entire workbook, but we will not be doing that at this time.

Step 3: You can choose the things you may still want users to continue to do after locking the sheets (Figure 4-11, Arrow C). In this example, we will leave this screen as it is, so that the user can **ONLY** select unlocked cells but nothing more. Do not add a password. If you do, and if you then forget it, you can't modify the design of this spreadsheet anymore.

Step 4: Click **OK**.

If you attempt to select cells that are locked, Excel will not permit it. You can only select and change data in the cells you unlocked in the earlier steps.

NOTE: If you wish to unlock the sheet again to make further changes to formulas, functions, the data table, or to labels, simply click the **Review** tab again in the **Home Ribbon** and then click **Unprotect Sheet**.

CHAPTER 4—ASSIGNMENT (SUMMARY OF TASKS)

1. Start a new spreadsheet.
2. In cell A1, type the words *Saving for College*, and make it bold with a font size of 22.
3. **Merge and center** those words across cells A1 through E1.
4. Make column A a width of *30*.
5. In cell A2, type the words *Annual Income*.
6. In cell A3, type the words *Annual Percent of Income Invested*.
7. In cell A4, type the words *Annual Investment*.
8. In cell A5, type the words *Expected Percent Returns*, and indent this text by five (5) characters.
9. In cell A6, type the words *Years Until Needed*, and indent this text by five (5) characters.
10. In cell A7, type the words *Future Value of Investment*, and indent this text by five (5) characters.
11. In cell A8, type the words *Total Dollar Investment*, and indent this text by five (5) characters.
12. Make the names of cells B2 through B8 the same as the labels at their left using the **Create from Selection** command in the **Formulas Ribbon**.
13. Make column B a width of *16*.

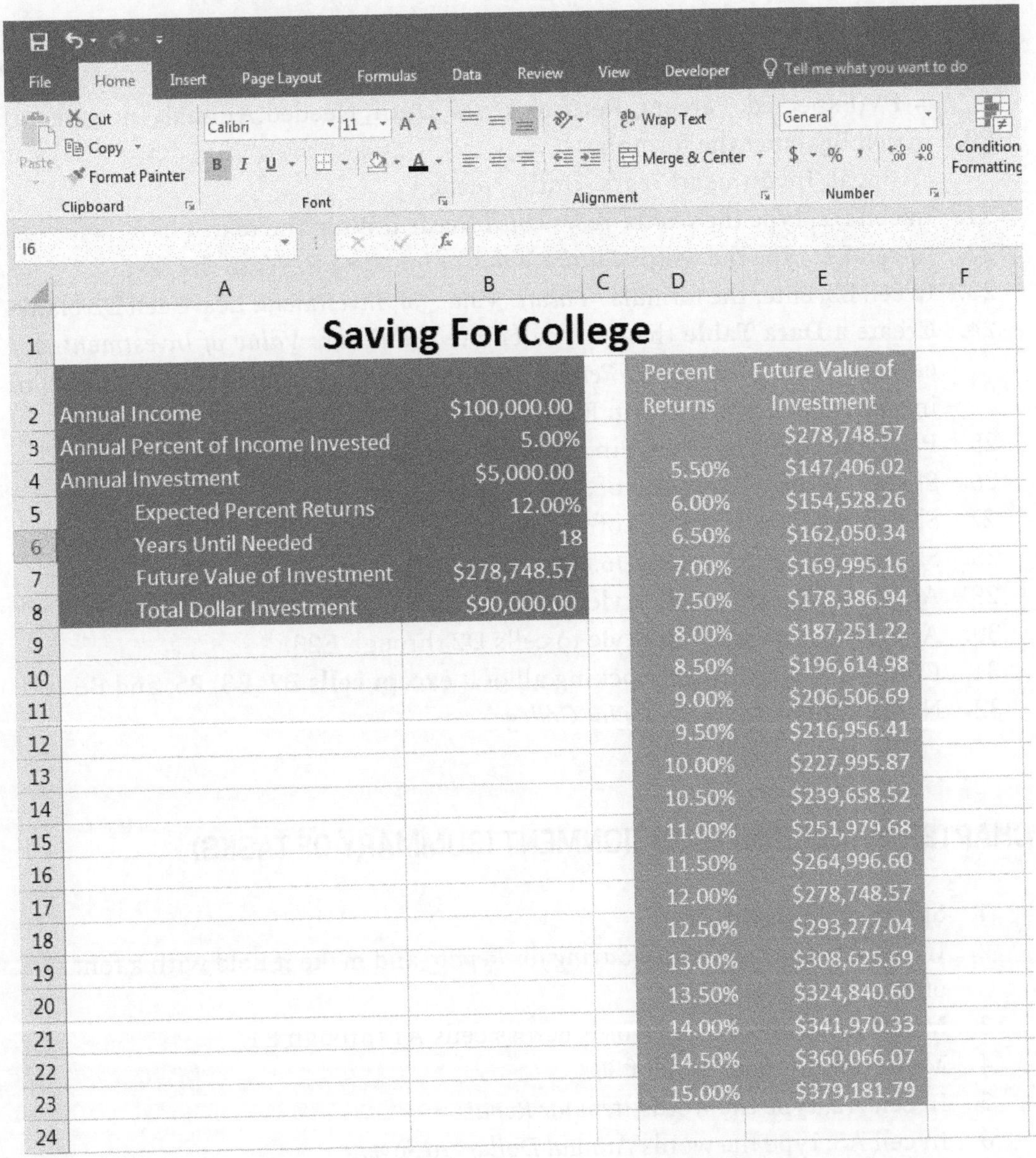

FIGURE 4-12

14. In cell B2, enter *100000*. Use the **Accounting Currency** format in cell B2.
15. In cell B3, enter *5.00%* with **TWO** places to the right of the decimal.
16. In cell B4, enter the **formula** *=Annual_Income*Annual_Percent_of_Income_Invested*. Use the **Accounting Currency** format for B4.
17. In cell B5, enter *12.00%* with **TWO** places to the right of the decimal.

18. In cell B6, enter *18* with **NO** places to the right of the decimal.
19. In cell B7, enter the **Future Value** (**FV**) function:
    =-FV(Expected_Percent_Returns,Years_Until_Needed,Annual_Investment)
20. In cell B8, enter the formula:
    =Annual_Investment*Years_Until_Needed
21. In cell D2, type the words *Years Until Needed*. Wrap the text.
22. In cell E2, type the words *Future Value of Investment*. Wrap the text.
23. In cell E3, enter the formula *=Future_Value_of_Investment*. Leave cell D3 blank.
24. Create a **Data Table** that will calculate the *Future Value of Investments* for each scenario of *Expected Percent Returns*, ranging from *5.50%* to *15.00%* in increments of *.50* as seen in Figure 4-12.
25. Paint the currency format in cell B7 to cells E3 through E23.
26. Shrink column C to a size of *3.00*.
27. Shrink column D to a size of *9.00*.
28. Shrink column E to a size *16.00*.
29. Apply the **Accent1 Cell Style** to cells A2 through B8.
30. Apply the **Accent6 Cell Style** to cells D2 through E23.
31. Protect the spreadsheet, locking all of it **except cells** B2, B3, B5, and B6.
32. Name the **sheet tab** *Saving for College*.

CHAPTER 4—ALTERNATE ASSIGNMENT (SUMMARY OF TASKS)

1. Start a new spreadsheet.
2. In cell A1, type the words *Saving for Repair*, and make it bold with a font size of 22.
3. **Merge and center** those words across cells A1 through E1.
4. Make column A a width of *30*.
5. In cell A2, type the words *Item for Repair*.
6. In cell A3, type the words *Annual Dollars to Save*.
7. In cell A4, type the words *Expected Rate of Returns*, and indent this text by five (5) characters.
8. In cell A5, type the words *Years Until Repair Needed*, and indent this text by five (5) characters.
9. In cell A6, type the words *Future Value of Savings*, and indent this text by five (5) characters.
10. In cell A7, type the words *Total Dollars Invested*, and indent this text by five (5) characters.

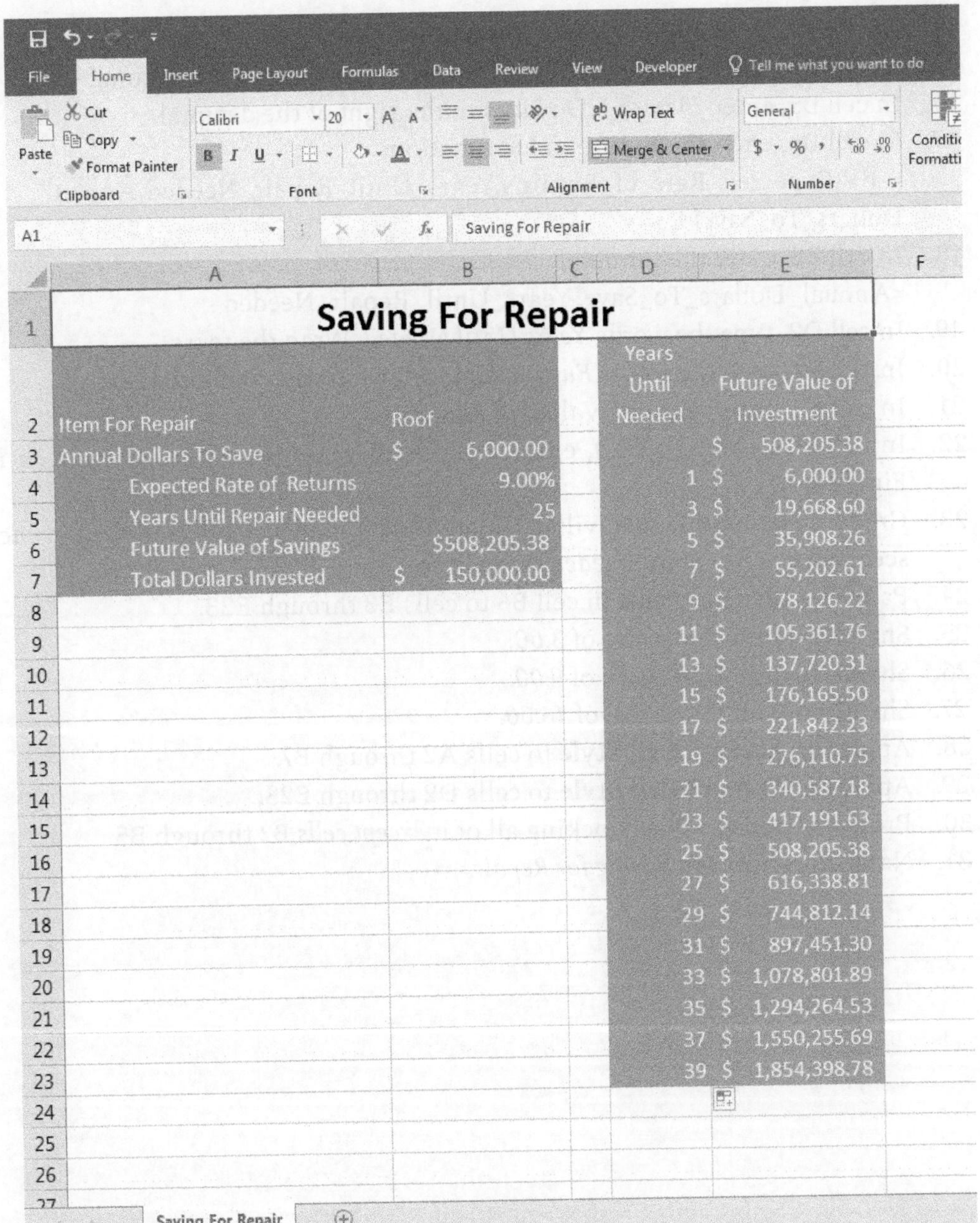

FIGURE 4-13

11. Now make the names of cells B2 through B7 the same as the labels at their left, using the **Create from Selection** command in the **Formulas Ribbon**.
12. Make column B a width of *16*.
13. In cell B2, enter the word *Roof*.

14. In cell B3, enter *6000*. Use the **Currency** format.
15. In cell B4, enter *9.00%* with **TWO** places to the right of the decimal.
16. In cell B5, enter *25* with **NO** places to the right of the decimal.
17. In cell B6, enter the Future Value function: =-FV(Expected_Rate_of_Returns,Years_Until_Repair_Needed,Annual_Dollars_To_Save)
18. In cell B8, enter the formula: =Annual_Dollars_To_Save*Years_Until_Repair_Needed
19. In cell D2, type the words *Years Until Needed*. Wrap the text.
20. In cell E2, type the words *Future Value of Investment*. Wrap the text.
21. In cell E3, enter =Future_Value_of_Savings.
22. In cells D4 through D23, enter a series of years from 1 to 39 as seen in Figure 4-13.
23. Create a **Data Table** that will calculate the *Future Value of Investment* of each scenario of *Years Until Needed*, as seen in Figure 4-13.
24. Paint the currency value in cell B6 to cells E3 through E23.
25. Shrink column C to a size of *3.00*.
26. Shrink column D to a size of *9.00*.
27. Shrink column E to a size of *16.00*.
28. Apply the **Accent2 Cell Style** to cells A2 through B7.
29. Apply the **Accent3 Cell Style** to cells D2 through E23.
30. Protect the spreadsheet, locking all of it *except cells* B2 through B5.
31. Name the **sheet tab** *Saving for Repair.*

CHAPTER FIVE

Managing Lists in a Spreadsheet

BY THE END of this chapter, it is our objective that the user will be able to:

1. Use filters.
2. Use the **COUNTIF** function.
3. Use the **SUMIF** function.
4. Use the **AVERAGEIF** function.
5. Sort.
6. Create **subtotals**.
7. Freeze panes.

CHAPTER 5—ANALYSIS OF STUDENT PLACEMENT SCORES

Sometimes large lists in a spreadsheet require some minimal level of database features, such as filtering out information that is not pertinent, looking up related values, and finding things such as the number of people or things with certain attributes. In our example, we are tracking scores on a student placement test of students with various college majors.

It is important to understand that spreadsheets are not a substitute for databases, but some things can be done to allow us to extract information from such lists.

Suppose you are a school administrator and you have given placement tests. Also, suppose you want to sort the data, extract certain data from the list, analyze the scores from those tests, and find things such as their letter grades and how the different students in different majors performed on the exam.

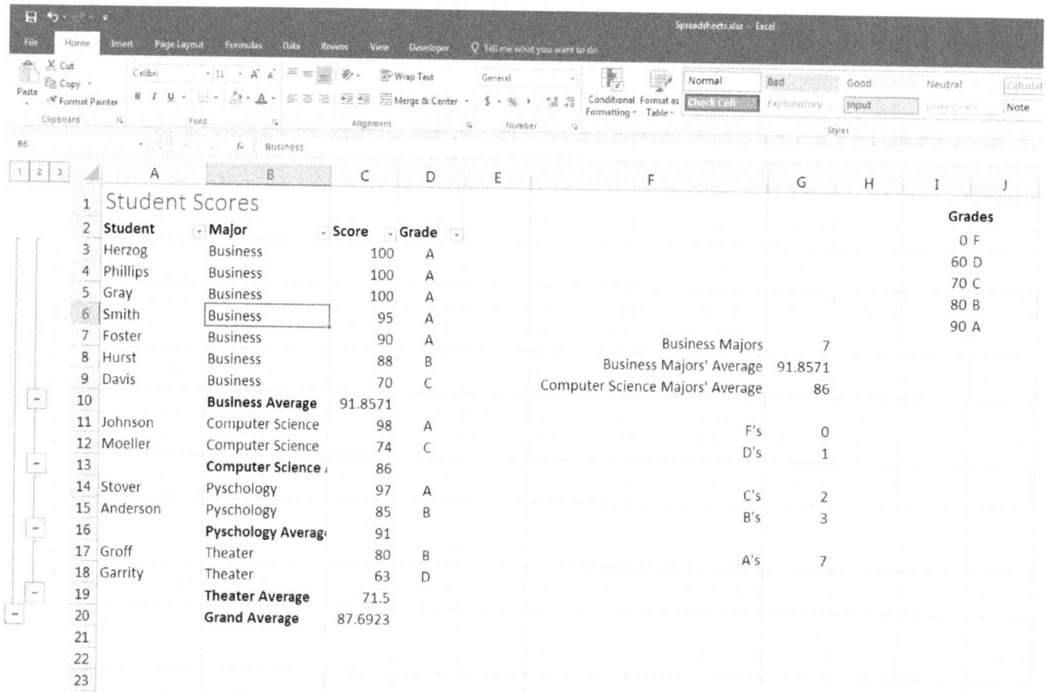

FIGURE 5-1

Some of these things we have already discussed in Chapter 2, but in this chapter, we will take such data to a higher level. Let us begin.

1. Open the file that is associated with this textbook named *Student Placement Scores.*
2. In cell D2, type the word *Grade.*
3. In cell D3, enter the **VLOOKUP** function that was discussed in Chapter 2, which will be *=VLOOKUP(C3,$I$1:$J$6,2)* (Figure 5-2, Arrow A).

Please notice that the second argument in this case has the absolute address applied to the range depicting the **lookup legend** (located in cells I1 through J6), as discussed in Chapter 2 (Figure 5-2, Arrow B). It would also have been acceptable to name that entire range (I1 through J6) as *Grades.* In that case, you could have then entered the function as:

=VLOOKUP(C3,Grades,2)

4. Pull the fill handle down to make sure that the grades are filled down to row 15 (Figure 5-2).
5. Center the grades inside column D (Figure 5-2).

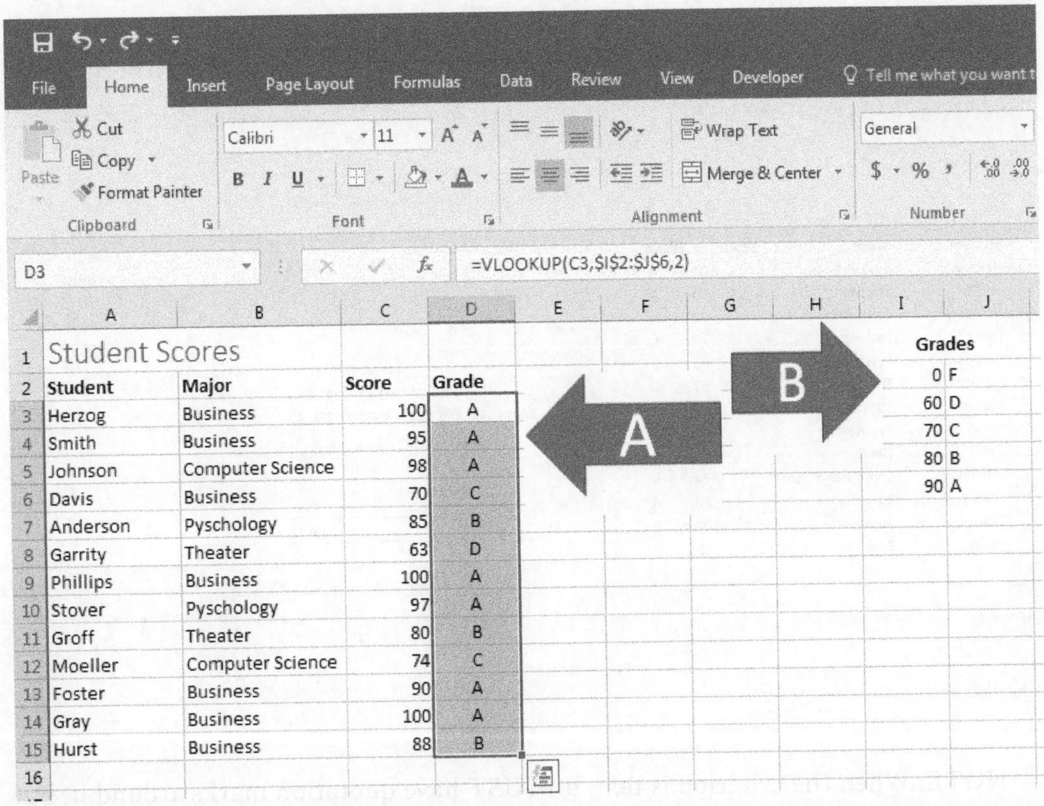

FIGURE 5-2

COUNTING RECORDS BASED UPON CERTAIN CRITERIA

If we had several hundred students in this example, it would take quite a bit of time to count the ones that meet certain criteria. For example, if we want to know how many *business* majors we have, we can use the **COUNTIF** function. It has only two arguments, described below:

=COUNTIF(Argument 1: The range of cells to be searched; Argument 2: The criteria in the search).

6. Widen column F to a width of 37.
7. In cells F7 through F14, type in the words as shown in Figure 5-3, which will describe each of the functions we will enter in a moment.
8. Suppose you want Excel to count the number of *business* majors listed in the *Student Scores* spreadsheet. In cell G7, type in the function: =COUNTIF(B3:B15,"Business")

It will give you the result you see in Figure 5-4.

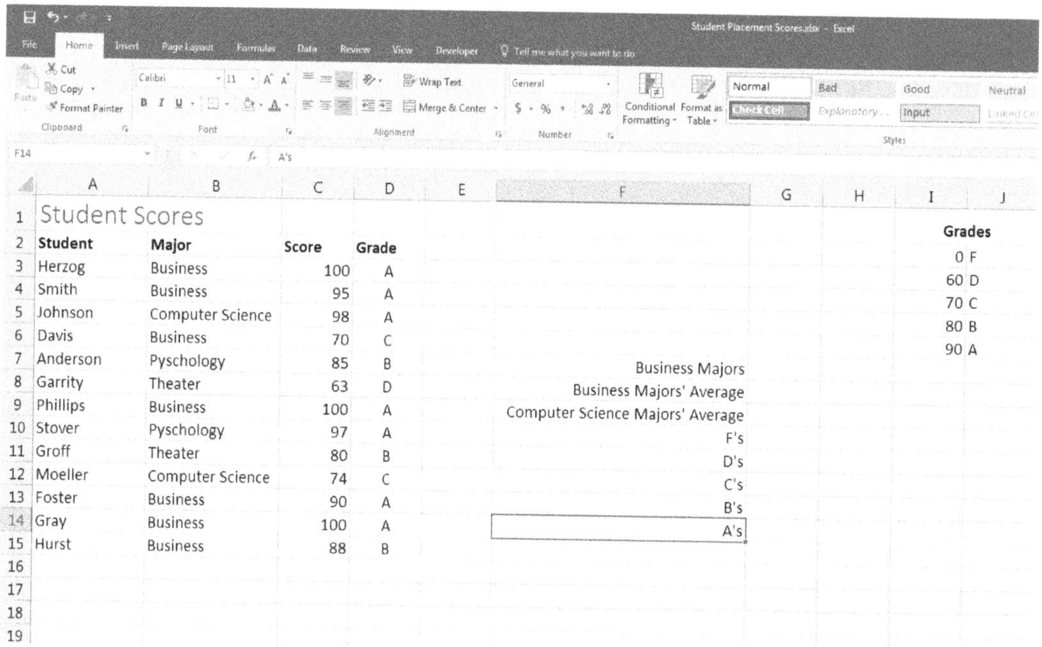

FIGURE 5-3

NOTE: When the criterion is text, it **MUST** have quotation marks around it, but **NOT** if data with numeric value is the criterion.

9. For now, let's skip cells G8 and G9. Suppose in cell G10 you wanted to count the number of students who received *F's* in these *test scores*. In this case, you can use the *F* in cell J2 as the criteria. In that case, you will NOT need to put parentheses around the argument because you will be using that cell reference instead of text. In cell G10, type the function:
=COUNTIF($D$3:$D$15,J2)

The reason for using the **absolute address** in this example is because in a moment we will fill this function down to the cells below it, and when we do, we want to use the same range for the search of data each time. However, we also want to use the **relative address** on the cell reference of J2, so it will change to J3, then J4, then J5, and then J6 in order to have it change the criteria to D, C, B, and then A, which are currently entered into those cells.

10. Fill the function you just entered in F10 all the way down to row 14. Your screen should look like what you see in Figure 5-4, and you will thus have a count of how many students received each different grade.

GETTING AVERAGES AND SUMS OF RECORDS BASED UPON CERTAIN CRITERIA

Suppose you want to know what the average placement score was for business majors or any of the other majors. This can be performed using the **AVERAGEIF** function. It works the same as the **COUNTIF** function, except that it adds a third argument. The third argument requires you to show the range you want averaged once the search criteria are found as shown below:

=AVERAGEIF (Argument 1: The range of cells to be searched; Argument 2: The criteria in the search; Argument 3: The range of cells to be averaged once the data with that criteria is found).

11. Let's now go back to cell G8 that we skipped a moment ago and type in the function:
    =AVERAGEIF(B3:B15"Business",C3:C15)
12. In cell G9, type the function:
    =AVERAGEIF(B3:B15,"Computer Science",C3:C15)

You will see the average scores of students who are *business majors* and *computer science majors* in Figure 5-4.

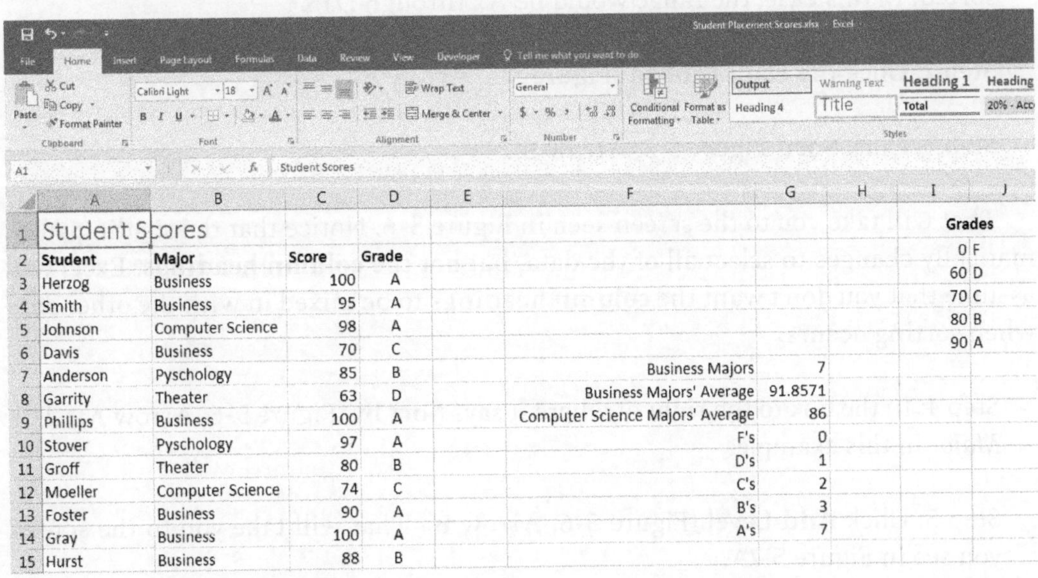

	A	B	C	D	E	F	G	H	I	J
1	Student Scores								Grades	
2	Student	Major	Score	Grade					0	F
3	Herzog	Business	100	A					60	D
4	Smith	Business	95	A					70	C
5	Johnson	Computer Science	98	A					80	B
6	Davis	Business	70	C					90	A
7	Anderson	Pyschology	85	B		Business Majors	7			
8	Garrity	Theater	63	D		Business Majors' Average	91.8571			
9	Phillips	Business	100	A		Computer Science Majors' Average	86			
10	Stover	Pyschology	97	A		F's	0			
11	Groff	Theater	80	B		D's	1			
12	Moeller	Computer Science	74	C		C's	2			
13	Foster	Business	90	A		B's	3			
14	Gray	Business	100	A		A's	7			
15	Hurst	Business	88	B						

FIGURE 5-4

NOTE: If you use these functions, you must spell the criteria **exactly** as the data reads, using spaces wherever they are needed. In this example, if you had spelled *Computer Science* incorrectly or if you had a space at the end of any of the entries, there would not be a perfect match, and Excel would not use those entries as part of the calculation.

The same thing would happen if you wanted to get totals. Of course, in this example, it would not be of any use to find the total of the scores. To do so, you would use the **SUMIF** function. It works very similarly to the **AVERAGEIF** function as follows:

=SUMIF (Argument 1: The range of cells to be searched; Argument 2: The criteria in the search; Argument 3: The range of cells to be totaled once the criteria are found). In this example, it would read:
=SUMIF(B3:B15,"Computer Science",C3:C15)

SORTING

In this example, suppose you want to sort the data by the different student majors and then sort the scores from the highest to the lowest score for each major. You can do this by using the following steps:

Step 1: Select or click inside any one of the cells within the range of cells to be sorted. In this case, the range would be A2 through D15.

Step 2: Click the **Data Ribbon** (Figure 5-5, Arrow A)

Step 3: Click **Sort** (Figure 5-5, Arrow B).

That will take you to the screen seen in Figure 5-6. Notice that the shading automatically changes to select all of the data, but not the **column headings**. Excel will assume that you don't want the **column headings** to be mixed in with the other data when sorting occurs.

Step 4: In the box to the right of where it says **Sort by** (Figure 5-6, Arrow A), click *Major* in this example.

Step 5: Click **Add Level** (Figure 5-6, Arrow B). That will take you to the screen you see in Figure 5-7A.

Step 6: In this example, click *Score* to the right of where it says **Then by** (Figure 5-7A, Arrow A).

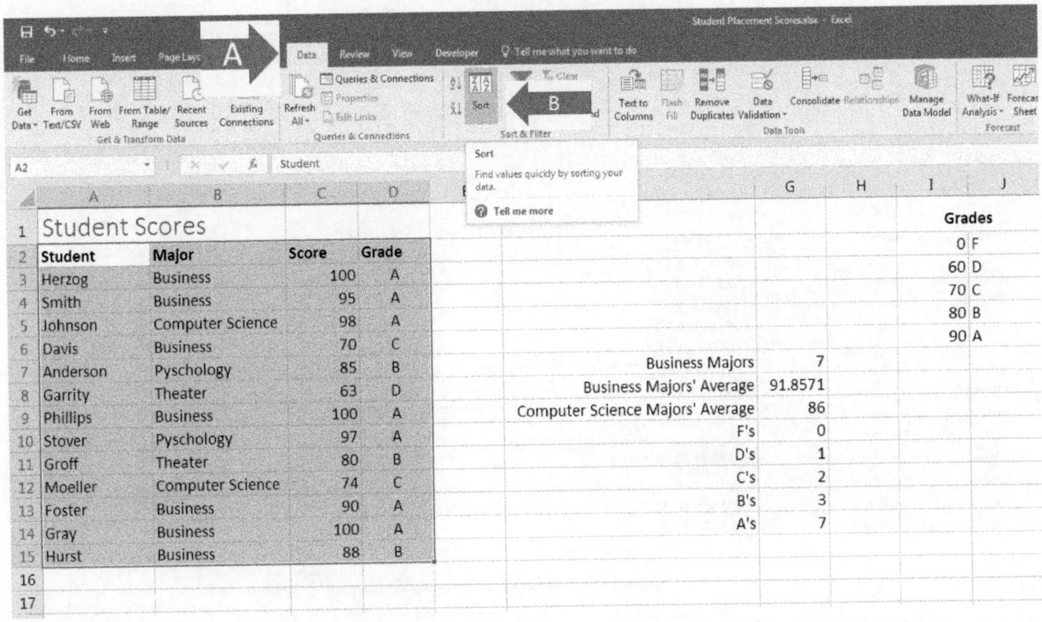

FIGURE 5-5

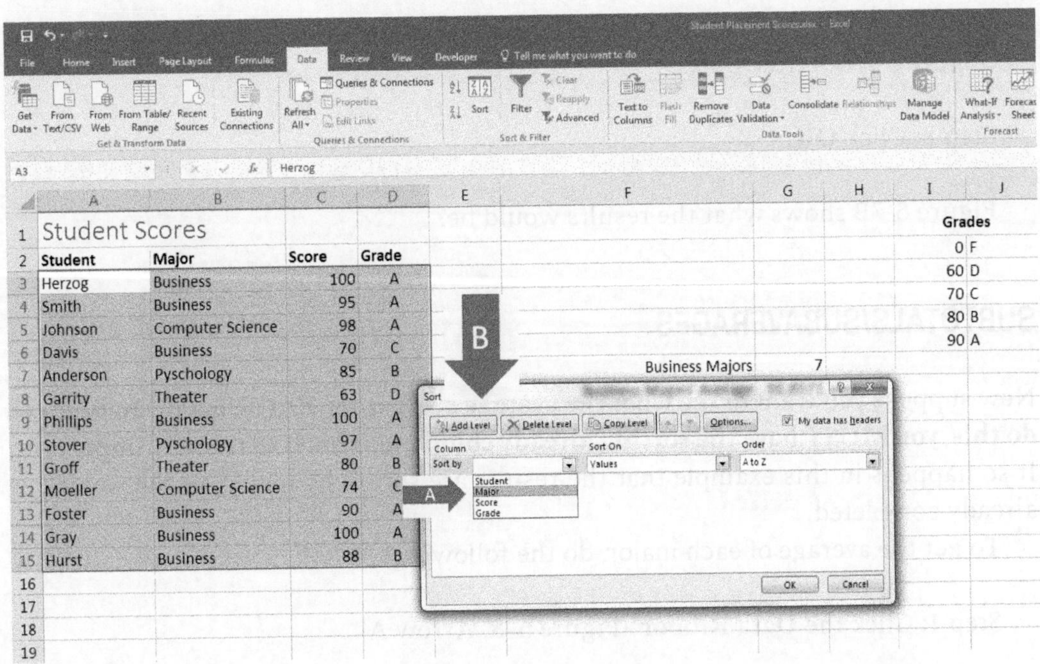

FIGURE 5-6

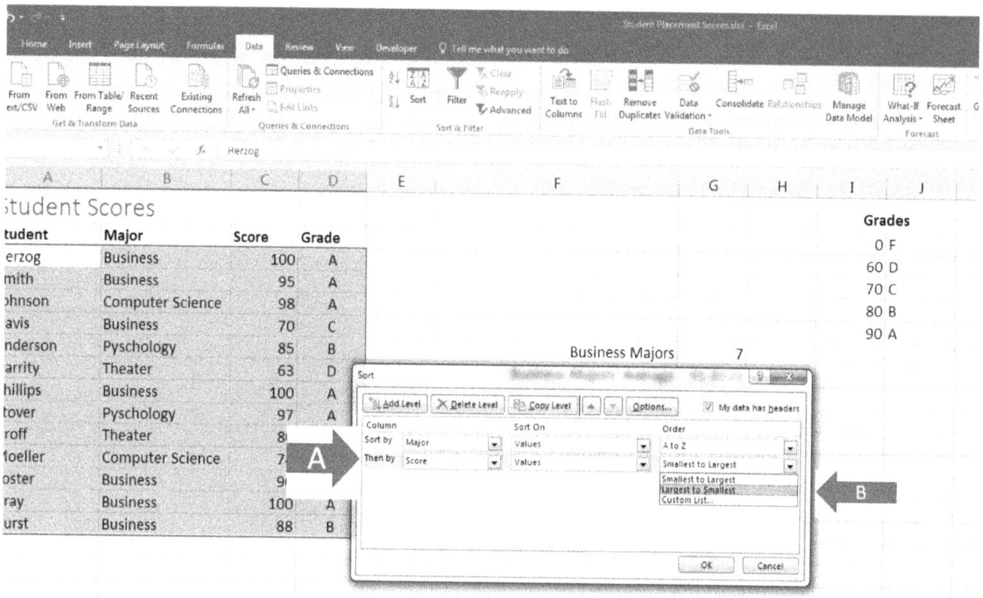

FIGURE 5-7A

Step 7: In the box under where it reads **Order**, click **Largest to Smallest** (Figure 5-7A, Arrow B).

Step 8: Click **OK**.

Figure 5-7B shows what the results would be.

SUBTOTALS/SUBAVERAGES

Now suppose you want to see what the average score is for **EACH** major. **In order to do this, you MUST have the sorting already complete. This is extremely important.** It so happens in this example that the results are sorted by major, so the sorting is already completed.

To get the average of each major, do the following:

Step 1: Click the **Data Ribbon** (Figure 5-8, Arrow A).

Step 2: Click **Subtotal** (Figure 5-8, Arrow B). NOTE: Please make sure you are still somewhere inside the cell range of A2 through D15 before you do this step!

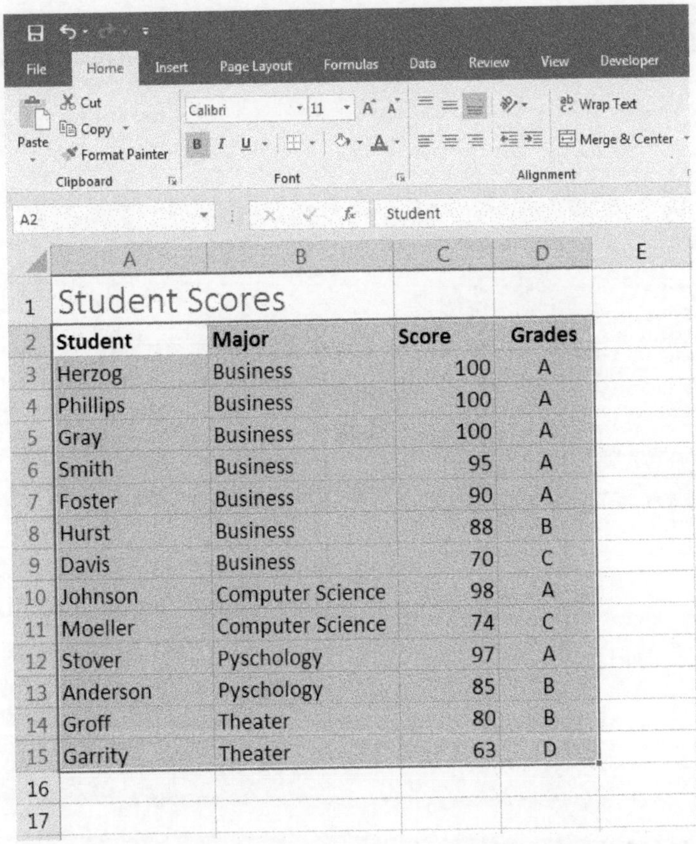

FIGURE 5-7B

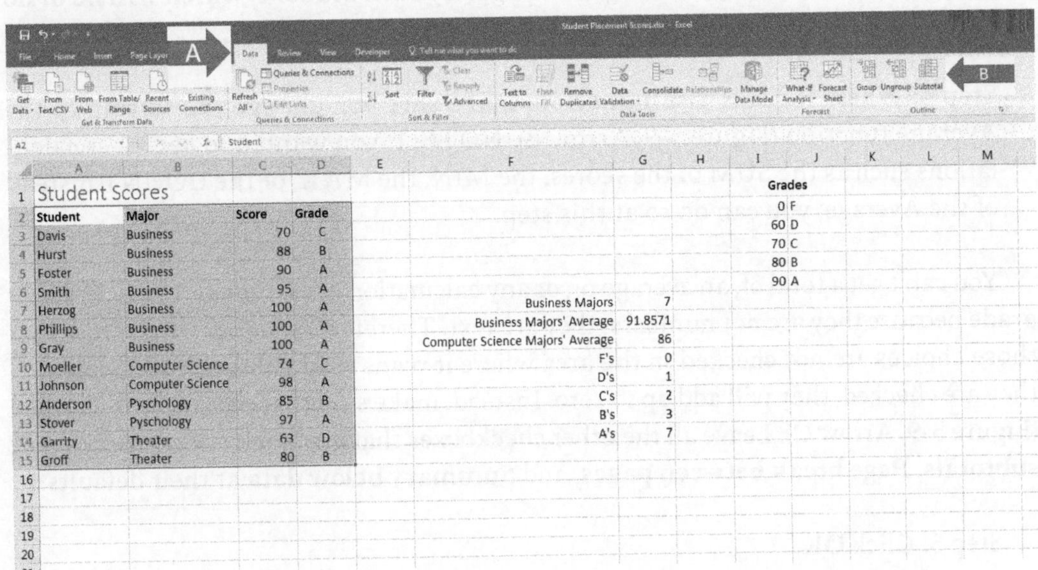

FIGURE 5-8

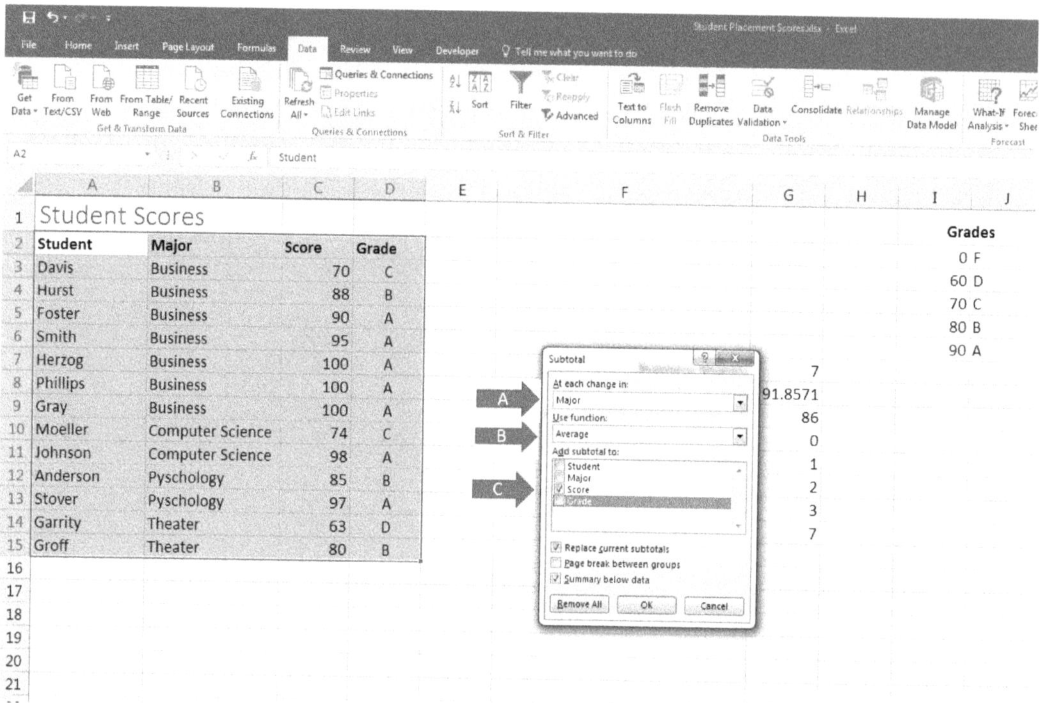

FIGURE 5-9

Step 3: At this point, you will see a pop-up menu that is shown in Figure 5-9. Where it reads **At each change in:** (Figure 5-9, Arrow A), change the choice from *Student* to *Major*. **If you don't, you will get averages by each student, which will be of no value because that is what you already have now.**

Step 4: Where it reads **Use Function**, make sure you change the setting to the **Average** function (Figure 5-9, Arrow B). NOTE: If you wanted to get other calculations such as the SUM of the scores, the MIN, the MAX, or the COUNT instead of the Average, you can do so at this step.

You can't get a total of, an average of, or any calculation of, a student, major, or letter grade because they are not numbers; they are text. Therefore, you must make sure that those choices are not checked in the area where it reads **Add subtotals to:** because if they are checked, they will add up to zero. Instead, make sure that only *Score* is checked (Figure 5-9, Arrow C). Leave all the other checkboxes that are labeled **Replace current subtotals, Page break between pages**, and **Summary below data** at their defaults.

Step 5: Click **OK**.

The results you will get are shown in Figure 5-10A.

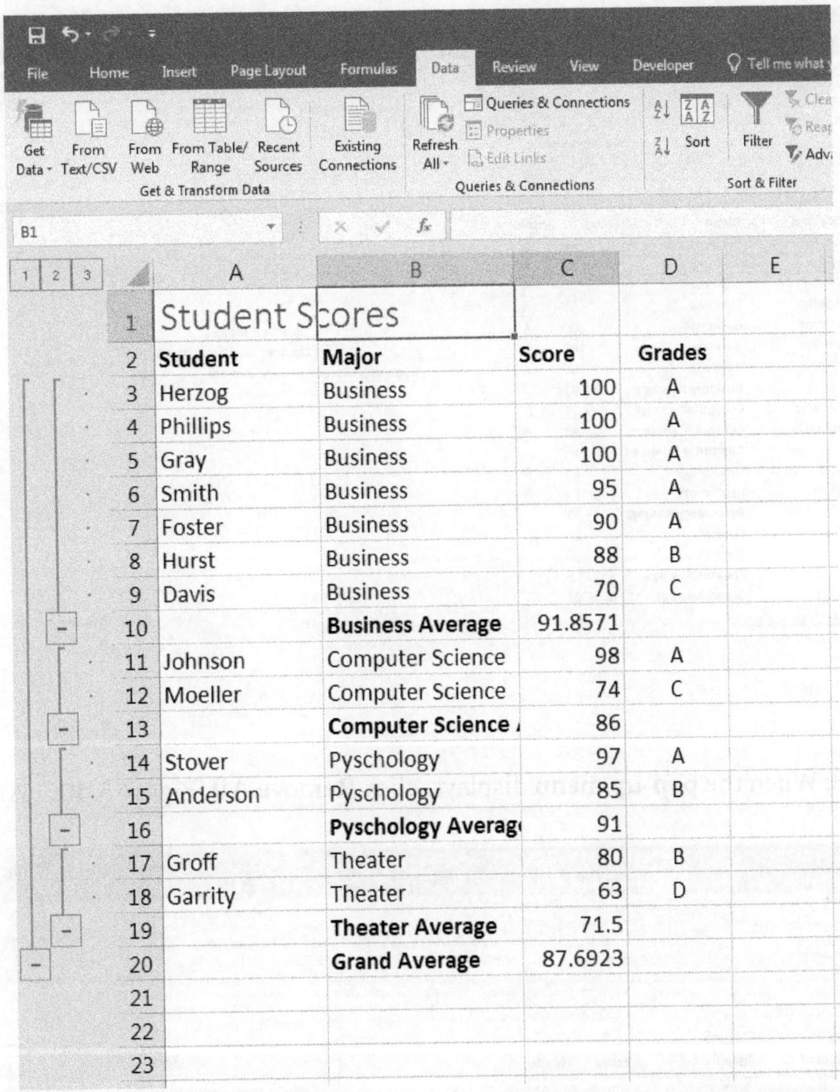

FIGURE 5-10A

TO REMOVE A SUBTOTAL

What should you do if you make a mistake and must redo the **subtotal**? This may happen if you forget to sort the data first or for various other reasons. This is how you can redo it.

Step 1: Click the **Data Ribbon** (Figure 5-10B, Arrow A).

Step 2: Click **Subtotal** (Figure 5-10B, Arrow B).

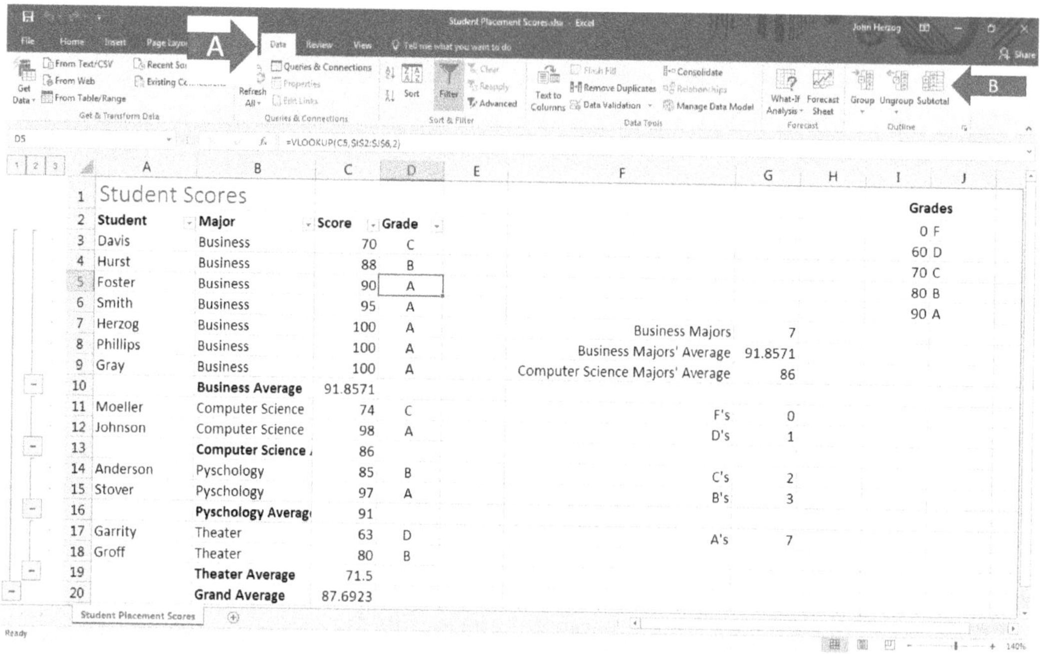

FIGURE 5-10B

Step 3: When the **pop-up menu** displays, click **Remove All** (Figure 5-10C, Arrow C).

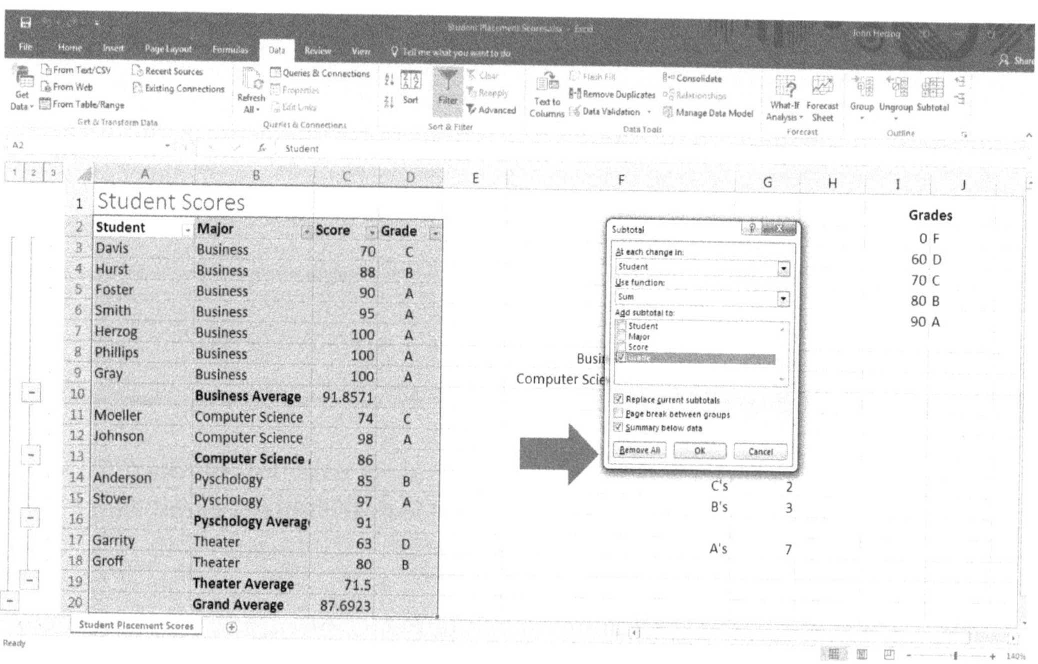

FIGURE 5-10C

The **subtotals** will vanish (Figure 5-10D)

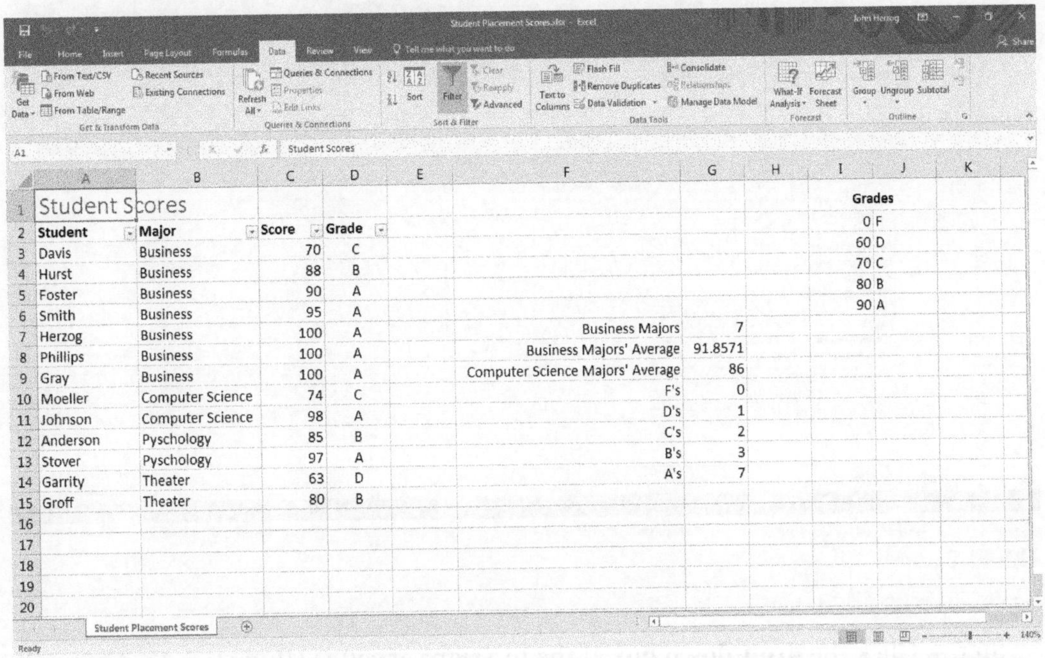

FIGURE 5-10D

USING FILTERS

Sometimes you will want to see a list of items that fit given criteria without having to sift through the data that does **NOT** fit the criteria. This can be done by using a **Filter**. For example, suppose you want to see the averages for only the students who received a *B* on their placement tests. Once again, with only a handful of records in a spreadsheet, you could do so by just looking over the list, but if you have several hundred or several thousand records, the filter will be a great tool. You can accomplish this by doing the following:

Step 1: Just like you did when you created **Subtotals**, make sure your cursor is anywhere inside the range of records in which you are working.

Step 2: Click the **Data Ribbon** as shown in Figure 5-11, Arrow A.

Step 3: Click **Filter** (Figure 5-11, Arrow B). When you do, you will see **drop-down boxes** displayed at the top right of each cell that Excel assumes you consider to be a column heading (Figure 5-11, Arrow C).

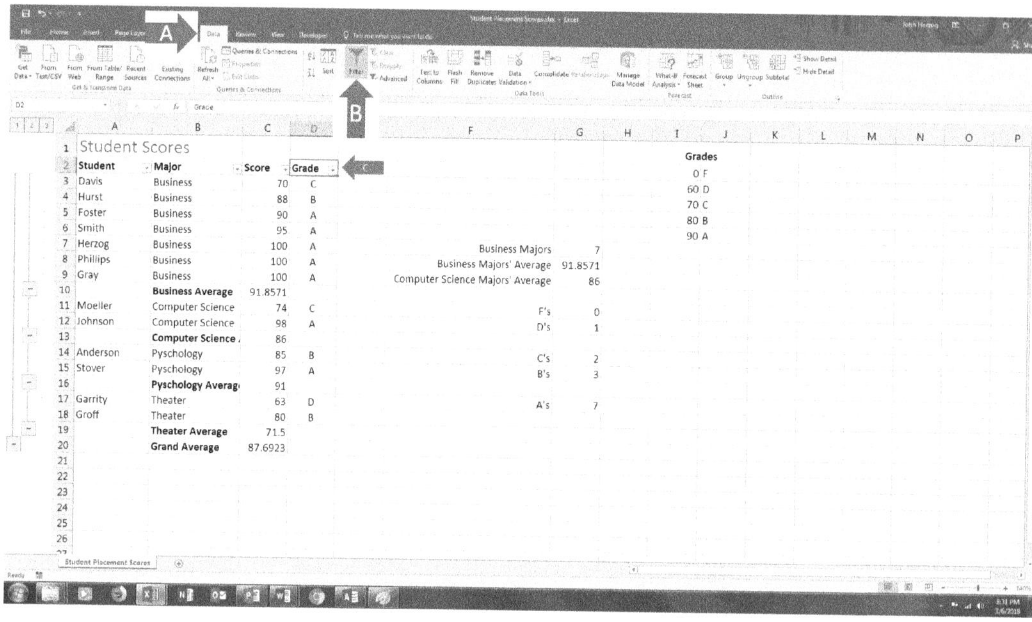

FIGURE 5-11

Step 4: Click the **drop-down box** at the top right of cell D2 (Figure 5-11, Arrow C).

Step 5: That will show all of the grades that are currently in the data that you have to choose from for criteria in this filter (Figure 5-12). Make sure that the only box that is checked is the one that is next to *B* (Figure 5-12, Arrow B). Rather than taking the time to uncheck all of the other *grades* that you don't want, you can uncheck the box that reads **Select All** (Figure 5-12, Arrow A). That will remove the checkmarks to all of the items listed. Then click only the *B* box (Figure 5-12, Arrow B).

Once you have taken these steps, your screen should look like Figure 5-13.

NOTE: If you wish, you can click numerous values. For example, if you wanted to view only the students who received a B or a C, you could click on the checkmarks for both of them.

Suppose that you are now interested in seeing only the *business* majors and *theater* majors with grades of *B*. You can then further filter out more data. To do so, do the following:

Step 1: Click the **drop-down box** at the top right of cell B2 (Figure 5-14A, Arrow A).

Step 2: Repeat the process you completed earlier by clicking and removing the checkmarks from all of the *majors* you don't need to see by unchecking the box

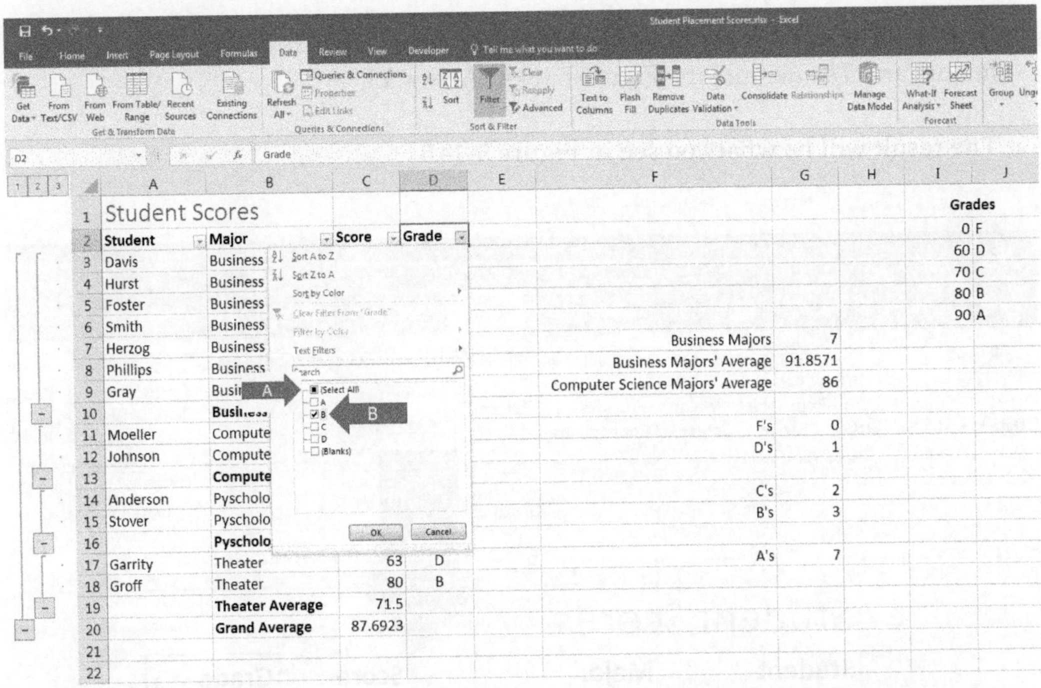

FIGURE 5-12

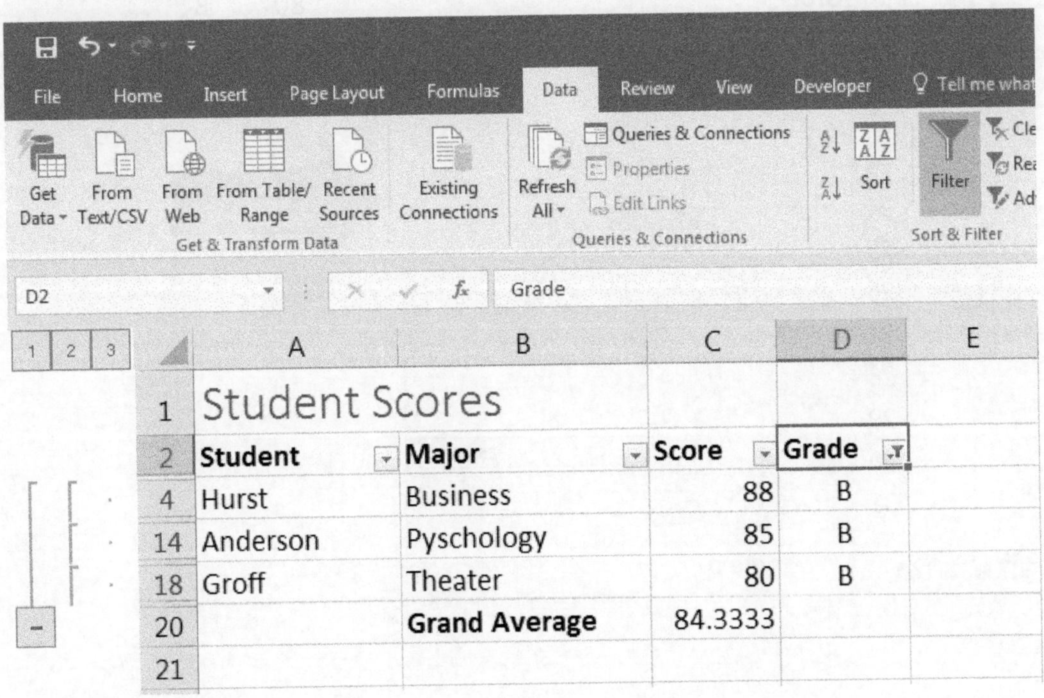

FIGURE 5-13

that reads **Select All** and then clicking only the checkboxes to the left of *Business* majors and *Theater* majors (Figure 5-14A, Arrow B).

The result will be what you see in Figure 5-14B.

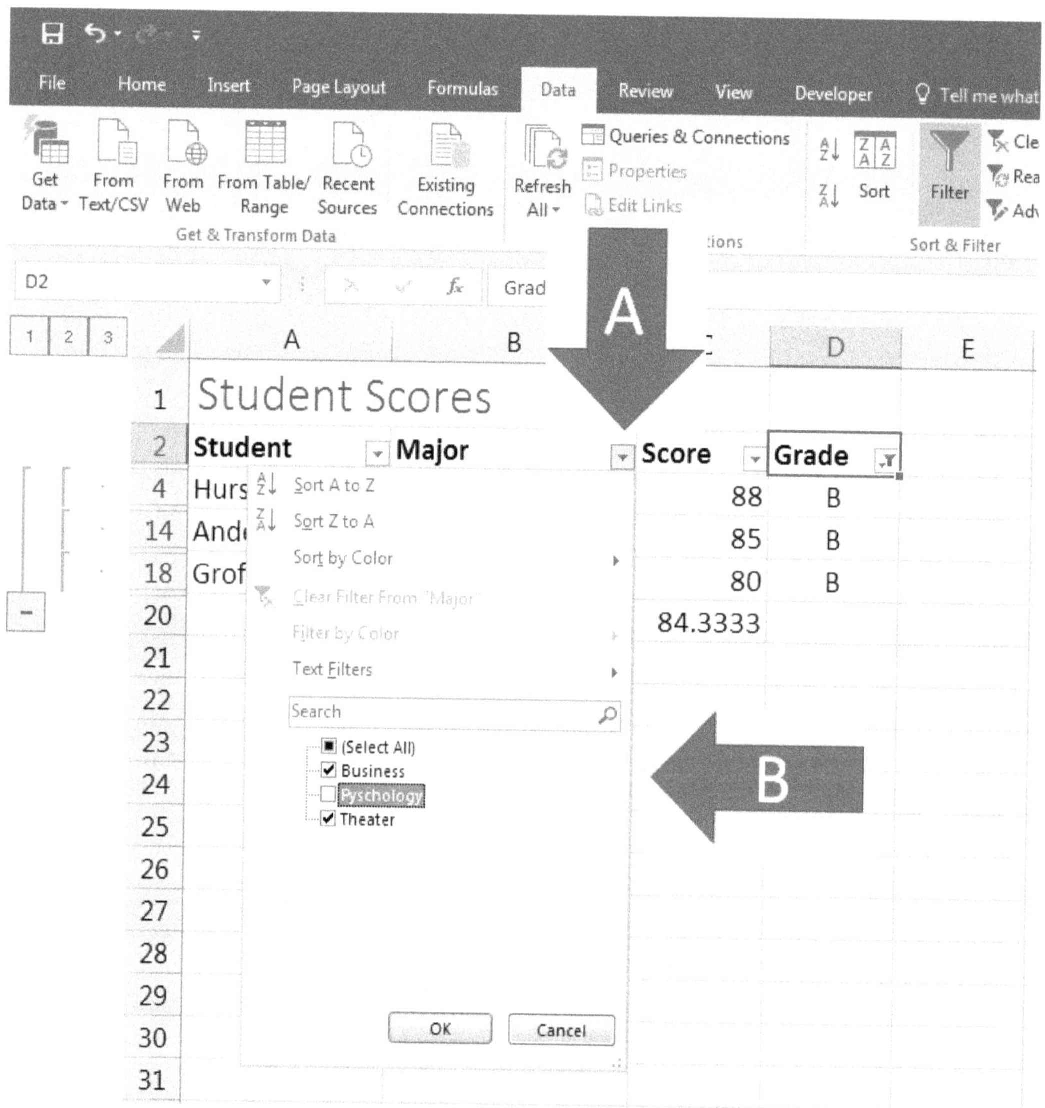

FIGURE 5-14A

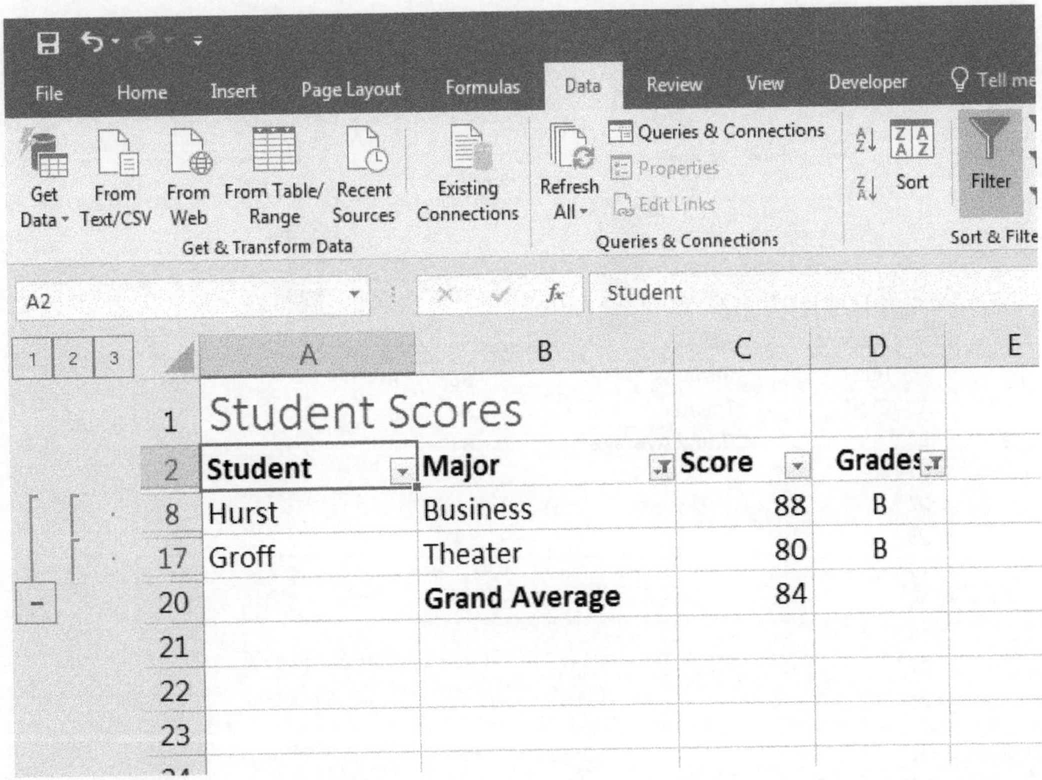

FIGURE 5-14B

TO REMOVE FILTERS

Sometimes we create a filter and it comes out wrong. Sometimes people create filters that they only need to view for a moment, but then they quickly want to remove them. If that happens, here is what you can do.

Step 1: Click anywhere inside the list.

Step 2: Click the **Data Ribbon** (Figure 5-15, Arrow A).

Step 3: Click **Filter** (Figure 5-15, Arrow B). Doing so will remove the filters.

NOTE: You can also remove the filters while keeping the **Filter Boxes** in place by going to the **Sort & Filter group** of the **Data Ribbon** and clicking **Clear** (Figure 5-15, Arrow C). All of the data will be redisplayed, but the **Filter Boxes** will remain intact.

NOTE: If data has changed after you have applied a filter, you can also click **Refresh All** (Figure 5-15, Arrow D) to see the latest results.

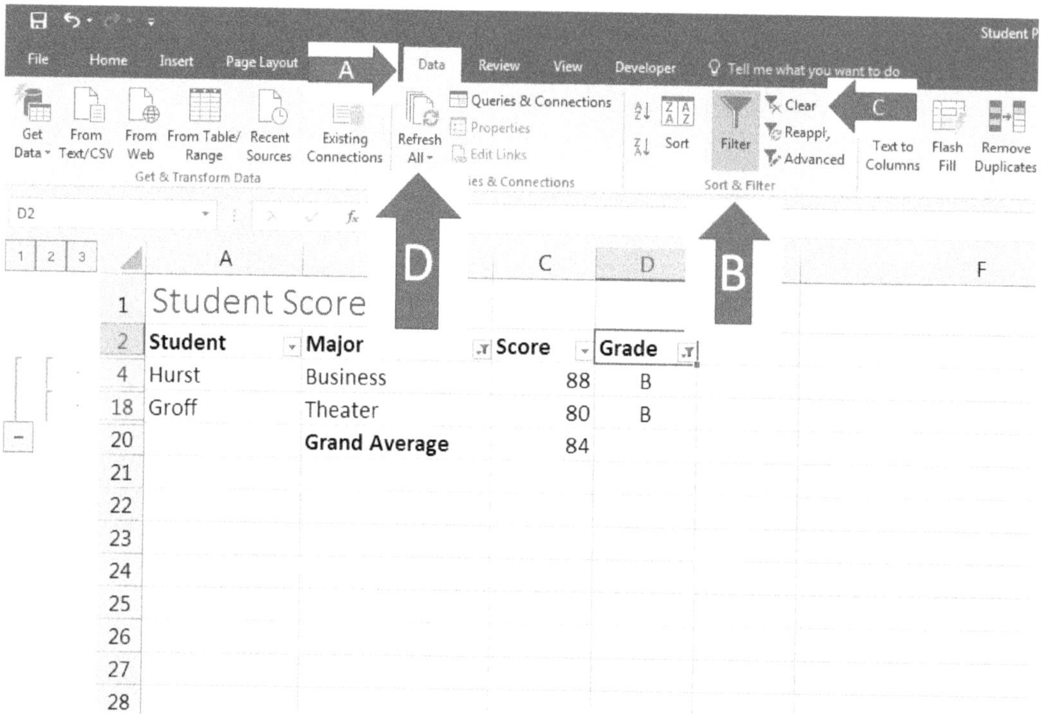

FIGURE 5-15

NOTE: You can also use filters beyond what you have already seen. In Figure 5-16, Arrow A shows many of those filter options. For example, you can see that you can find words that are **equal** to a given word, words that are **NOT equal** to certain words, words that **begin** with certain letters, words that **end** with certain letters, words that **contain** certain letters or words or **do NOT contain** certain letters or words. You can also use the **Filter menu** to **sort** your data as well (Figure 5-16, Arrow B). In that same list, you can also create a **Custom Filter**.

With a **Custom Filter**, you can use the **and feature** or the **or feature** to filter out unneeded data (Figure 5-17). With the **and feature**, two different criteria must **both** be true in order for the data to be displayed, while with the **or feature**, one **or** the other of two criteria need to be true.

With the number filters, you can do similar things (Figure 5-18, Arrow). You can display numbers that are **equal** to a given value, numbers that are **NOT equal** to certain values, numbers that are **greater than** certain values, numbers that are **greater than or equal to** given values, numbers that are **less than** certain values, numbers that are **less than or equal to** given values, or numbers that are **between** certain values. You can also display the **top 10** of values or items that are **above average** and **below average**. You can also create a **Custom Filter** here as well.

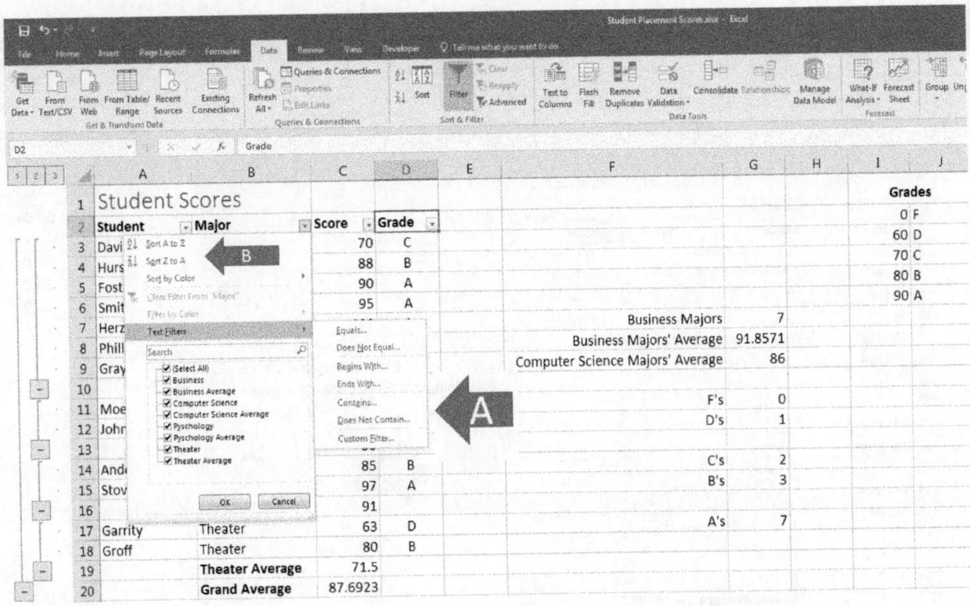

FIGURE 5-16

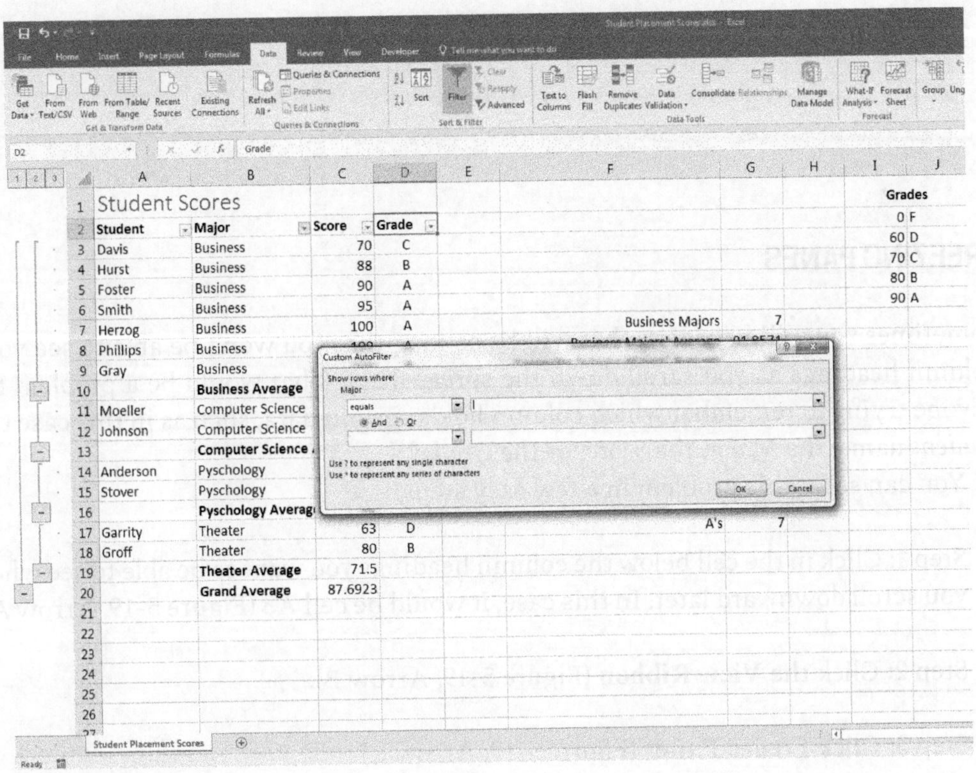

FIGURE 5-17

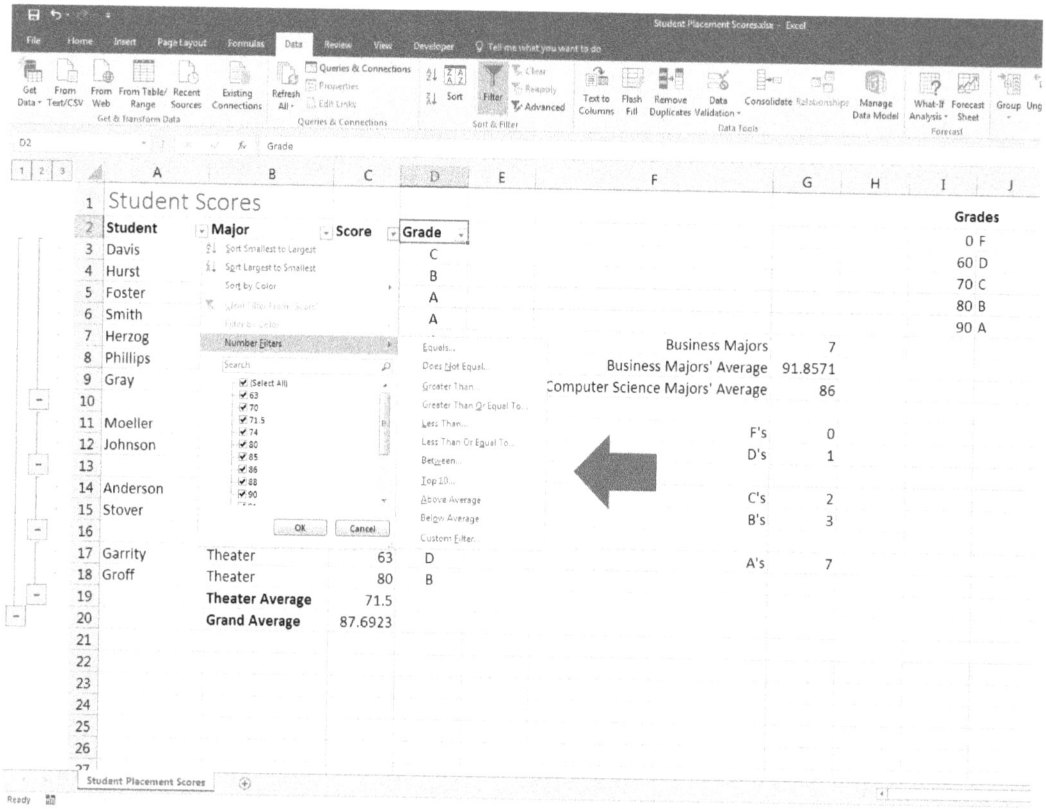

FIGURE 5-18

FREEZING PANES

Sometimes you will have lists that will be so long that you won't be able to see your column headings as you scroll down the spreadsheet. That would be a problem for anyone trying to remember which column holds certain data, such as in this case the *Student* name, the *Major*, the *Score*, or the *Grade*.

You can solve this problem in a few easy steps.

Step 1: Click in the cell below the column headings you want to be able to see when you scroll downward later. In this case, it would be cell A3 (Figure 5-19, Arrow A).

Step 2: Click the **View Ribbon** (Figure 5-19, Arrow B).

Step 3: Click **Freeze Panes** (Figure 5-19, Arrow C).

Step 4: Click **Freeze Panes** in the **drop-down list** (Figure 5-19, Arrow D).

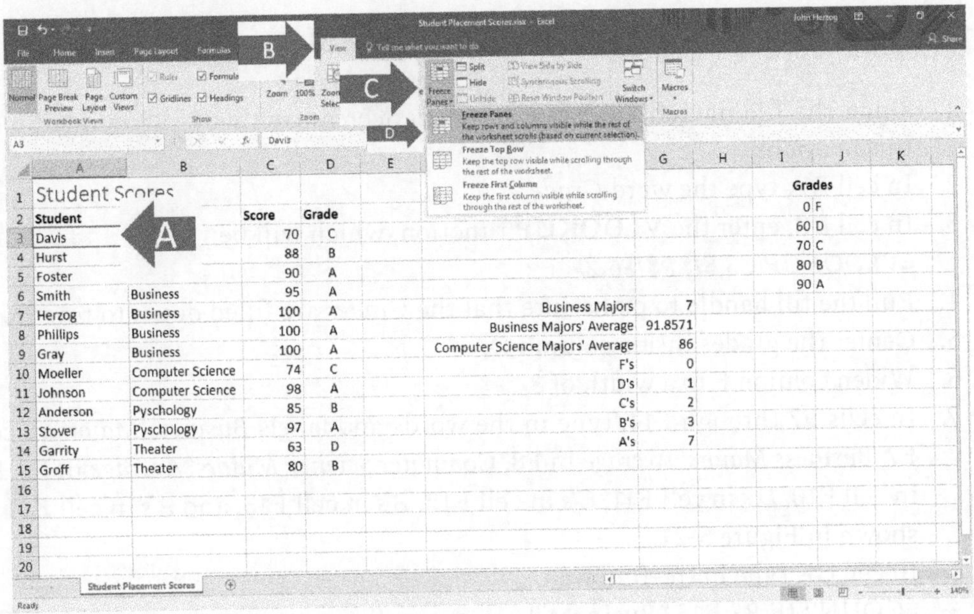

FIGURE 5-19

When you do, you will see a thin line that marks where the freezing of the panes begins (Figure 5-20, Arrow, Thin Line). You can then scroll down and still see the column headings. Notice that in Figure 5-20, the spreadsheet is scrolled down, showing rows 11 through 28, but the headings in rows 1 and 2 are still clearly visible.

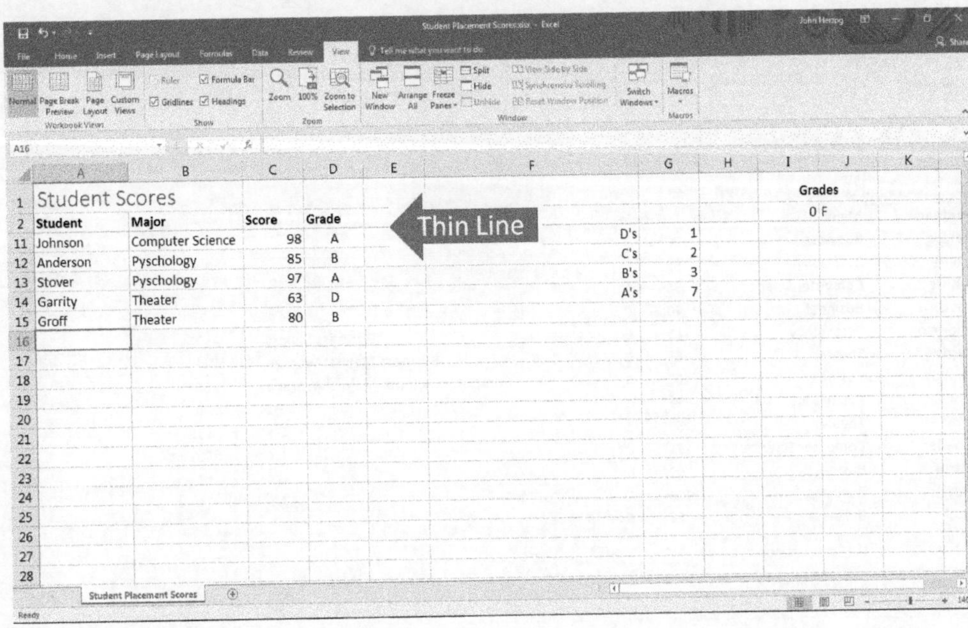

FIGURE 5-20

CHAPTER 5—ASSIGNMENT (SUMMARY OF TASKS)

1. Open the file named *Student Placement Scores.xlsx* that is associated with this textbook.
2. In cell D2, type the word *Grade.*
3. In cell D3, enter the **VLOOKUP** function, which will be *=VLOOKUP(C3,$I$1:$J$6,2)*
4. Pull the fill handle to make sure that the *Grades* are filled down to row 15.
5. Center the grades inside column D.
6. Widen column F to a width of *37.*
7. In cells F7 through F14, type in the words and labels *Business Majors* in cell F7, *Business Majors' Average* in F8, *Computer Science Majors' Average* in F9, *F's* in cell F10, *D's* in cell F11, *C's* in cell F12, *B's* in cell F13, and *A's* in cell F14 as shown in Figure 5-21.
8. In cell G7, type in the function: =COUNTIF(B3:B15,"Business")
9. In cell G10, type in the function: =COUNTIF($D$3:$D$15,J2)
10. Fill the function down to row 14.

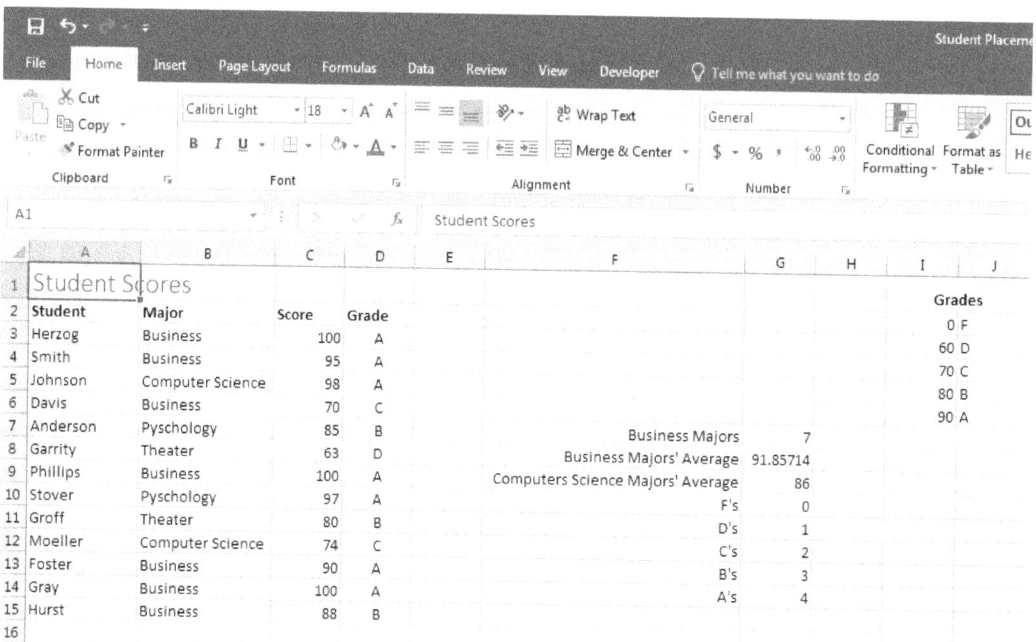

FIGURE 5-21

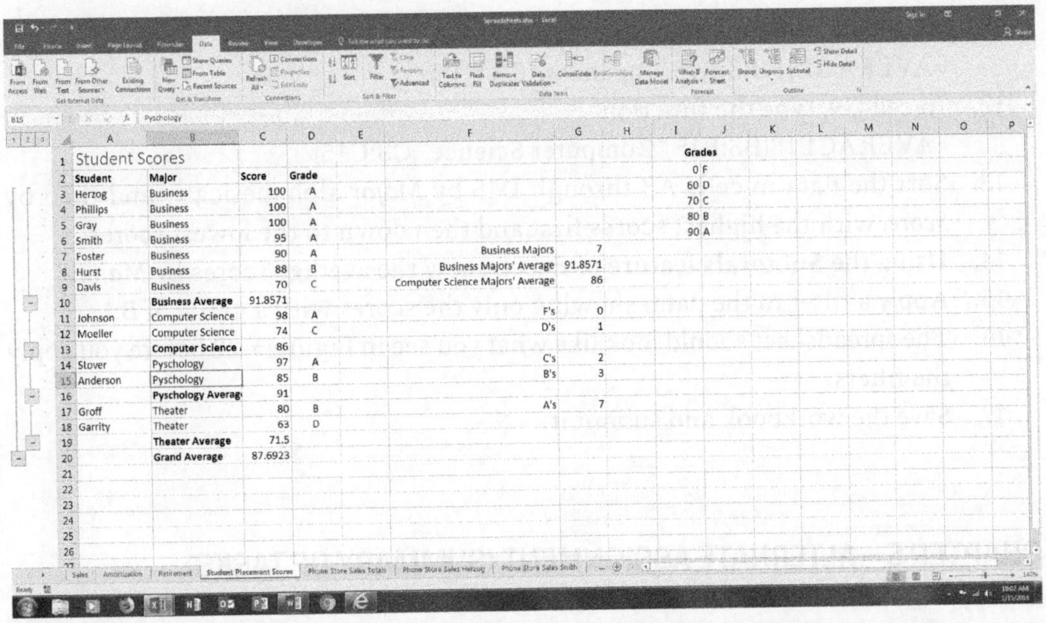

FIGURE 5-22

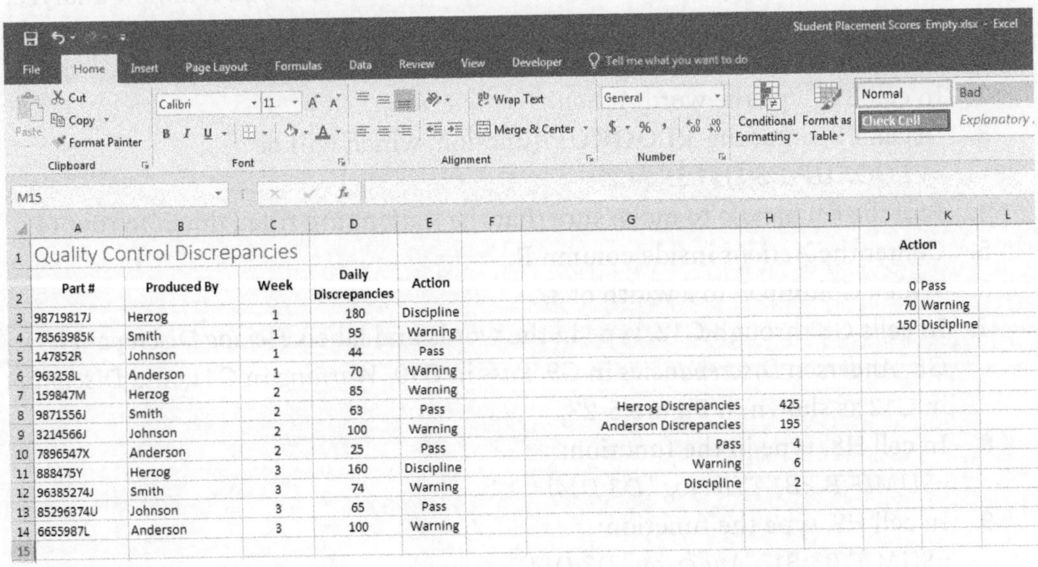

FIGURE 5-23

11. In cell G8, type in the function: =AVERAGEIF(B3:B15,"Business",C3:C15)
12. In cell G9, type the function: =AVERAGEIF(B3:B15,"Computer Science",C3:C15)
13. Sort the data in cells A3 through D15 by *Major* alphabetically and then by *Score*, with the highest scores first and then down to the lowest scores.
14. Using the **Subtotals** feature of Excel, show the average scores by *Major*.
15. Apply a filter to the data, showing only the scores with a grade of *B*.
16. The spreadsheet should look like what you see in Figure 5-22 before you apply the filters.
17. Save the workbook and submit it.

CHAPTER 5—ALTERNATE ASSIGNMENT (SUMMARY OF TASKS)

A manufacturing firm tests the production of tens of thousands of parts they have produced in a given week. Workers who have produced those parts are permitted only 70 discrepancies or defects per week before receiving a warning, and only 150 before receiving discipline. They are tracked in the spreadsheet that you will now analyze.

1. Open the file named *Quality Control.xlsx* that is associated with the textbook.
2. In cell E2, type the word *Action*.
3. In cell E3, enter the **VLOOKUP** function, which will be *=VLOOKUP(D3,$J$1:$K$4,2)*
4. Pull the fill handle to make sure that the *Actions* are filled down to row 14.
5. Center the *Actions* inside column E.
6. Widen Column G to a width of *37*.
7. In cells G8 through G12, type in the words and labels *Herzog Discrepancies* in G8, *Anderson Discrepancies* in G9, *Pass* in G10, *Warning* in G11, and *Discipline* in G12 as shown in Figure 5-23.
8. In cell H8, type in the function: *=SUMIF(B3:B17,"Herzog",D3:D14)*
9. In cell H9, type the function: *=SUMIF(B3:B17,"Anderson",D3:D14)*
10. In cell H10, type in the function: *=COUNTIF($E$3:$F$14,K2)*
11. Fill the function down to row 12.
12. Sort the data in cells A3 through E14 by *Produced By alphabetically*, and then by *Discrepancies* with the highest number first down to the lowest.

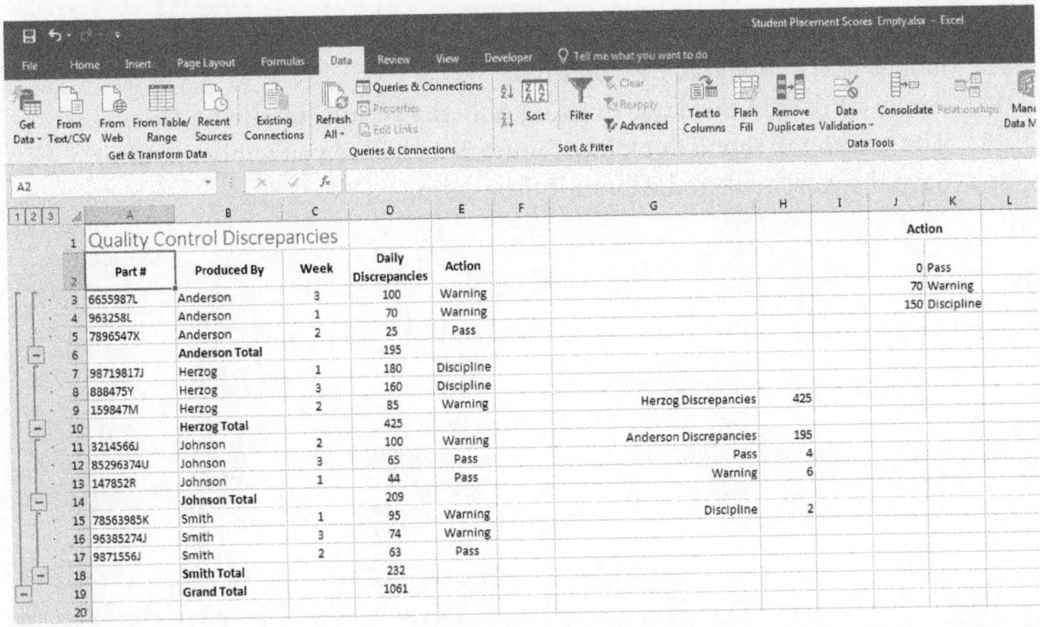

FIGURE 5-24

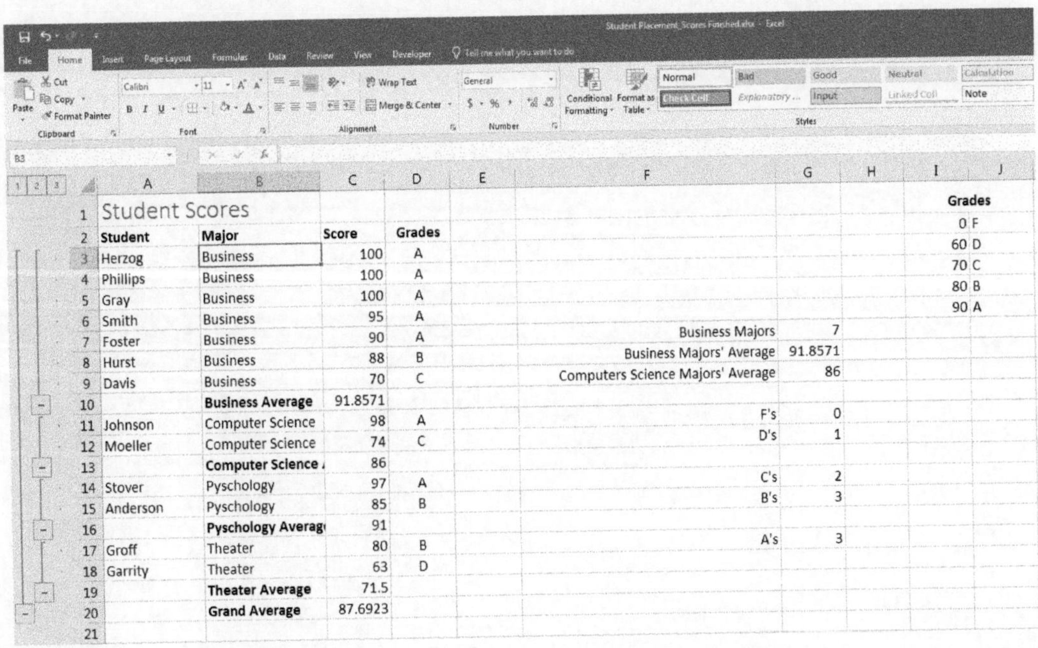

FIGURE 5-25

13. Using the **Subtotals** feature of Excel, show the total *Daily Discrepancies* by the names in the *Produced By* column.
14. Apply a **Filter** to the data, showing only the actions requiring *Discipline*.
15. The workbook should look like Figure 5-24 before you apply the filter.
16. Save the workbook and submit it for grading.

CHAPTER SIX

3-D Cell Reference, Multiple Data Column Charts, and Creating a Template

BY THE END of this chapter, it is our objective that the user will be able to:

1. Use the **3-D cell reference.**
2. Create a **column chart** with multiple data sets.
3. Create a **template.**
4. **Color-code sheet tabs.**
5. Use the **NOW** function.
6. Use **Auto Fill Options**.
7. Create **comments.**

CHAPTER 6 TASK: PHONE STORE SALES ANALYSIS

Herzog Phones Sales stores are trying to get grand totals of the sales generated by their employees. Each employee's sales are already entered on each of their own spreadsheets. The company needs to make sure that the spreadsheets don't have to be re-created from scratch at the end of each year. They also desire to get totals of their sales on one spreadsheet with a column chart added with as little work as possible.

1. Open the file that is associated with this textbook named *Phone Store Sales*.
2. Click on the first sheet tab titled *Phone Store Sales Totals* (Figure 6-2, Arrow).

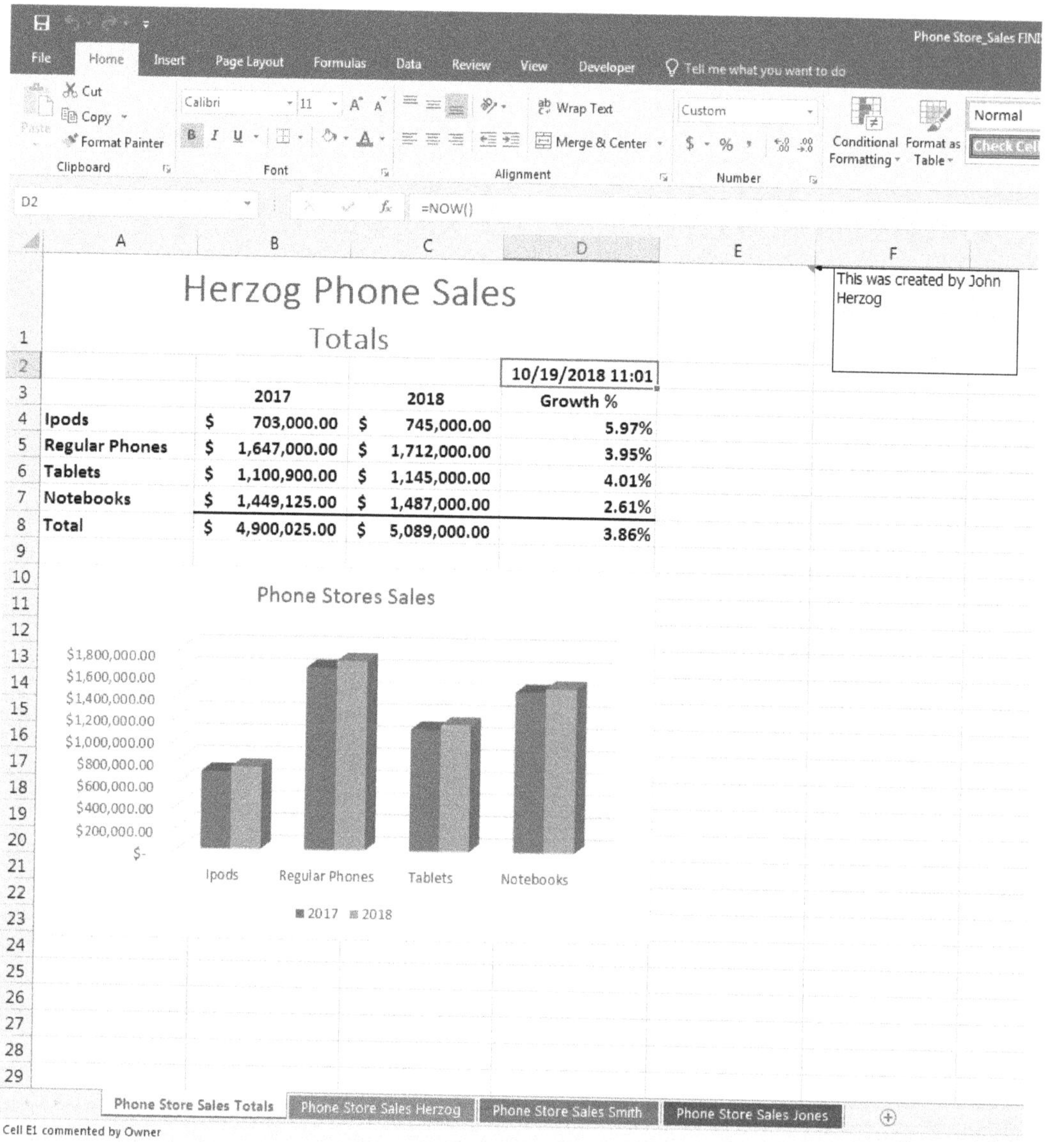

FIGURE 6-1

CREATING A 3-D CELL REFERENCE FOR TOTALS

In our example, we want to place the dollar total of all of the *iPods* sold by *Herzog, Smith*, and *Jones* into cell B4 of the *Phone Store Sales Totals* spreadsheet. Their sales are already entered in their respective spreadsheets. If you wanted to do this the slow way, you could click on cell B4 of the *Phone Store Sales Totals* spreadsheet and type = and then click on cell B4 in the *Phone Store Sales Herzog* sheet; then press +, and repeat this process for each of the remaining sheets. However, using that approach could

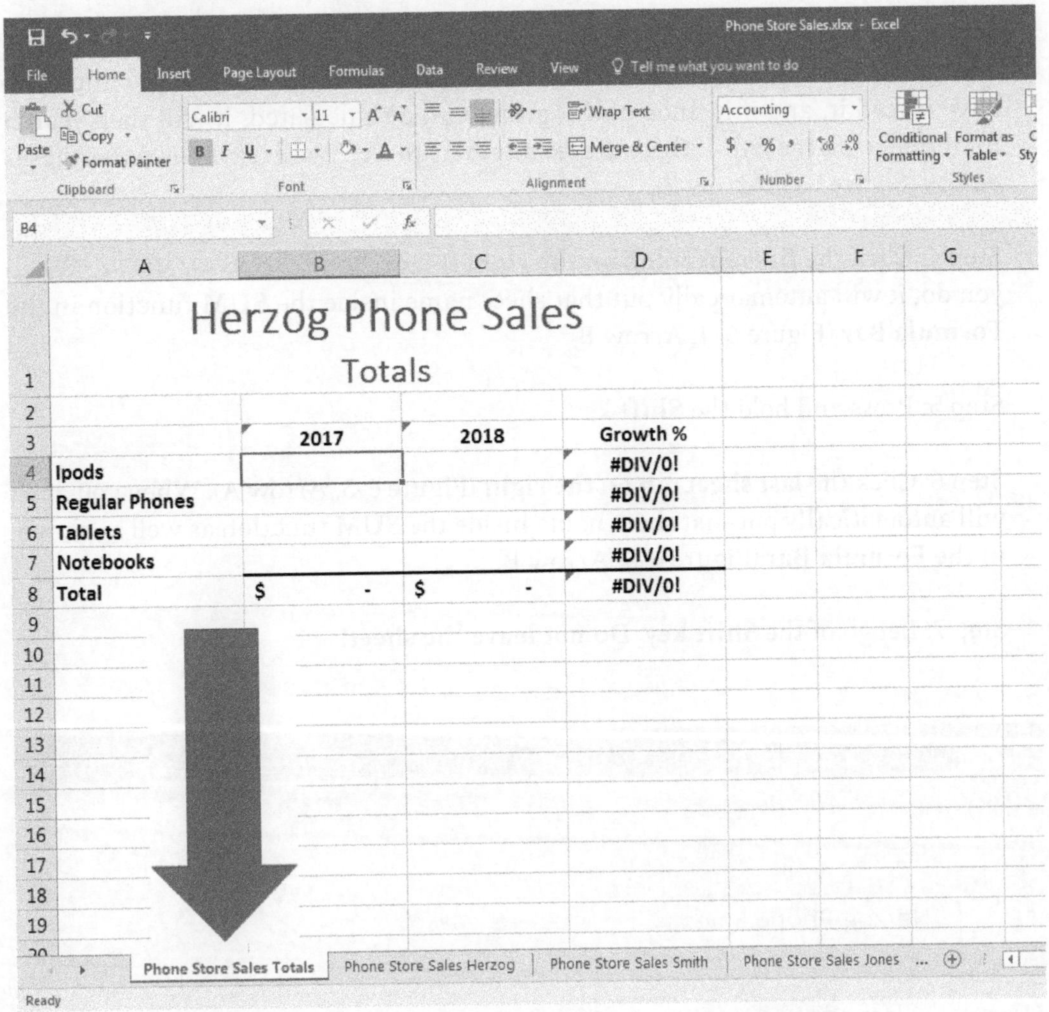

FIGURE 6-2

take a very long time if you have dozens or hundreds of sales staff with individual spreadsheets. That is why we use the **3-D cell reference** instead. It can be applied to this example using the following steps:

Step 1: Make sure the *Phone Store Sales Totals* sheet is at the beginning or the end of the series of sheets (and not in the middle of them). That is the sheet that you have just selected, and it should already be the first sheet in the workbook.

Step 2: Go to the upper-leftmost cell of that spreadsheet that has numbers that can be calculated. In this case, that would be cell B4 (also seen in Figure 6-2).

Step 3: Click **AutoSum** (Figure 6-3, Arrow A). NOTE: If you remember, when you click **AutoSum**, it will attempt to add cells straight above the cell in which you have placed it, and if it finds nothing that can be calculated, it will then search for numbers to the left. If there are no numbers there, it will leave the parentheses empty (Figure 6-3, Arrow B).

Step 4: Click the **first sheet tab on the right** (Figure 6-4, Arrow A). NOTE: When you do, it will automatically put that sheet name inside the **SUM** function in the **Formula Bar** (Figure 6-4, Arrow B).

Step 5: Press and hold the **Shift** key.

Step 6: Click the **last sheet tab on the right** (Figure 6-5, Arrow A). When you do, it will automatically put that sheet name inside the **SUM** function as well as shown in the **Formula Bar** (Figure 6-5, Arrow B).

Step 7: Let go of the **Shift key. Do not leave the sheet!**

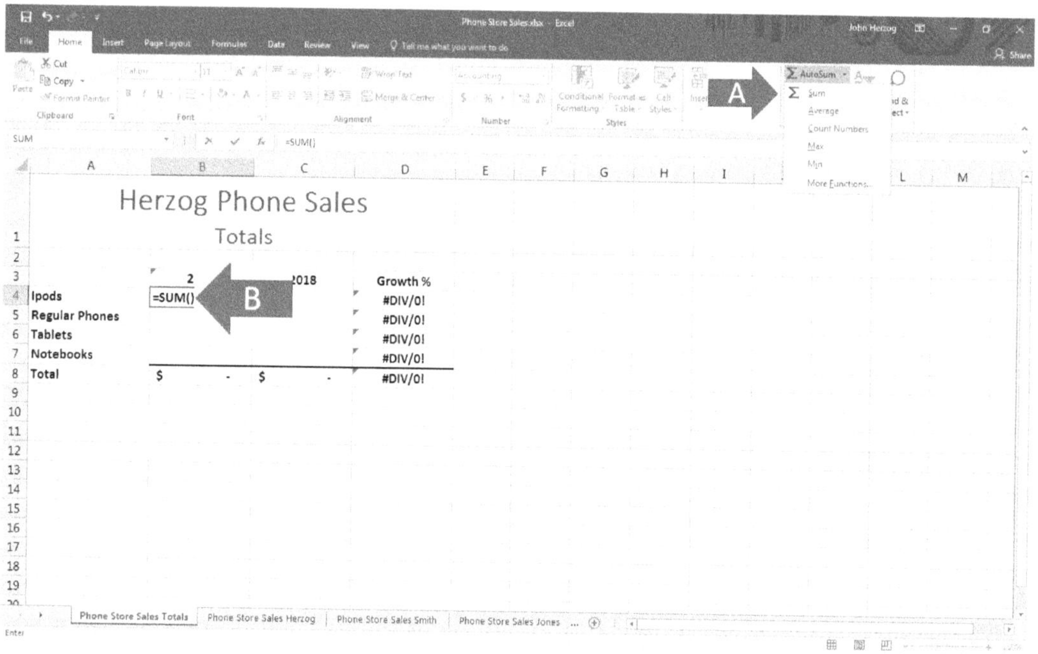

FIGURE 6-3

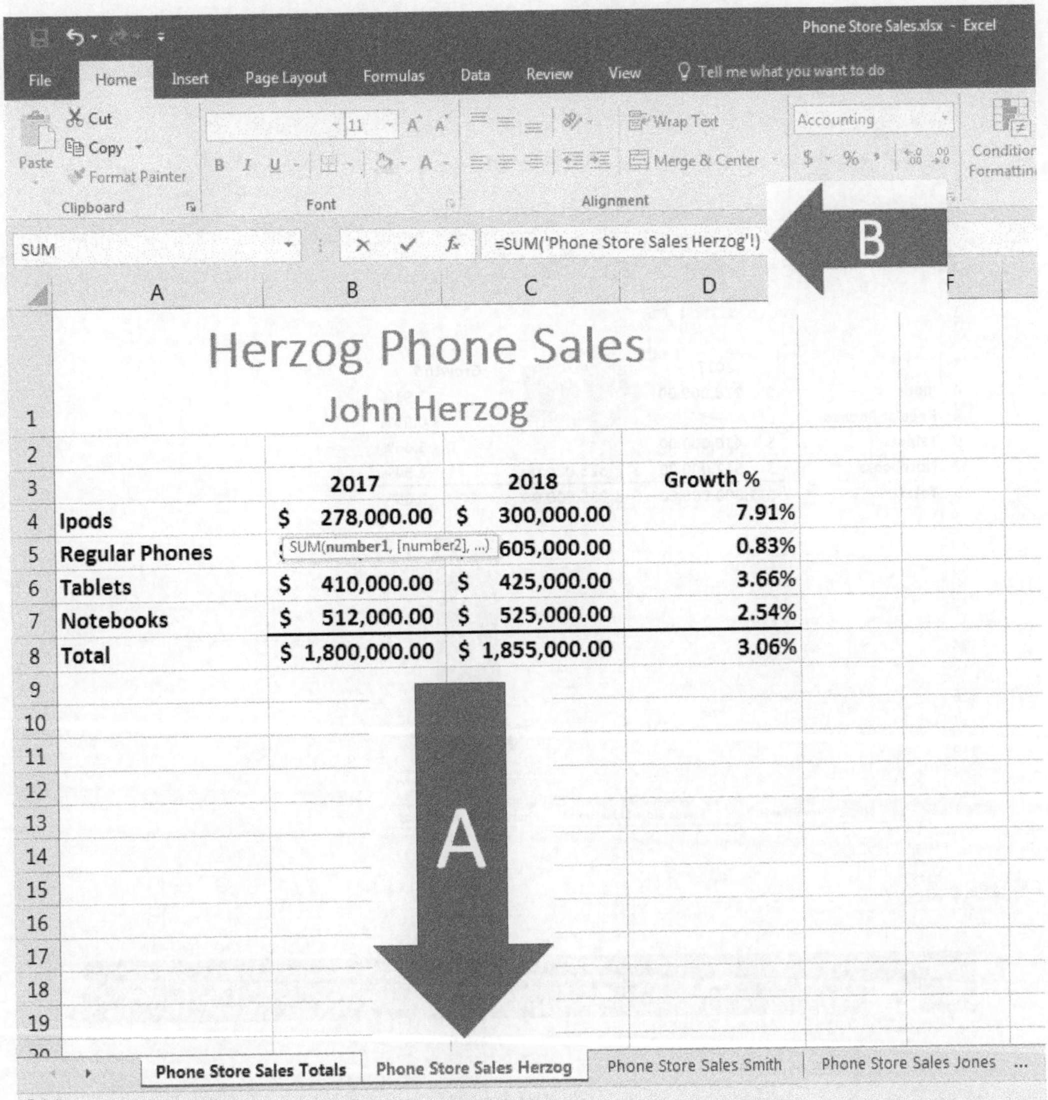

FIGURE 6-4

Step 8: **Click on the upper-leftmost cell that has calculable numbers in the sheet YOU ARE NOW IN.** In this case, it is B4 again, and in the sheet named *Phone Store Sales Jones* (Figure 6-5, Arrow C). When you do, it will add B4 to the end of the function range (Figure 6-6, Arrow). The function is basically saying "give me the **SUM** of cell B4 in the sheets from *Phone Store Sales Herzog* through *Phone Store Sales Jones* **inclusive**."

Step 9: Press **Enter**.

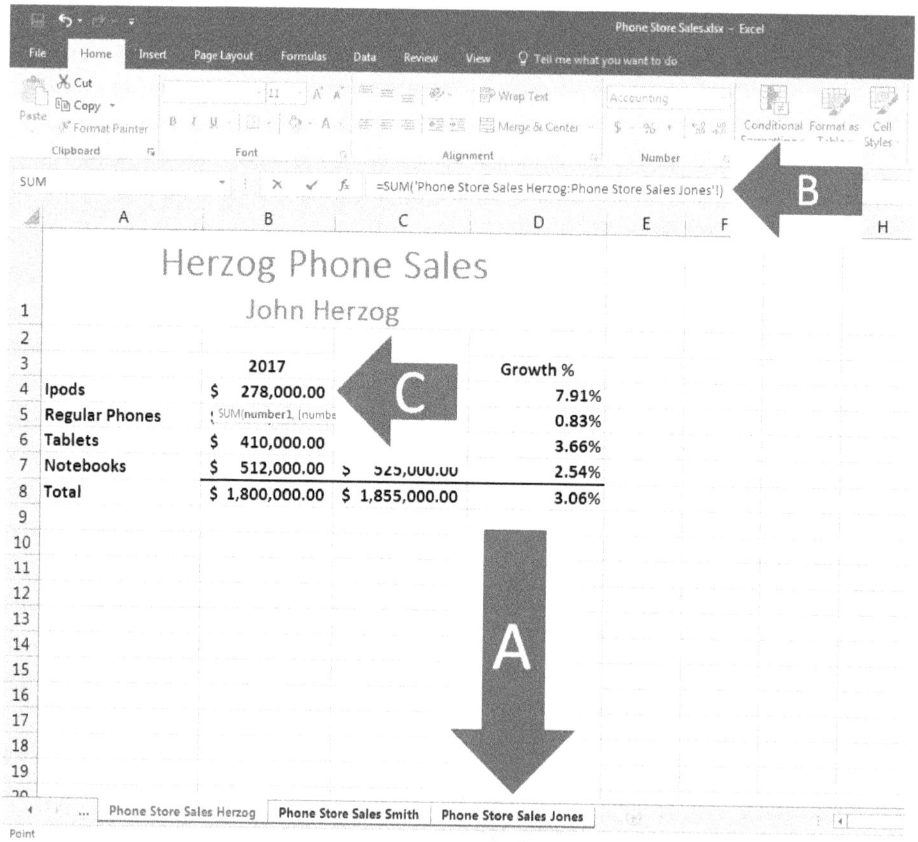

FIGURE 6-5

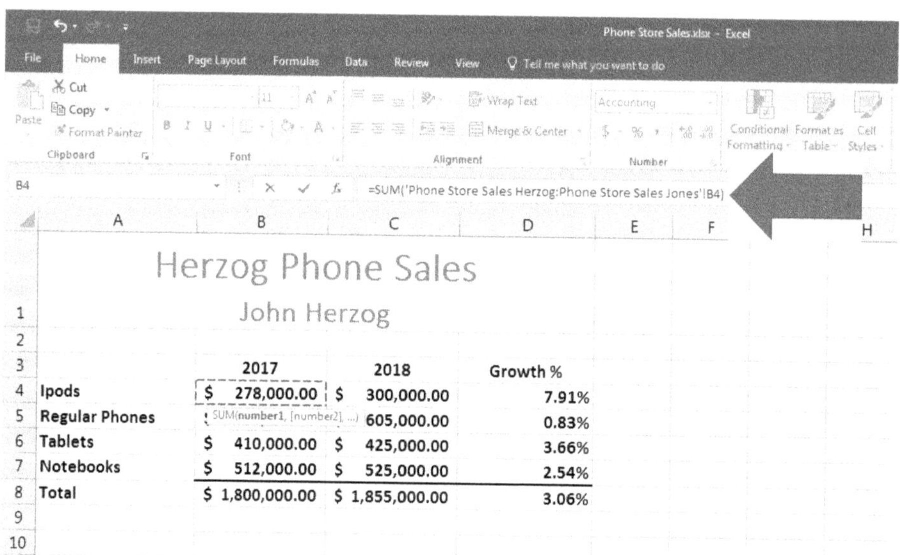

FIGURE 6-6

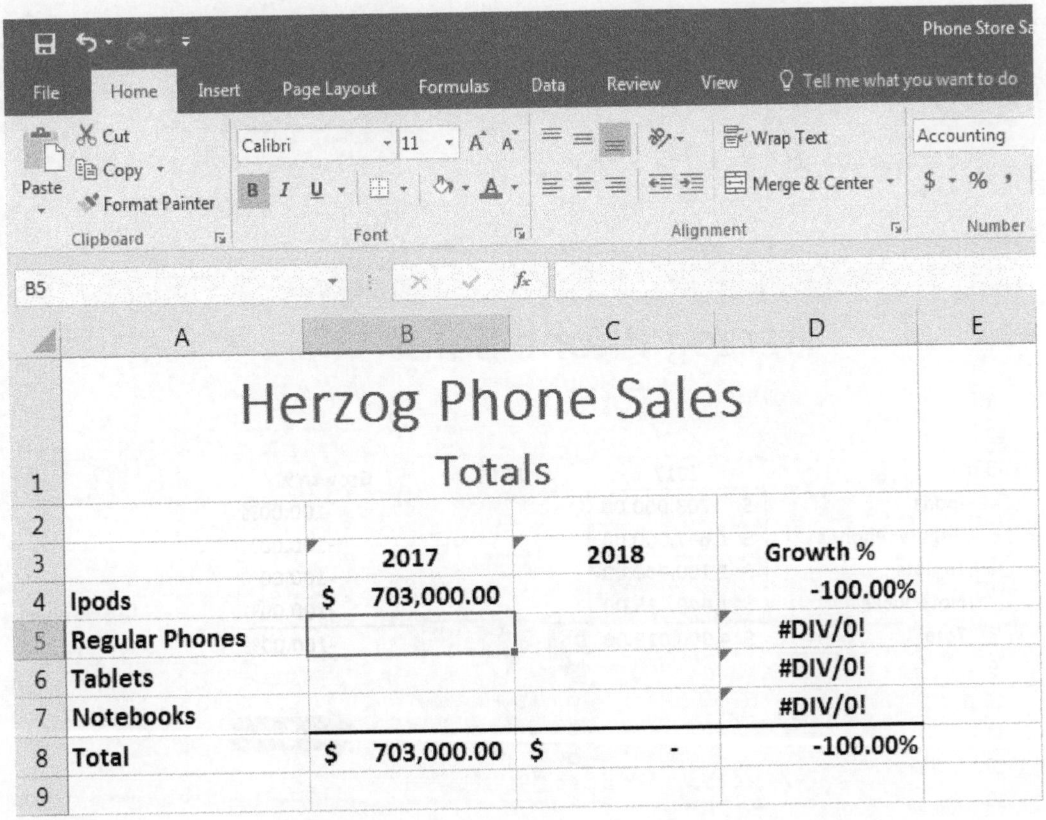

FIGURE 6-7

When you do, the result will be what you see in Figure 6-7.

You don't need to repeat this process to put the **3-D cell reference** in the remaining cells. All you need to do now is fill that function down to row 7 (or down to cell B7).

NOTE: When you fill this function down, you don't want to have the format of B4 overwrite the format of cell B7 because B7 has a border that you don't want to remove. After you fill the function down, you will see the **Auto Fill Options** (Figure 6-8, Arrow A). Click on it. When it displays this menu, in order to protect the border in B7, click **Fill Without Formatting** (Figure 6-8, Arrow B). Whenever you fill or copy cells, they will always copy their formats to their **destination cells** unless you click the **Fill Without Formatting** command.

Now fill the functions in cells B4 through B7 to cells C4 through C7 by pulling the fill handle on the lower right of cell B7 to the right, by one column.

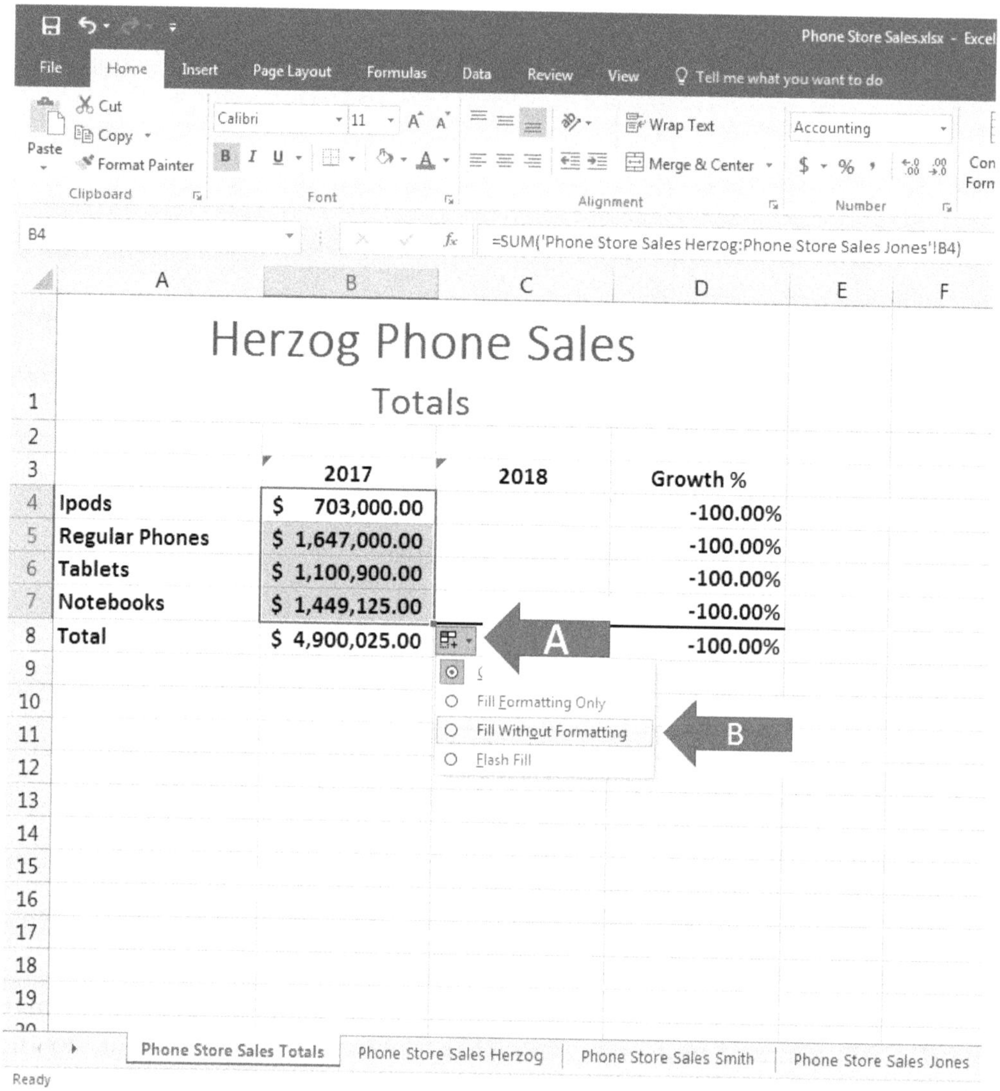

FIGURE 6-8

THE NOW FUNCTION

Anytime you want to date and time-stamp your spreadsheets, the NOW() function will allow you to do that. In this case, suppose that *Herzog Phone Sales* wants only to display the date and time that the data was most recently entered. You can achieve this by doing the following:

Step 1: In cell D2, type *=NOW()*.

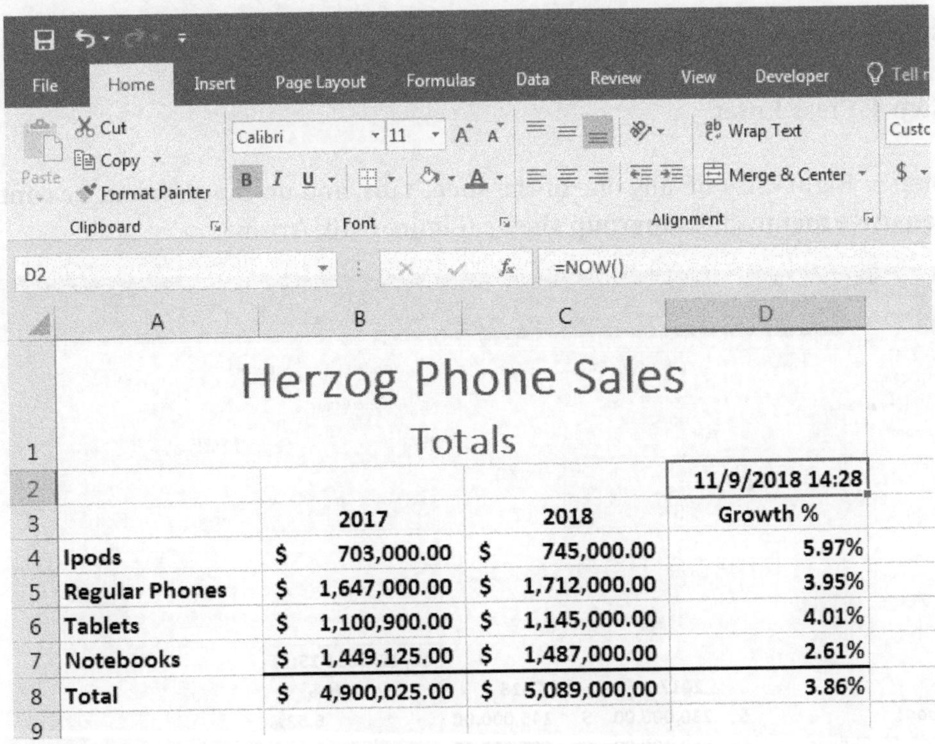

FIGURE 6-9

Step 2: Press **Enter** (Figure 6-9). The current date will be displayed with the current time. Each time there is any change in the spreadsheet, the time will change. Tomorrow, when you open the file, the cell will then display tomorrow's date and the time you opened it.

DRILLTHROUGHS

If we had wanted the NOW function to be in cell D2 on all of the sheets in the workbook, we can do so by creating what is called a **Drillthrough**. These can be created by a few easy steps.

Step 1: Select all of the sheets in your workbook by clicking on the first sheet tab to the left (named *Phone Store Sales Totals*), pressing and holding down the **shift** key, and then clicking the last sheet tab on the right (named *Phone Store Sales Jones*). All of those sheets are now selected.

Step 2: Delete the **NOW** function you entered there a moment ago so we can retype it as a **Drillthrough**.

Step 3: Retype =*NOW()* into cell D2.

Step 4: Press **Enter**.

Step 5: Right-click on any one of the sheet tabs, and on the **shortcut** or **context-sensitive menu**, click **ungroup sheets** (Figure 6-10, Arrow).

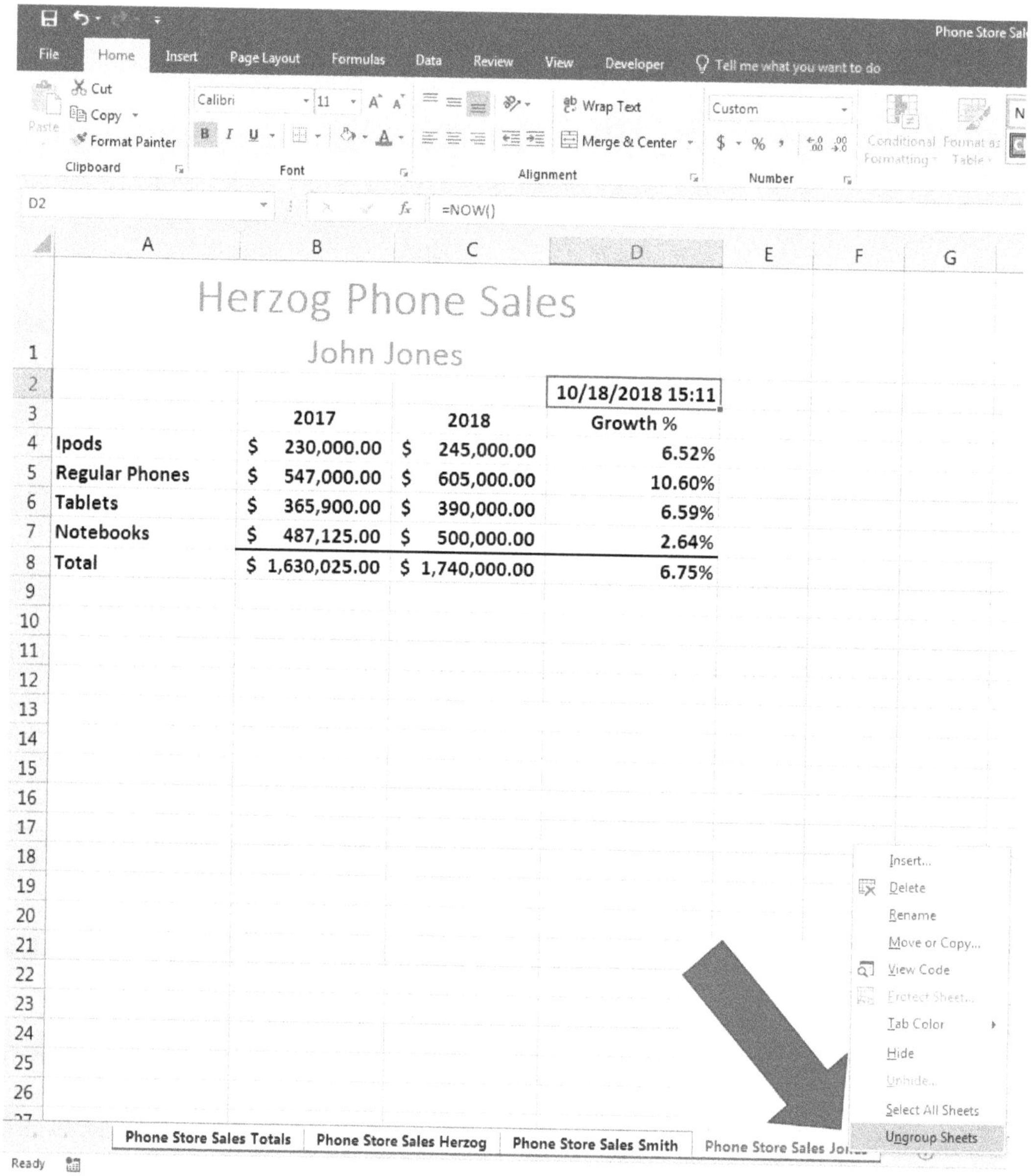

FIGURE 6-10

If you click on each of the sheets, you will see that each of them will have the **NOW** function in cell D2.

NOTE: Be very careful to **ungroup** the sheets immediately after you create a **drill-through**. If you forget to ungroup them, anything you type in any one of the cells will be placed in that same cell in each sheet, and what was previously in those cells will be overwritten!

CREATING THE EMBEDDED COLUMN CHART WITH MULTIPLE DATA SETS

If you create a column chart as we did in Chapter 1, you need to know that column charts are not limited to one data set. An example of one data set would be as in Chapter 1, when we discussed a *store name* and its *sales*. In this chapter, we have two data sets: one for a given year and one for a different year. When you make a column chart of this data therefore, you will have two columns (not one) representing the different items being tracked; in this case, the types of products that *Herzog Phone Sales* offers (Figure 6-10). This chart will also be embedded.

The procedure is the same as the one used for creating the chart in Chapter 1.

Step 1: Select the range to be charted. In this example, it will be cells A3 through C7 as shown in Figure 6-11. **Do not include the totals!**

Step 2: Click the **Insert Ribbon** (Figure 6-12, Arrow A), **Chart** (Figure 6-12, Arrow B), and **3-D Clustered Column** (Figure 6-12, Arrow C). When you do, the program will give you the screen that you see in Figure 6-12.

Step 3: Click and drag the border of the **embedded chart** to move it to its desired location. In this example, it will be cells A10 through D23 (Figure 6-13). Make sure it has a title that reads *Phone Stores Sales*, using the procedures discussed in Chapter 1.

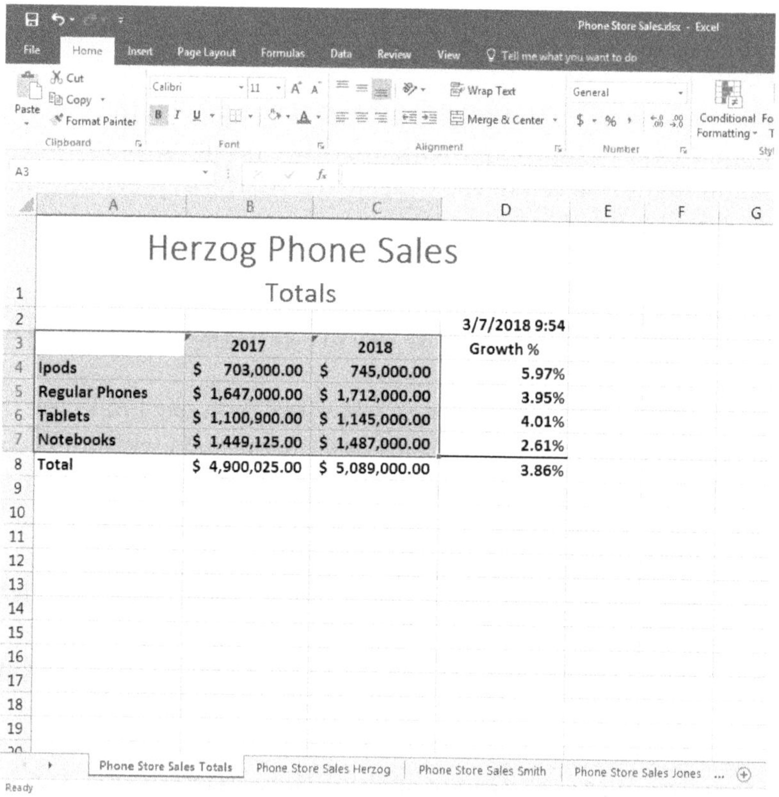

FIGURE 6-11

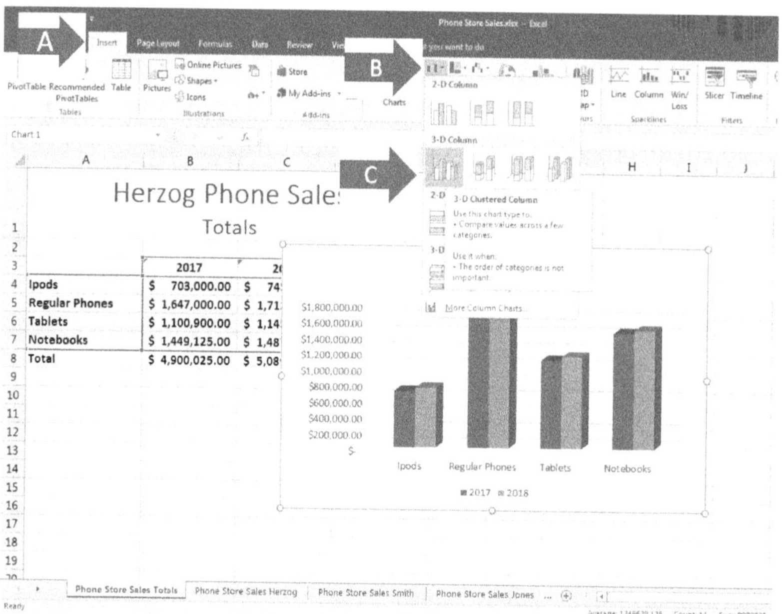

FIGURE 6-12

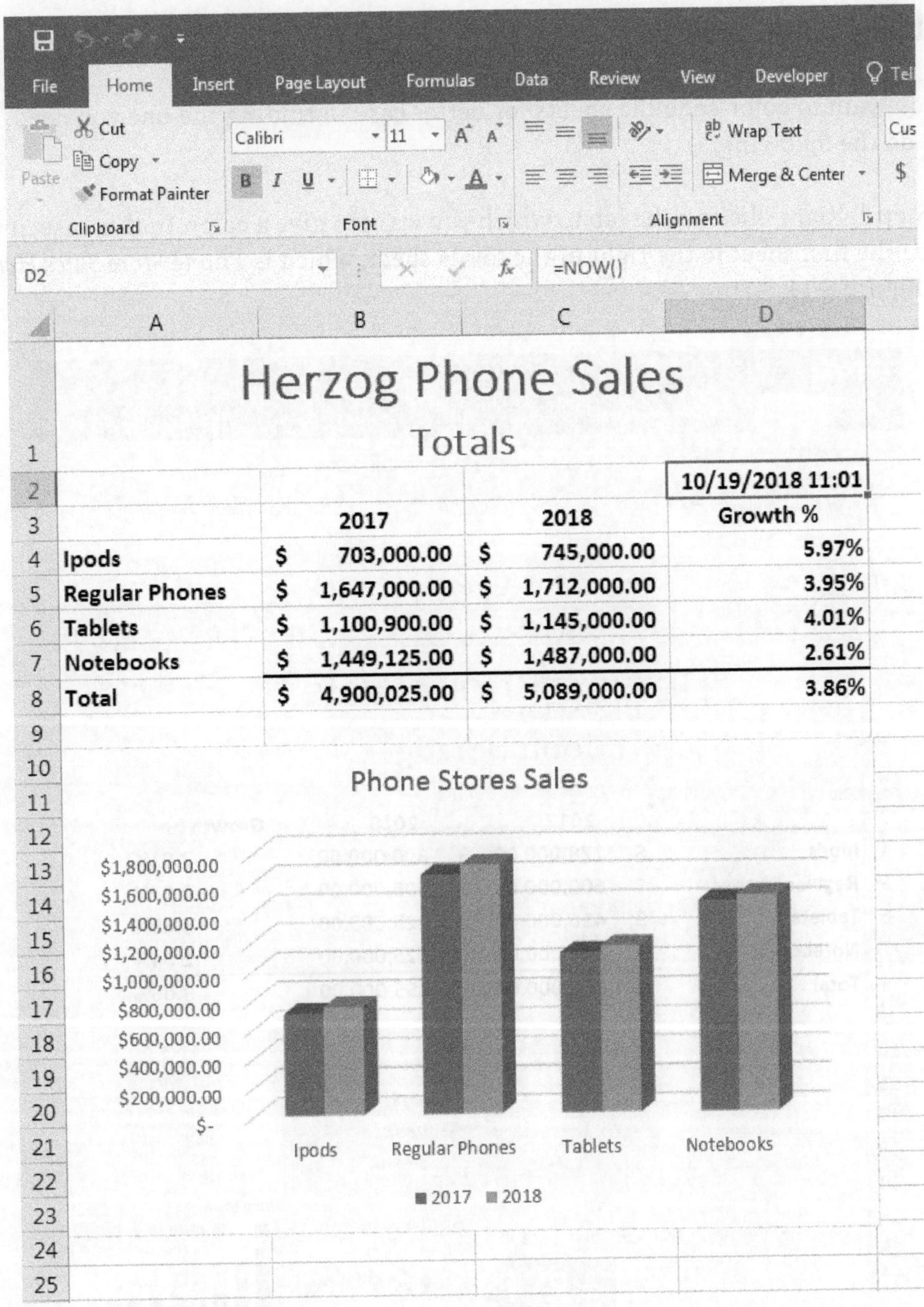
File
Home
Insert
Page Layout
Formulas
Data
Review
View
Developer
Cut
Copy
Format Painter
Paste
Clipboard
Calibri
11
Font
Wrap Text
Merge & Center
Alignment
D2
=NOW()
Herzog Phone Sales
Totals
10/19/2018 11:01
2017
2018
Growth %
Ipods $ 703,000.00 $ 745,000.00 5.97%
Regular Phones $ 1,647,000.00 $ 1,712,000.00 3.95%
Tablets $ 1,100,900.00 $ 1,145,000.00 4.01%
Notebooks $ 1,449,125.00 $ 1,487,000.00 2.61%
Total $ 4,900,025.00 $ 5,089,000.00 3.86%
Phone Stores Sales
$1,800,000.00
$1,600,000.00
$1,400,000.00
$1,200,000.00
$1,000,000.00
$800,000.00
$600,000.00
$400,000.00
$200,000.00
$-
Ipods
Regular Phones
Tablets
Notebooks
2017
2018

FIGURE 6-13

COLOR-CODING SHEETS

If you want to color-code the sheets for better ease in finding the one you need, you can do the following:

> Step 1: Right-click on the tab to which you want to give a color. In this case, it will be the first sheet to the right of the totals sheet, which is *Phone Store Sales Herzog* (Figure 6-14, Arrow A).

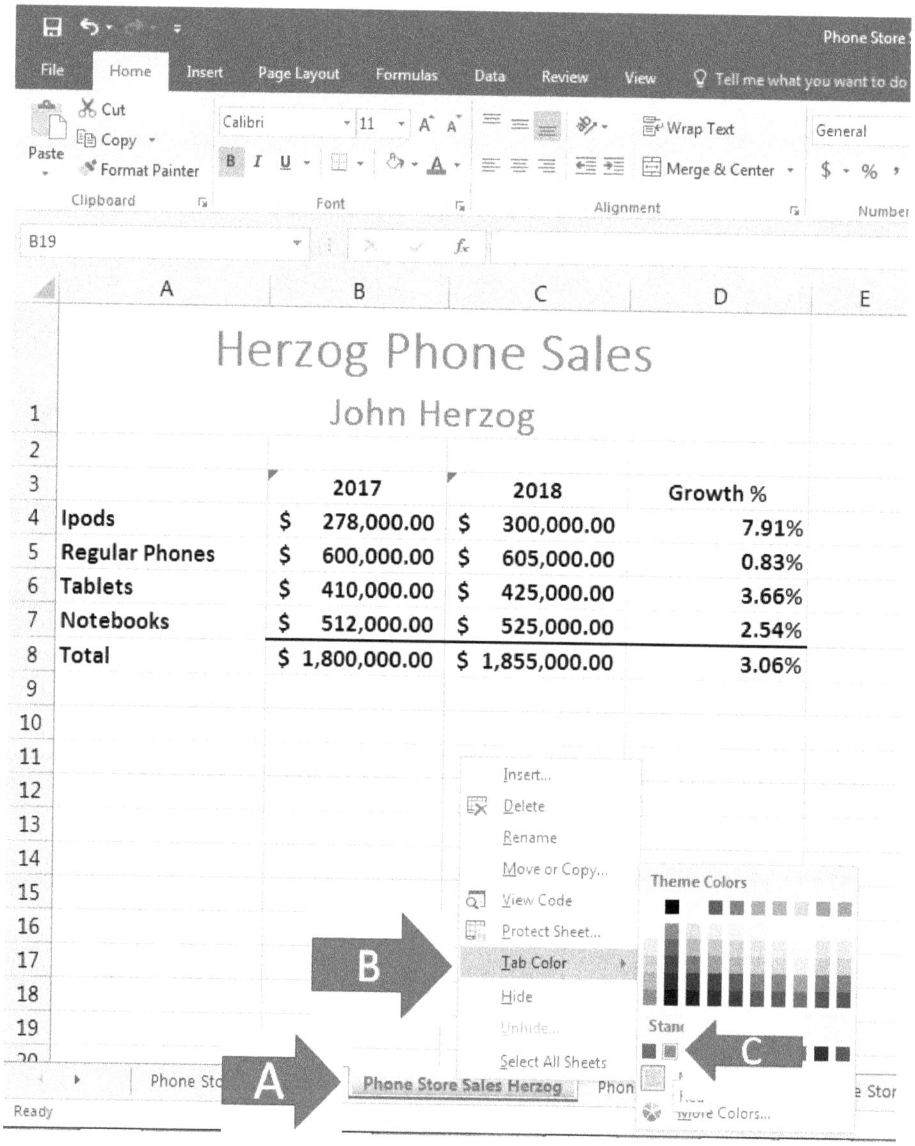

FIGURE 6-14

Step 2: In the **pop-up menu** that appears, click **Tab Color** (Figure 6-14, Arrow B).

Step 3: Choose the color you want for the tab. In this case, we will choose **red** (Figure 6-14, Arrow C). NOTE: When you choose the color, the sheet tab will not show the full color until you select a different sheet.

In this example, make sure all of the sheets have a different color.

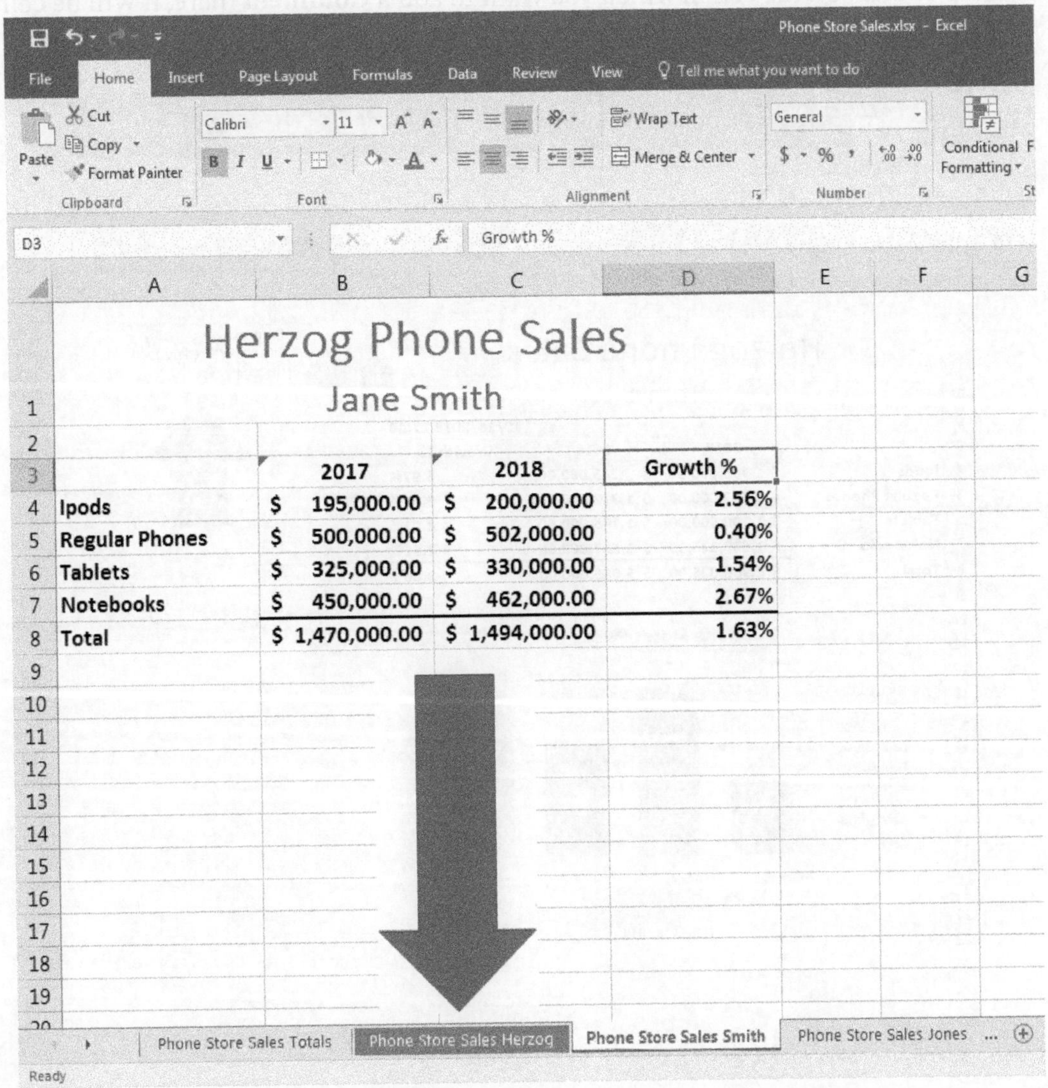

FIGURE 6-15

TO ADD A COMMENT

Sometimes you may want to make notes or comments in a spreadsheet for yourself or for future users. The problem is that you may not want it to be taking up space in a printout or you may not want them visible all of the time. Sometimes the comments you make will be lengthy.

One way to create a comment is by doing the following:

Step 1: Click on the cell in which you want to add a **Comment** (here, it will be cell E1 on the *Phone Store Sales Totals* sheet, Figure 6-16A).

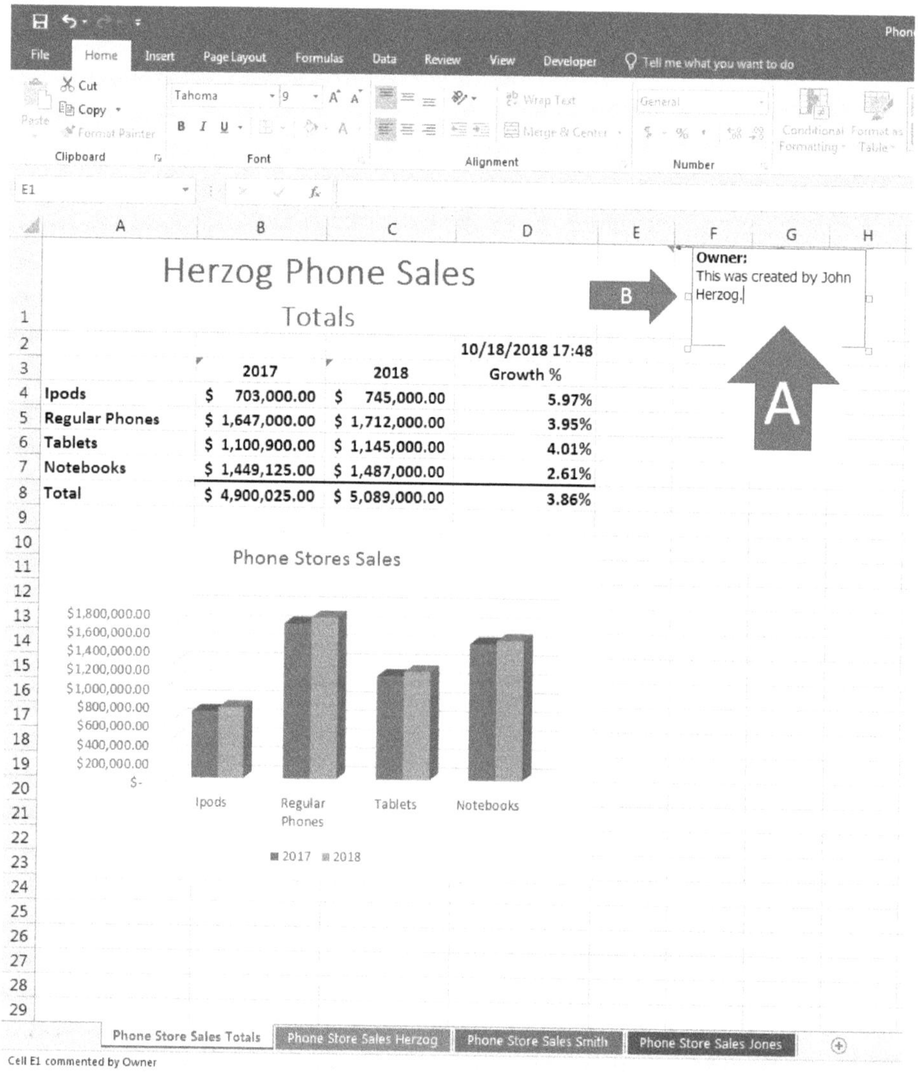

FIGURE 6-16A

Step 2: When you do, it will give you a box with a place to add your comment (Figure 6-16A, Arrow A). It will often put the word *Owner* there or the name of the person who is logged in. Sometimes it will put the name there to whichever person or organization this particular version of Excel is licensed. Whatever the case, you can use it, or you can delete it by pressing and holding the backspace. Once you do, you can then write your comment in the box (Figure 6-16A, Arrow A).

To see the comment, you need to only place your mouse over the cell that contains the comment, and it will be displayed (Figure 6-16B).

NOTE: If the box is not big enough to enter your comment, you can resize the **Comment box** (as you would by resizing any chart or object in any Microsoft product) by placing your mouse on one of the handles on the side of it or in one of the corners (Figure 6-16A, Arrow B) and then clicking and dragging the double arrow until the box becomes the size you need.

Printing spreadsheets and using **Page Setup** options will be discussed in depth in Chapter 7. The **Print Menu** is found in the **File Ribbon**.

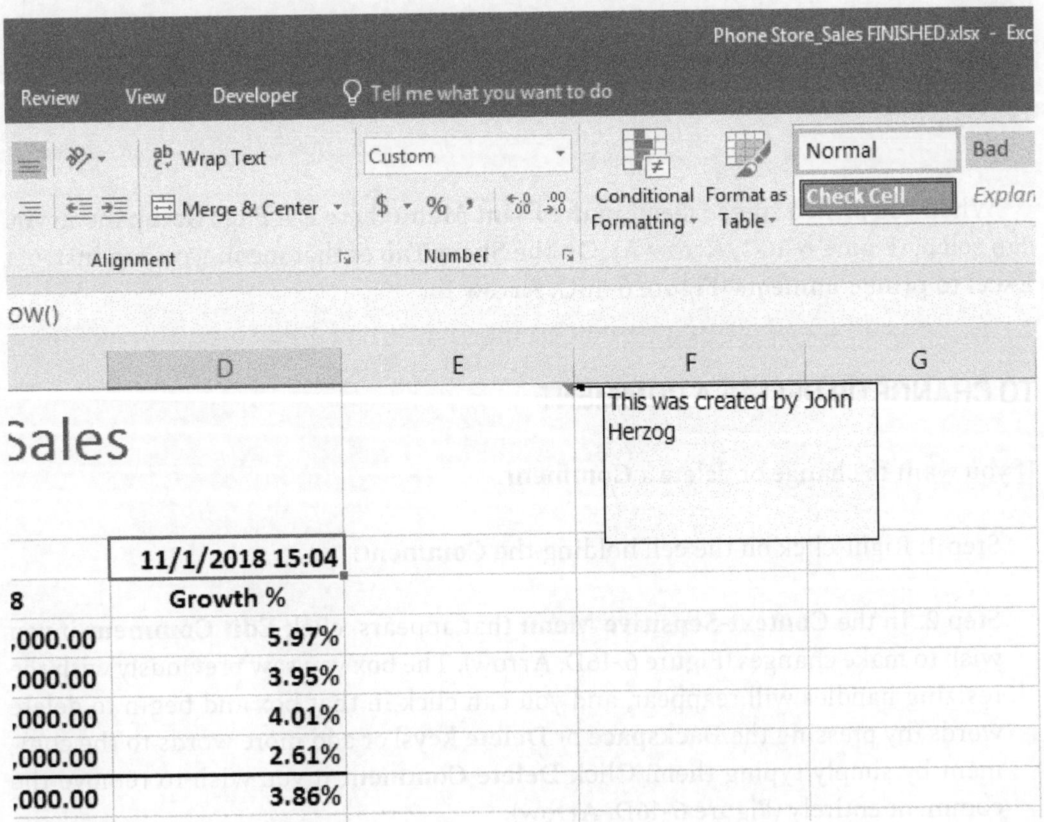

FIGURE 6-16B

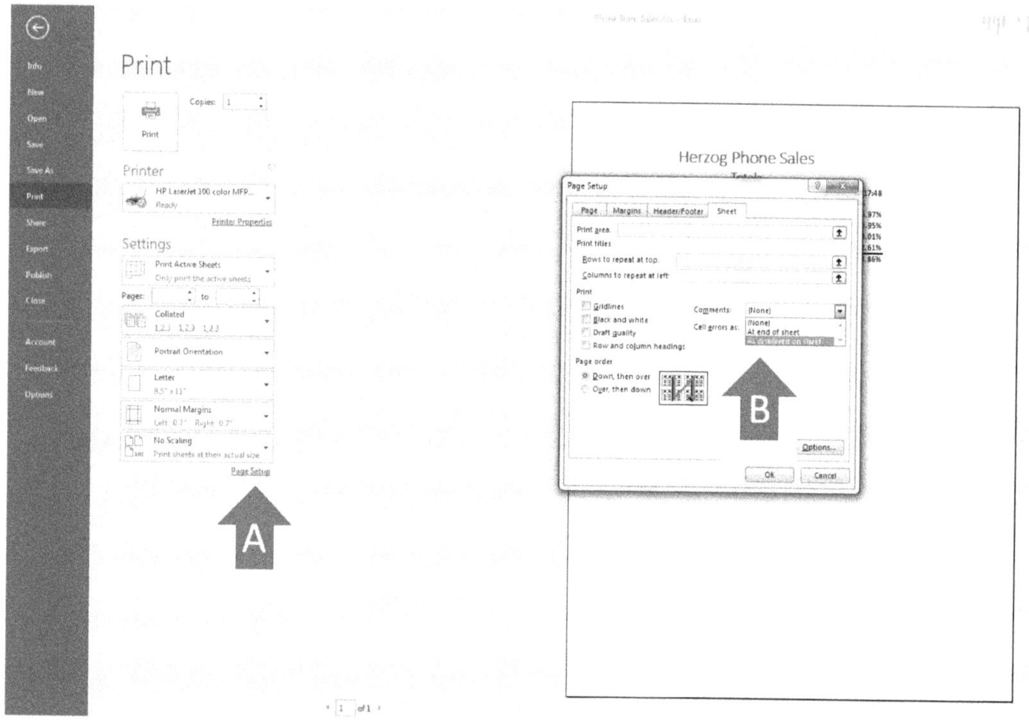

FIGURE 6-16C

When you print a spreadsheet, in the **Print Menu** there is a **Page Setup** menu you can select (Figure 6-16C, Arrow A). On the **Sheet Tab** of that menu, you can instruct Excel to print comments (Figure 6-16C, Arrow B).

TO CHANGE OR DELETE A COMMENT

If you want to change or delete a **Comment**:

Step 1: Right-click on the cell holding the **Comment**.

Step 2: In the **Context-Sensitive Menu** that appears, click **Edit Comment** if you wish to make changes (Figure 6-16D, Arrow). The box you saw previously with the resizing handles will reappear, and you can click in that box and begin to delete words (by pressing the **Backspace** or **Delete keys**) or add more words to the comment by simply typing them. Click **Delete Comment** if you wish to remove the comment entirely (Figure 6-16D, Arrow).

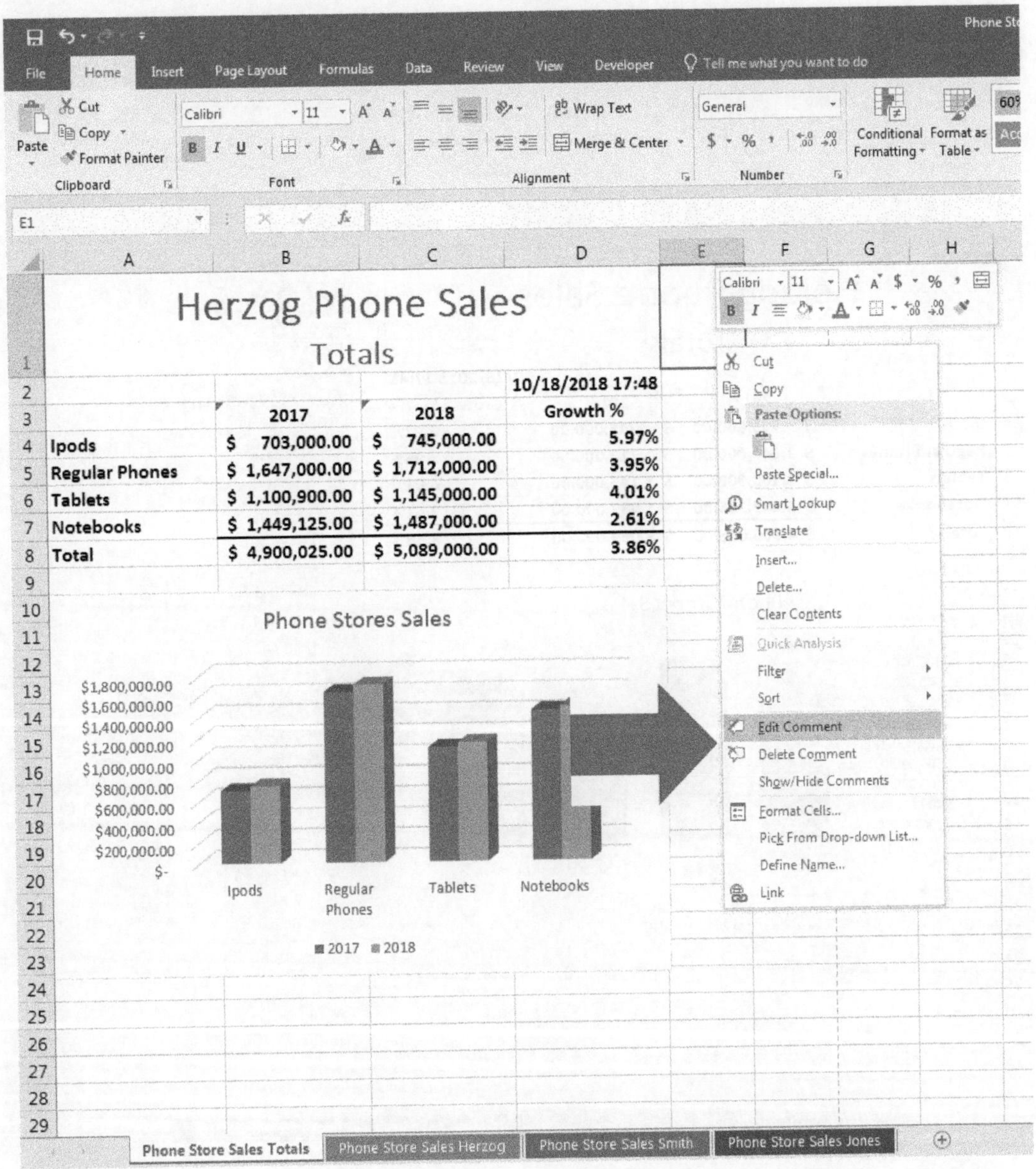

FIGURE 6-16D

NOTE: Another way to insert a **Comment** is by right-clicking on the cell in which you want to insert it. On the **Context-Sensitive Menu,** you will see an option labeled **Insert Comment** (Figure 6-16E, Arrow). At that point, you will see the same menu you saw previously when you pressed **Shift F2**. You can then simply begin to type the comment as before. Notice that when you began to edit the comment earlier, in the **Context-Sensitive Menu,** the words **Insert Comment** were replaced by the words **Edit Comment**.

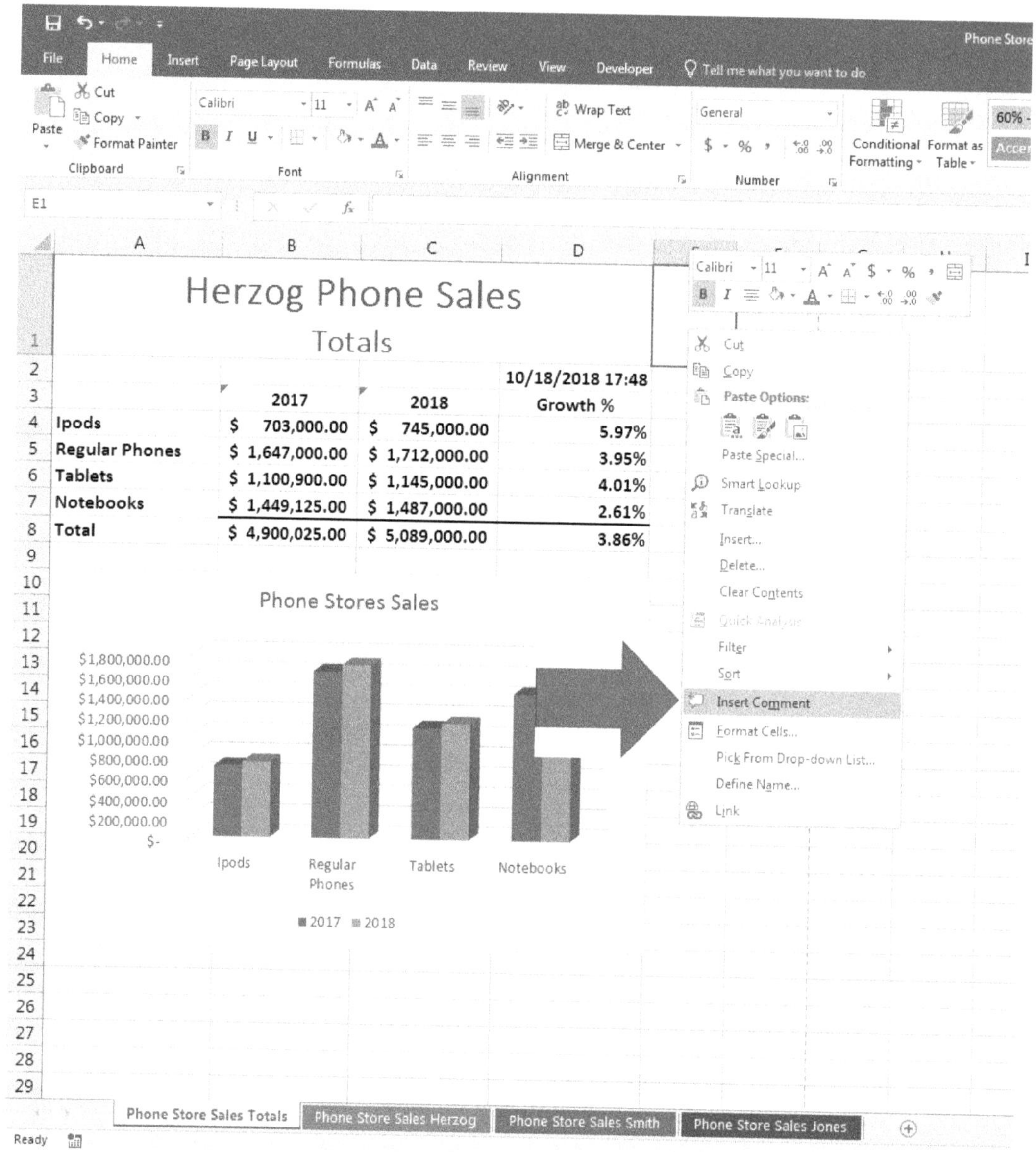

FIGURE 6-16E

MAKING A TEMPLATE

When you create a new spreadsheet, you can do it from scratch (or as a blank document). You can also use previously created spreadsheets as **templates** if they would fit your needs. For example, you could create an academic calendar, a back-to-school planner, and much more. You can even create a template of your own that fits your particular needs that could be used in the future to keep you from having to reinvent the wheel each time you put a spreadsheet together. Some users will simply copy a previously created spreadsheet and remove its data or replace it with current data. Unfortunately, that is often time consuming, and sometimes users forget to delete some of the older data. That is why a template is better in that it will automatically remove all of your old data for you and will allow you to start fresh with new data whenever you open one.

TO MAKE THIS FILE A TEMPLATE FOR FUTURE USE

Suppose that we want to create a **template** of the workbook we have just modified. To do so, use the following steps:

Step 1: Remove all data, but do not remove any formulas, functions, or labels from your spreadsheet (Figure 6-17).

Step 2: Click the **File Ribbon** (Figure 6-17, Arrow).

Step 3: In the next menu, click **Save As** (Figure 6-18, Arrow A).

Step 4: Click **Browse** (Figure 6-18, Arrow B).

Step 5: At the next menu, click the **Save as Type drop-down list** (Figure 6-19, Arrow A).

Step 6: Click **Excel Template** (***.xltx**) (Figure 6-19, Arrow B).

Step 7: Give the template the name you would like, then click **OK**.

When you do these steps, your template will then be displayed in the **New Document** menu that was discussed in the Introduction chapter of this textbook (Figure 6-20).

NOTE: There are some organizations that will not permit you to add templates to their systems without proper network permissions.

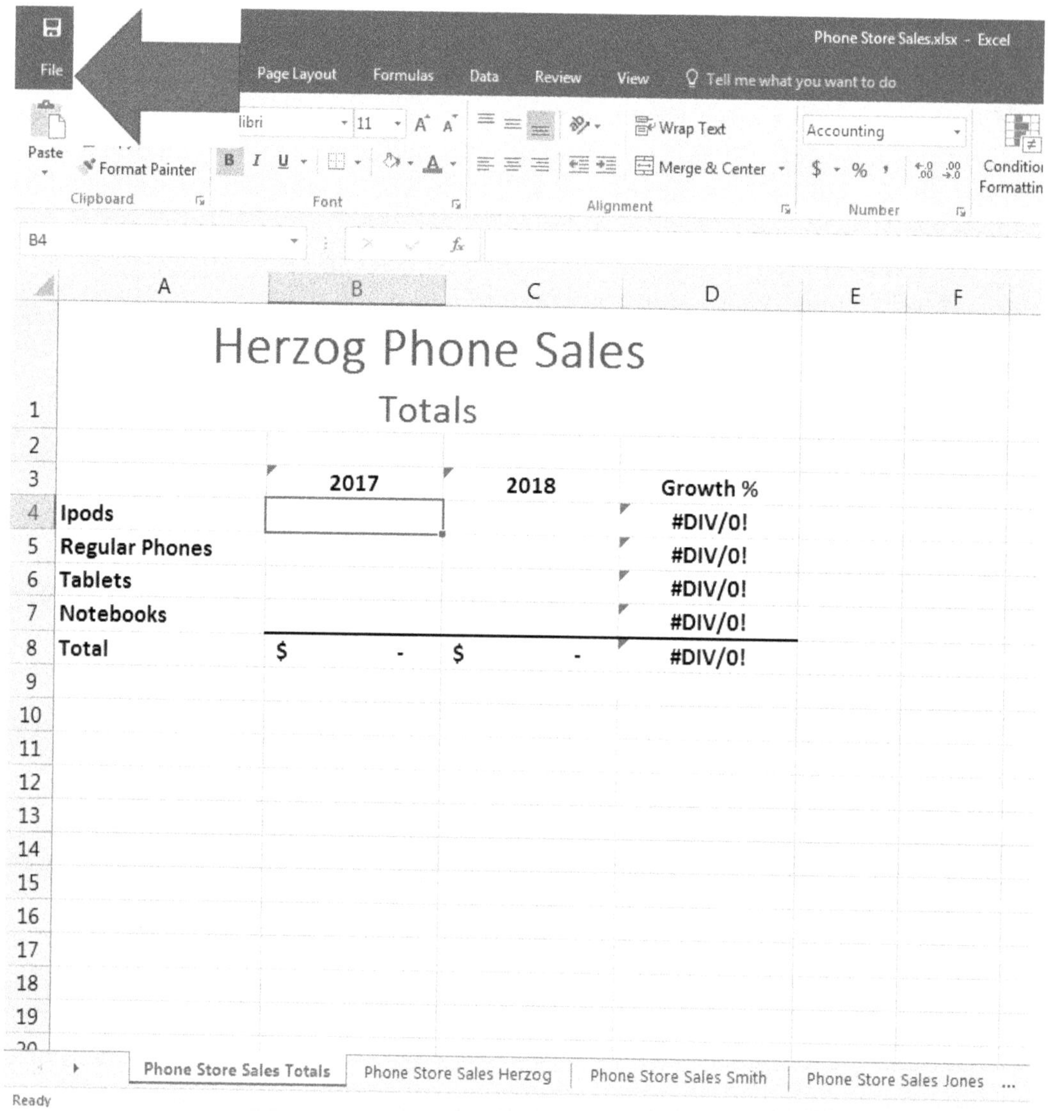
File
Page Layout
Formulas
Data
Review
View
Tell me what you want to do
Phone Store Sales.xlsx - Excel
Paste
Format Painter
Clipboard
Font
Wrap Text
Merge & Center
Alignment
Accounting
Number
B4
Herzog Phone Sales
Totals
2017
2018
Growth %
Ipods
Regular Phones
Tablets
Notebooks
Total
#DIV/0!
Phone Store Sales Totals
Phone Store Sales Herzog
Phone Store Sales Smith
Phone Store Sales Jones
Ready

FIGURE 6-17

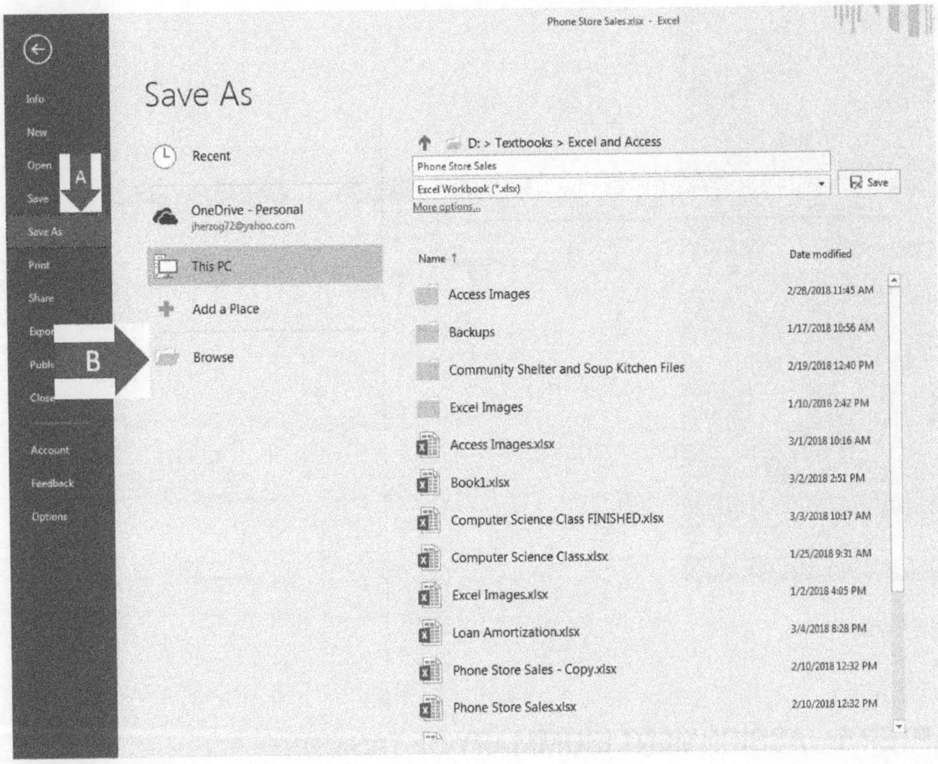

FIGURE 6-18

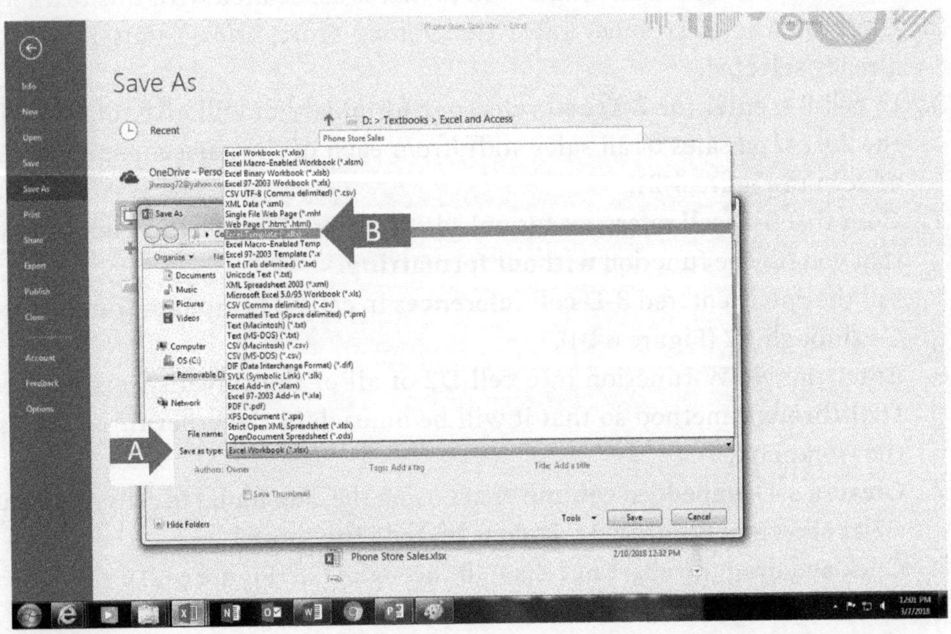

FIGURE 6-19

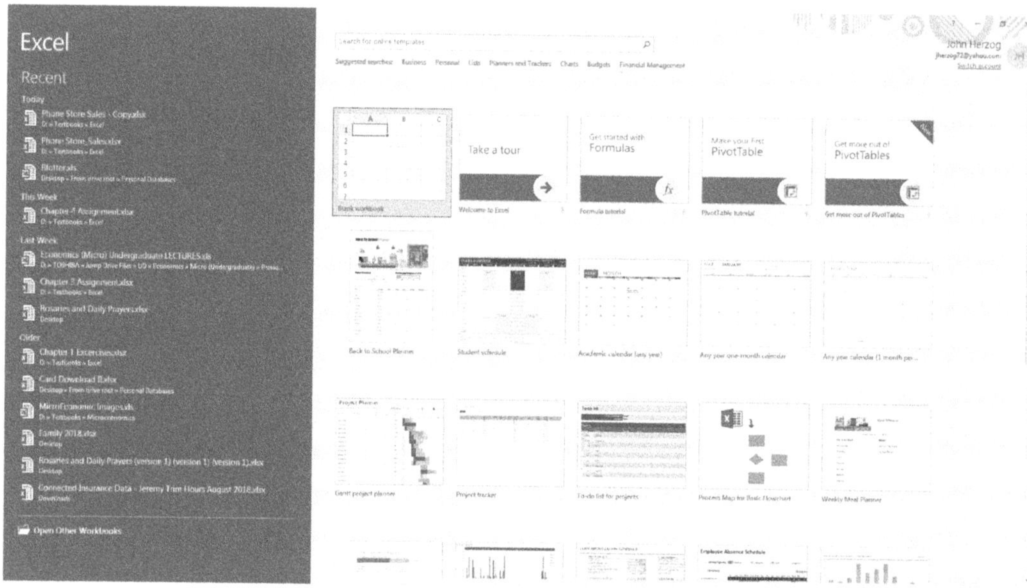

FIGURE 6-20

CHAPTER 6—ASSIGNMENT (SUMMARY OF TASKS)

1. Open the file named *Phone Store Sales* that is associated with this textbook.
2. Click on the first sheet tab titled *Phone Store Sales Totals* if it is not already selected.
3. In cell B4, enter the **3-D cell reference** formula that will give totals of all of the *2017 iPod* sales of all sales staff (from each of the adjacent spreadsheets) as seen in Figure 6-21.
4. Copy that **3-D cell reference** formula down to row 7 (or cell B7), and make sure that you fill the function **without formatting**.
5. Fill the newly entered **3-D cell references** in cells B4 through B7 over to cells C4 through C7 (Figure 6-21).
6. Enter the **NOW** function into cell D2 of all of the spreadsheets using the **Drillthrough** method so that it will be on each of the other three sheets in the workbook.
7. Create a **3-D embedded column chart** using the data found in cells A3 through C7 as shown in Figure 6-11. **Do not include the totals!**
8. Click and drag the chart to cells A10 through D23 (Figure 6-21).
9. Add a **chart title** to the chart that reads *Phone Store Sales*.

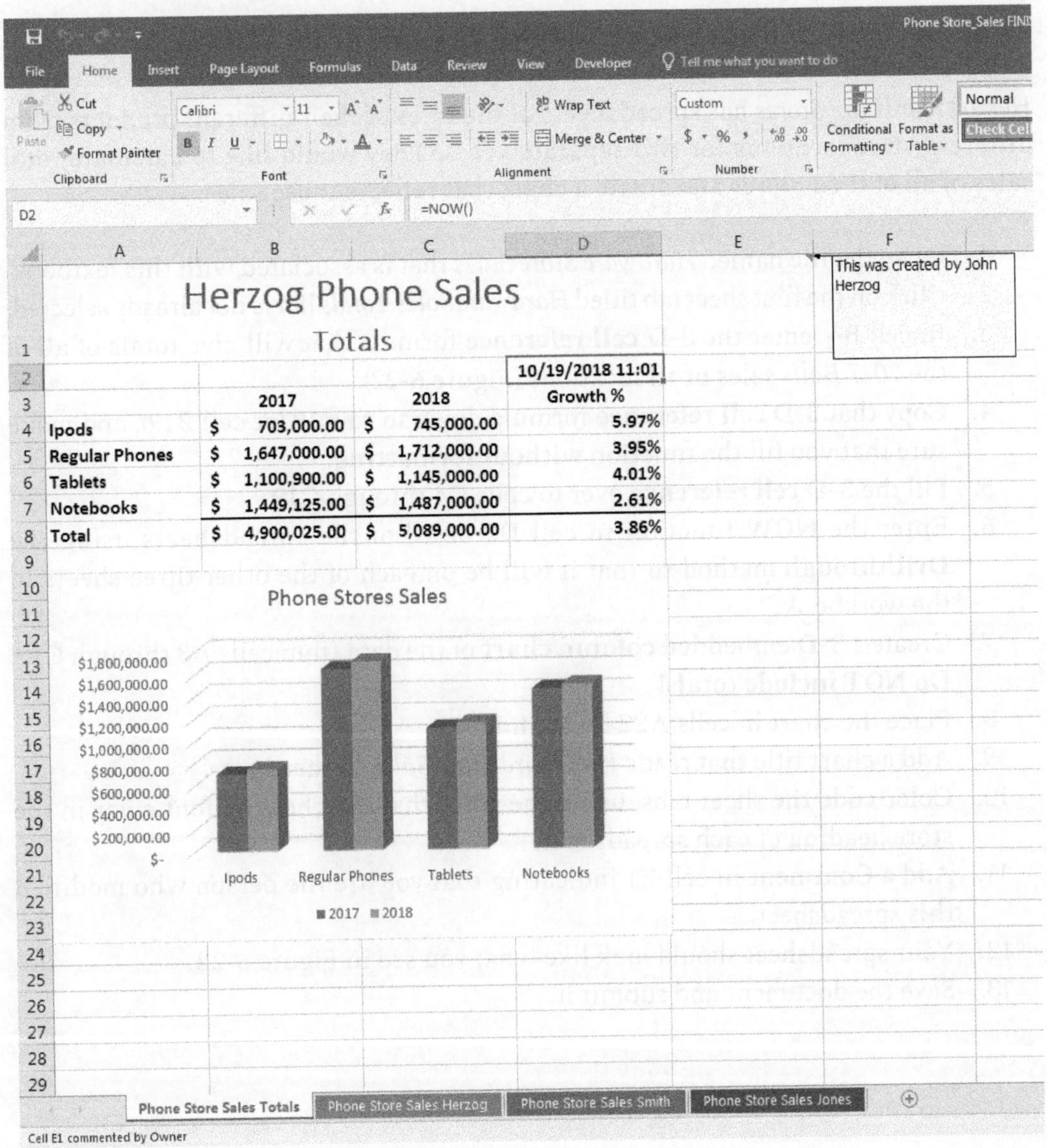

FIGURE 6-21

10. Color-code the sheet tabs, using the color that matches the font color in the heading of each spreadsheet.
11. Add a **Comment** in cell E1 of the *Phone Store Sales Totals* spreadsheet to indicate that you are the one who modified this spreadsheet.
12. The spreadsheet should look like Figure 6-21.
13. Save the workbook and submit it.

CHAPTER 6—ALTERNATE ASSIGNMENT (SUMMARY OF TASKS)

Herzog Hardware Stores has spreadsheets of the sales of their different products from different store locations for two separate years. They would like to calculate total sales of all of their stores and create a chart that reflects those sales.

1. Open the file named *Hardware Store Sales* that is associated with this textbook.
2. Click on the first sheet tab titled *Hardware Sales Totals* if it is not already selected.
3. In cell B4, enter the **3-D cell reference** formula that will give totals of all of the *2017 Bolts* sales of all locations (Figure 6-22).
4. Copy that **3-D cell reference** formula down to row 19 (or cell B19), and make sure that you fill the function **without formatting**.
5. Fill the **3-D cell reference** over to cells C4 through C19.
6. Enter the **NOW** function in cell D2 of all of the spreadsheets using the **Drillthrough** method so that it will be on each of the other three sheets in the workbook.
7. Create a **3-D embedded column chart** of the data from cells A3 through C19. **Do NOT include totals!**
8. Place the chart in cells A22 through D36.
9. Add a chart title that reads *Hardware Store Sales* to the chart.
10. Color-code the sheet tabs, using the color that matches the font color in the store heading of each spreadsheet.
11. Add a **Comment** in cell E1 indicating that you are the person who modified this spreadsheet.
12. Your spreadsheet should look like what you see in Figure 6-22.
13. Save the document and submit it.

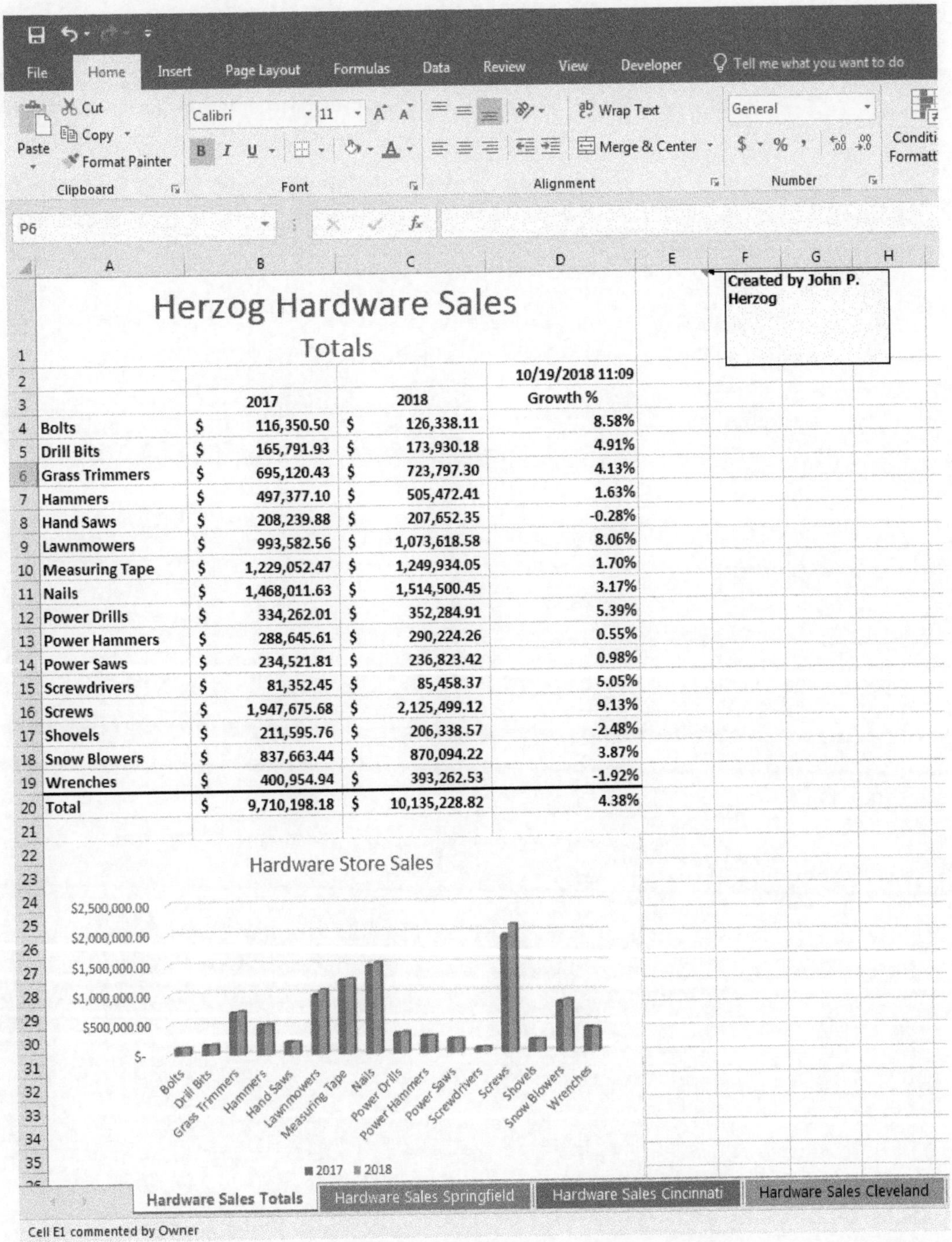

Herzog Hardware Sales

Totals

	2017	2018	Growth %
			10/19/2018 11:09
Bolts	$ 116,350.50	$ 126,338.11	8.58%
Drill Bits	$ 165,791.93	$ 173,930.18	4.91%
Grass Trimmers	$ 695,120.43	$ 723,797.30	4.13%
Hammers	$ 497,377.10	$ 505,472.41	1.63%
Hand Saws	$ 208,239.88	$ 207,652.35	-0.28%
Lawnmowers	$ 993,582.56	$ 1,073,618.58	8.06%
Measuring Tape	$ 1,229,052.47	$ 1,249,934.05	1.70%
Nails	$ 1,468,011.63	$ 1,514,500.45	3.17%
Power Drills	$ 334,262.01	$ 352,284.91	5.39%
Power Hammers	$ 288,645.61	$ 290,224.26	0.55%
Power Saws	$ 234,521.81	$ 236,823.42	0.98%
Screwdrivers	$ 81,352.45	$ 85,458.37	5.05%
Screws	$ 1,947,675.68	$ 2,125,499.12	9.13%
Shovels	$ 211,595.76	$ 206,338.57	-2.48%
Snow Blowers	$ 837,663.44	$ 870,094.22	3.87%
Wrenches	$ 400,954.94	$ 393,262.53	-1.92%
Total	$ 9,710,198.18	$ 10,135,228.82	4.38%

FIGURE 6-22

CHAPTER SEVEN

Macros, Pivot Tables, and Pie Charts

BY THE END of this chapter, it is our objective that the user will be able to:

1. Create a **pivot table**.
2. Create a **pie chart**.
3. Create a **macro**.
4. Create **macro buttons**.
5. Use **data validation**.
6. Use **table features** where cells are formatted as a table.
7. Add a **header or footer** to a spreadsheet.
8. **Print** a spreadsheet.

CHAPTER 7 TASK: TERRITORIAL SALES ANALYSIS

Suppose a firm wants to analyze its sales data in another dimension and save steps in doing so. In this chapter, we will use the *Territorial Sales* spreadsheet that is associated with the textbook as an example that shows sales by *Salesperson* and by *Sales Area*.

CREATING A PIVOT TABLE

Pivot tables show totals, averages, and other statistics in more depth and dimensions than has been done so far in this text, with more than one variable. You can create one using the following steps.

Step 1: Open the workbook named *Territorial Sales* that is associated with this textbook.

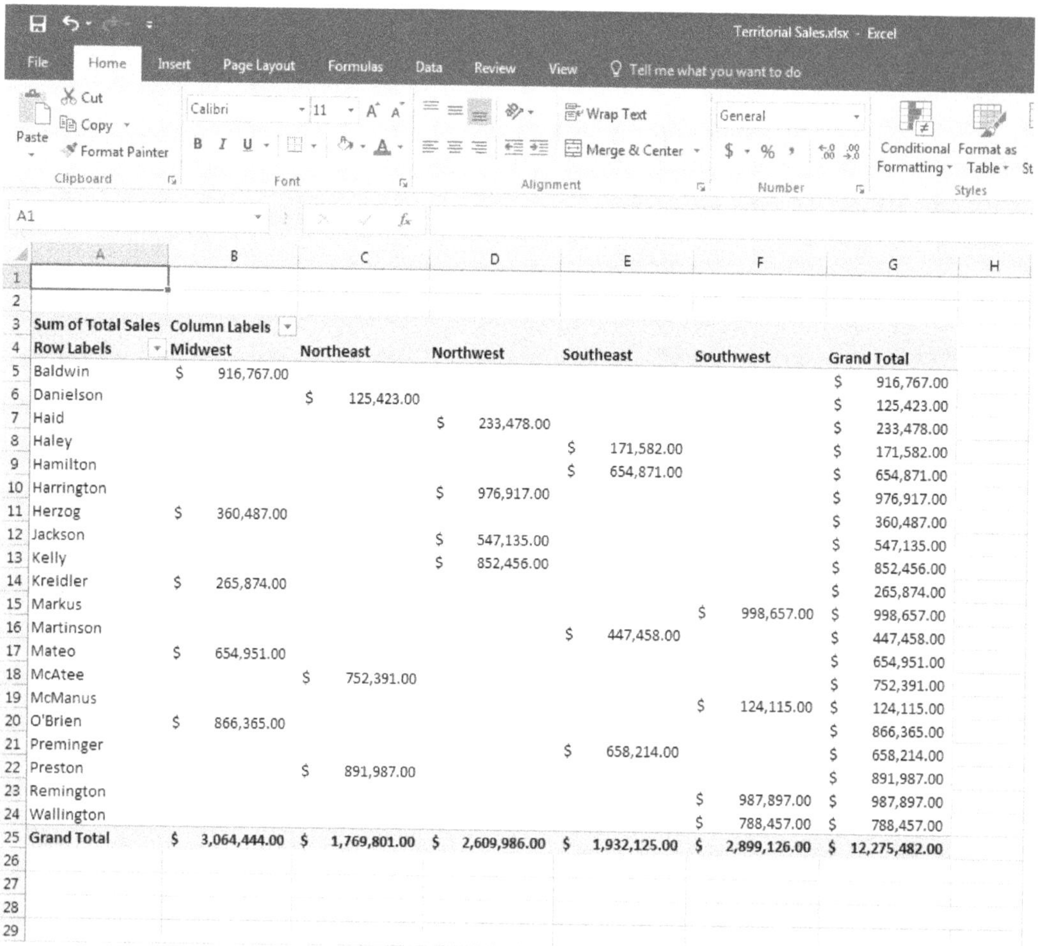

Sum of Total Sales	Column Labels					
Row Labels	Midwest	Northeast	Northwest	Southeast	Southwest	Grand Total
Baldwin	$ 916,767.00					$ 916,767.00
Danielson		$ 125,423.00				$ 125,423.00
Haid			$ 233,478.00			$ 233,478.00
Haley				$ 171,582.00		$ 171,582.00
Hamilton				$ 654,871.00		$ 654,871.00
Harrington			$ 976,917.00			$ 976,917.00
Herzog	$ 360,487.00					$ 360,487.00
Jackson			$ 547,135.00			$ 547,135.00
Kelly			$ 852,456.00			$ 852,456.00
Kreidler	$ 265,874.00					$ 265,874.00
Markus					$ 998,657.00	$ 998,657.00
Martinson				$ 447,458.00		$ 447,458.00
Mateo	$ 654,951.00					$ 654,951.00
McAtee		$ 752,391.00				$ 752,391.00
McManus					$ 124,115.00	$ 124,115.00
O'Brien	$ 866,365.00					$ 866,365.00
Preminger				$ 658,214.00		$ 658,214.00
Preston		$ 891,987.00				$ 891,987.00
Remington					$ 987,897.00	$ 987,897.00
Wallington					$ 788,457.00	$ 788,457.00
Grand Total	**$ 3,064,444.00**	**$ 1,769,801.00**	**$ 2,609,986.00**	**$ 1,932,125.00**	**$ 2,899,126.00**	**$ 12,275,482.00**

FIGURE 7-1

Step 2: Click somewhere inside cells A1 through C21.

Step 3: Click the **Insert Ribbon** (Figure 7-2, Arrow A)

Step 4: Click **Pivot Table** (Figure 7-2, Arrow B).

Step 5: A **pop-up menu** will be displayed confirming the range of cells you will be using as your source for the **Pivot Table** (Figure 7-2, Arrow C). By clicking within that range of cells A1 through C21, Excel assumes that your data is part of a **Table** or an **Array**. **Tables** and **Arrays** are data sets with **column headings** and **rows** of data. The **rows** of data would be called **Records**. Excel then assumes that the range of cells A1 through C21 is what you want to use as your **Pivot Table** source. At this point, you will want to make sure that the **Pivot Table** is going to

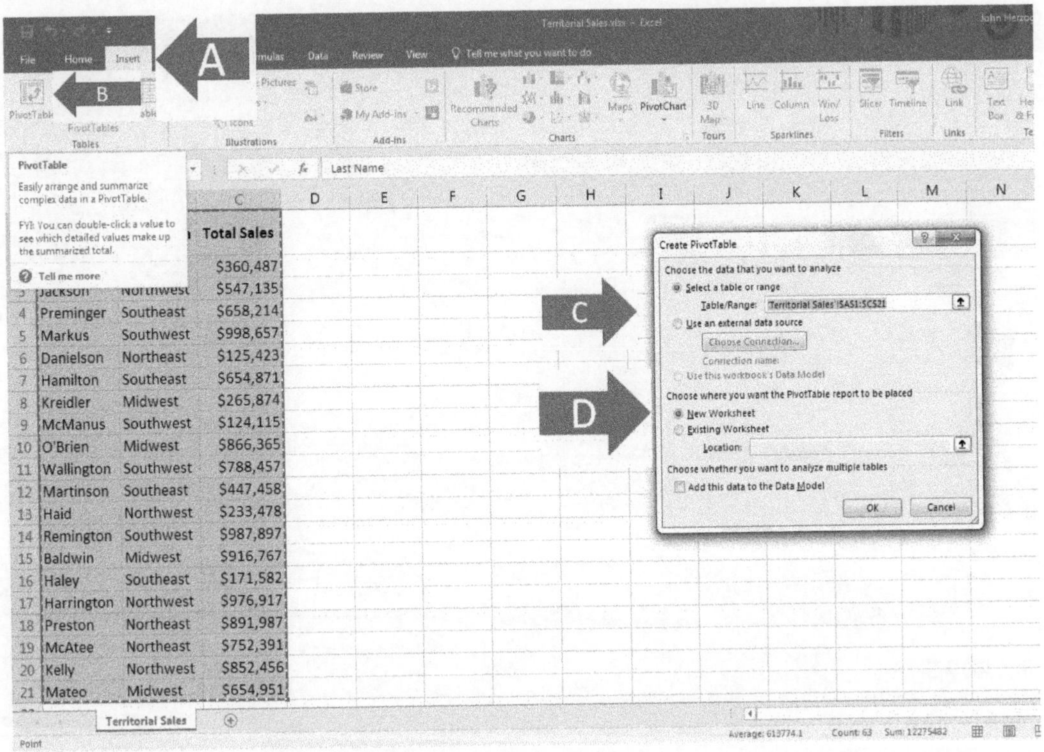

FIGURE 7-2

be sent to a new sheet once you have created it. As you can see, the default setting has already declared that that **Pivot Table** will in fact be sent to another sheet (Figure 7-2, Arrow D), thus nothing needs to be changed in the menu at this time; therefore, click **OK**.

Step 6: At the next screen, you will see the **PivotTable Fields Option Box** on the right of your screen (Figure 7-3). Click and drag the **Pivot Table Field** called **Last Name** (Figure 7-3, Arrow A) down to the **field area** called **Rows** (Figure 7-3, Arrow B).

Step 7: Notice that when you move the **Last Name** field data to the **Rows** area, all of the **Last Names** from the **Territorial Sales** spreadsheet will be displayed on the left of your new spreadsheet (Figure 7-4, Arrow A). Click and drag the **Pivot Table Field** called **Sales Area** (Figure 7-4, Arrow B) down to the field area called **Columns** (Figure 7-4, Arrow C).

Step 8: Click and drag the **Pivot Table Field** called **Total Sales** (Figure 7-5, Arrow A) down to the box that is called **Values** (Figure 7-5, Arrow B).

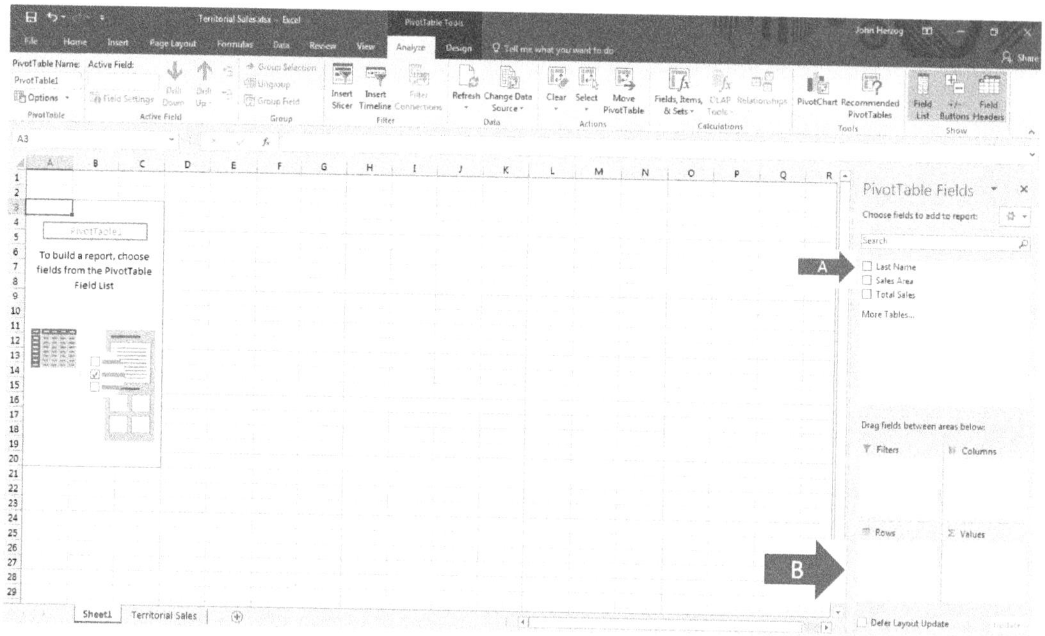

FIGURE 7-3

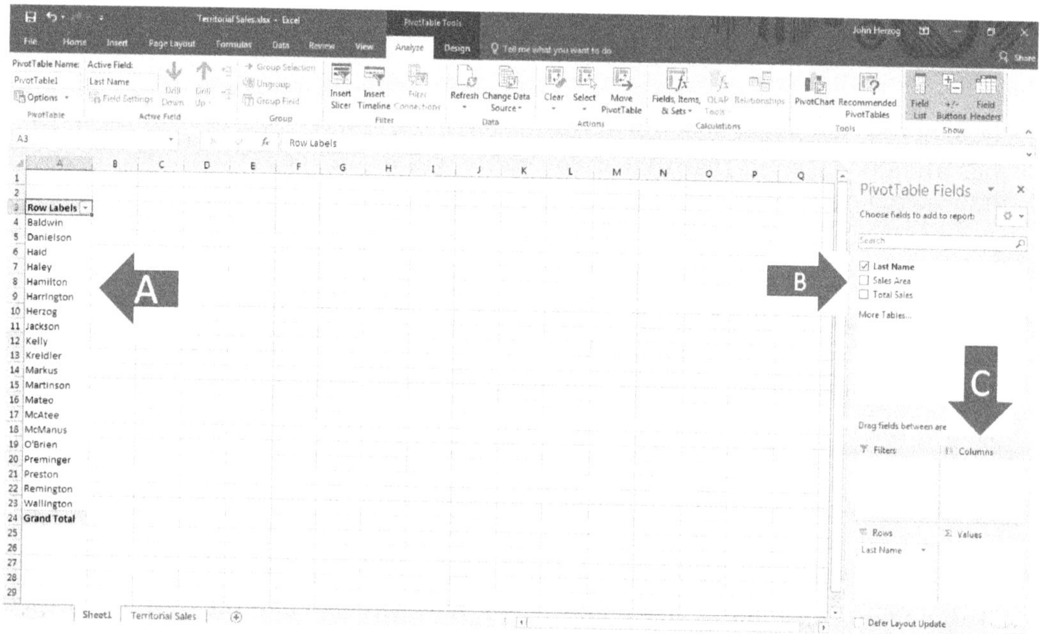

FIGURE 7-4

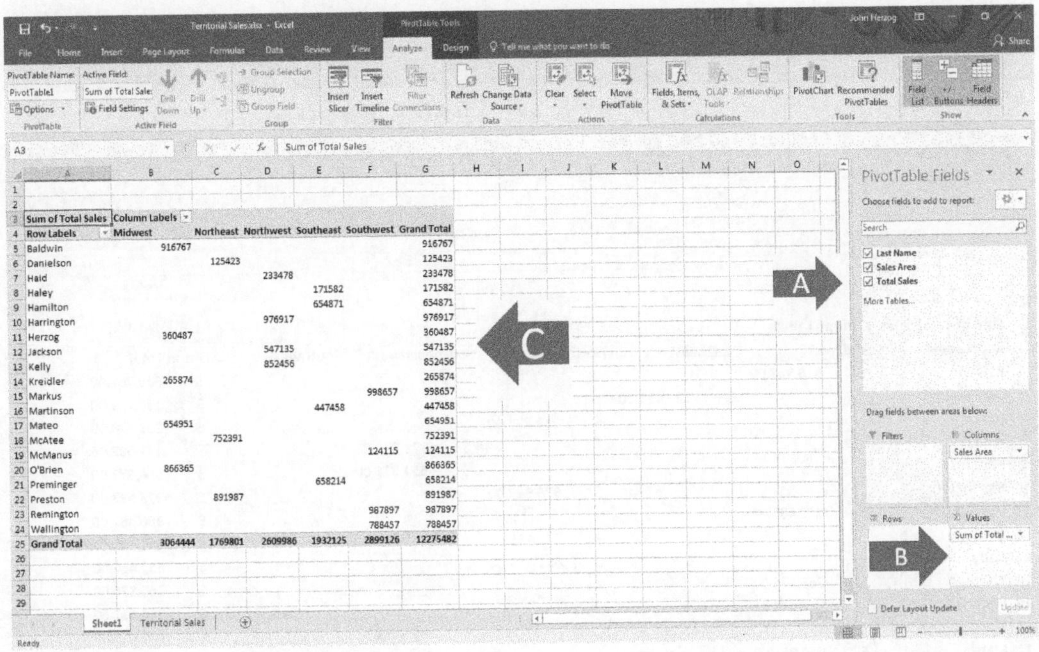

FIGURE 7-5

Your results will be seen at the left of your new spreadsheet as you can see in Figure 7-5, Arrow C.

The **Pivot Table** is in a new sheet of its own called Sheet1. In this case, rename it as **Sales Pivot Table**.

Step 9: If you would like to make the numbers appear as currency, you can do so as you have done to any other dollar data in the past. In this case, we will use the **Accounting** format (Figure 7-6).

Step 10: You can also adjust the column widths. Widen the columns to a width of *16* characters (as seen in Figure 7-6).

Sometimes you may want to filter out some of the data in a **pivot table**. In this case, suppose we want to see only sales from the *Sales Areas* in the *Northwest* and the *Southeast*. You can do so by doing the following:

Step 1: Click on the **Column Labels** drop-down box as seen in Figure 7-7, Arrow A.

Step 2: As you can do with any filter, click the **Select All** checkbox (Figure 7-7, Arrow B) to remove all of the checkmarks in the checkboxes at the left.

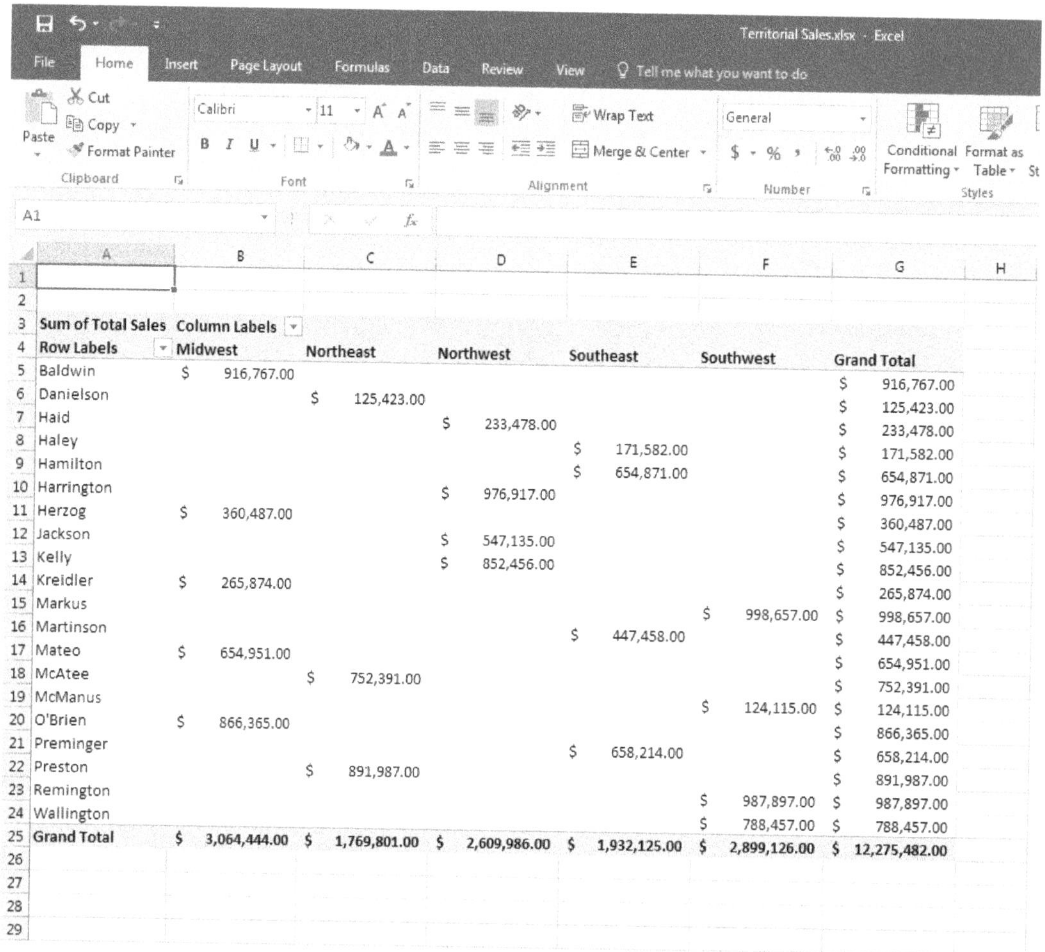

FIGURE 7-6

Step 3: Click only on the checkboxes for *Northwest* and *Southeast* (Figure 7-7, Arrow C).

Step 4: Click **OK**. You can see the result in Figure 7-8.

NOTE: The **Pivot Table** will remove rows that have no sales.

THE AVERAGE FUNCTION

Now let's return to the original spreadsheet. Suppose you want to display an average of the sales at the bottom of the sales column (with a label) as we did in a previous chapter. You can do so by using the following steps.

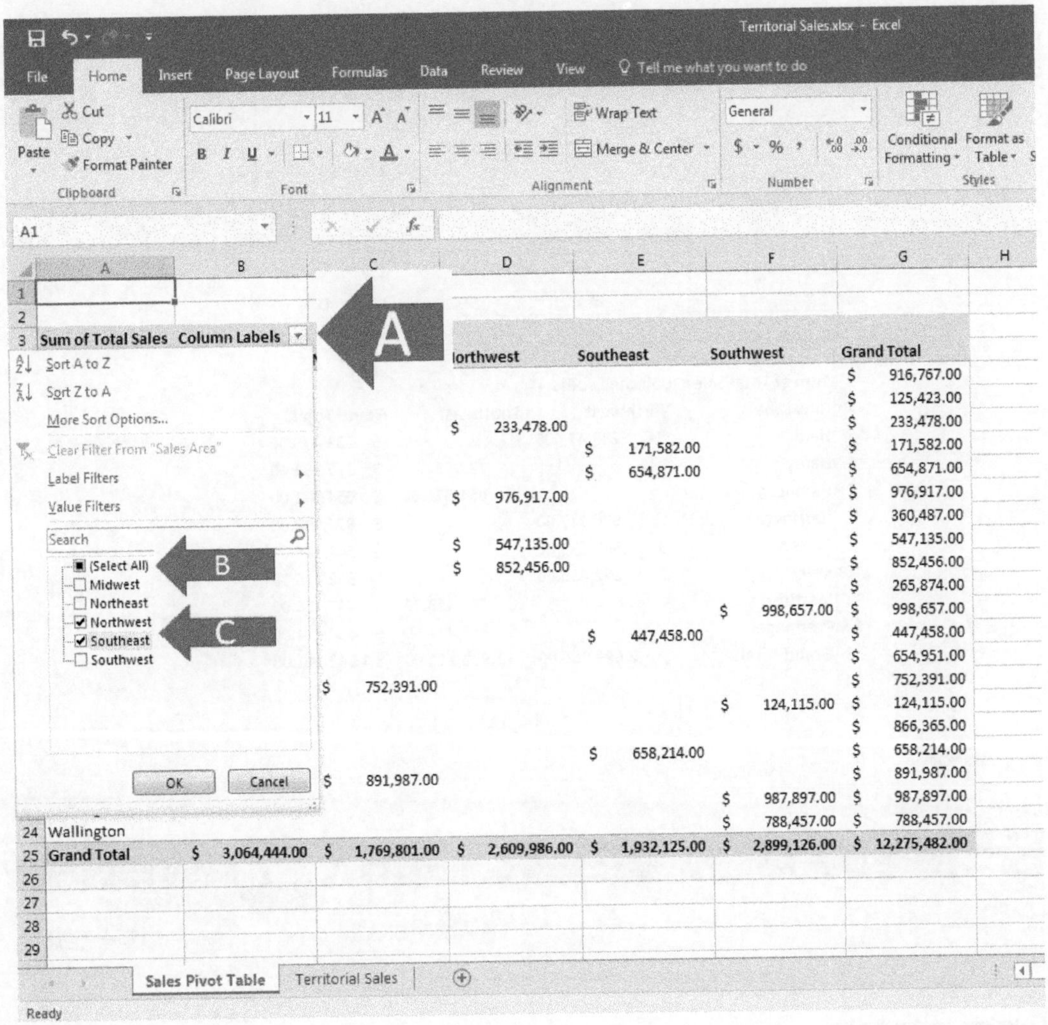

FIGURE 7-7

Step 1: Click in cell B22 and type the word *Average* (Figure 7-9, Arrow A) and make it **bold**.

Step 2: Click on cell C22.

Step 3: Click the tiny arrow to the right of the **AutoSum** button in the **Home Ribbon** (Figure 7-9, Arrow B).

Step 4: Click **Average** (Figure 7-9, Arrow C).

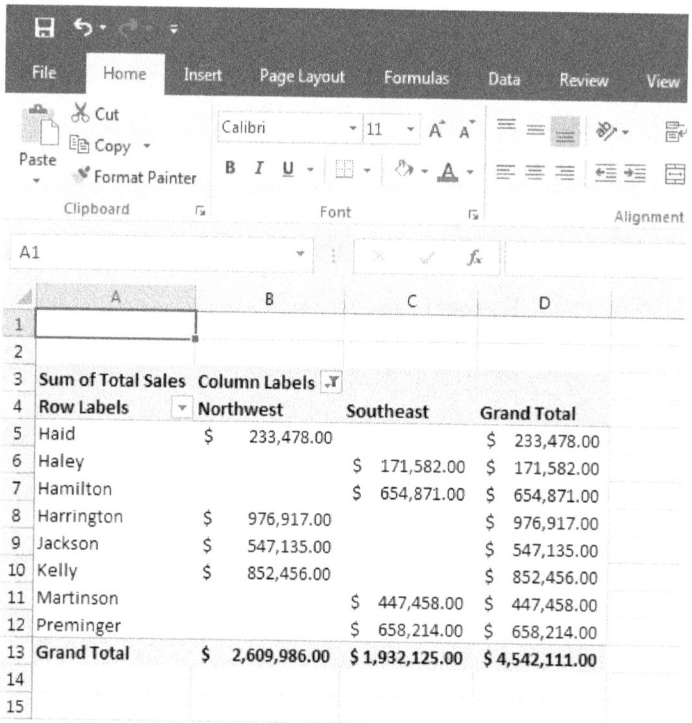

FIGURE 7-8

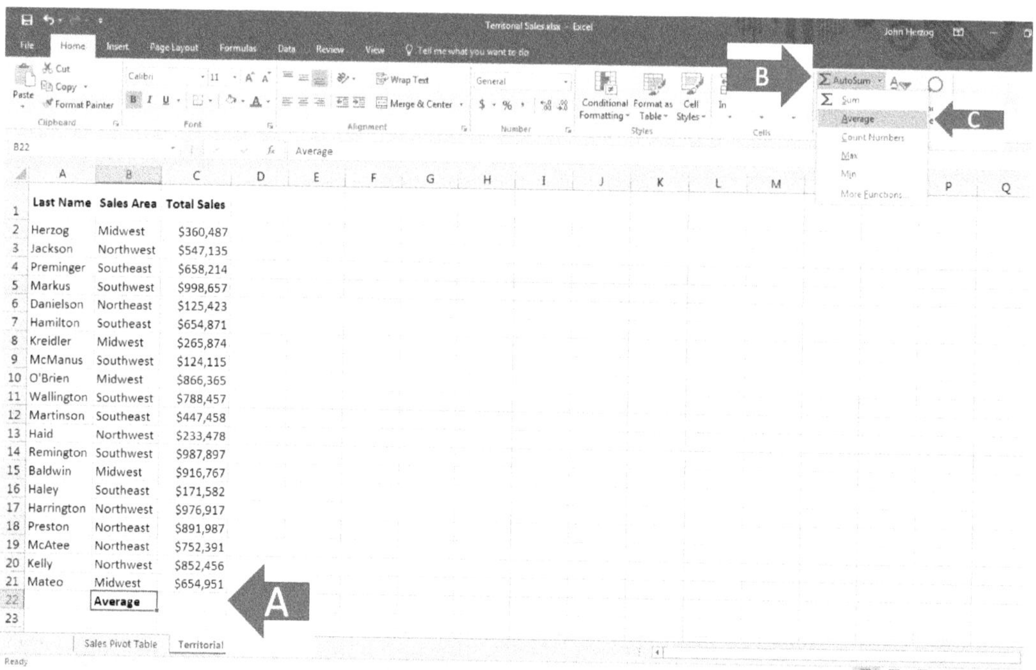

FIGURE 7-9

The result will be as you see in Figure 7-10. Notice that it will give you the same rules as the **AutoSum**. The other alternative is just to type in the **Average** function with the cell range you want to have averaged inside the parentheses by typing *=AVERAGE(C2:C21).*

Step 5: Press **Enter**. The average will appear.

FIGURE 7-10

CREATING A PIE CHART

Pie charts are good for showing numbers as a proportion of the total. **However, they can only display data in two dimensions, such as *Sales Area* and *Total Sales*, as in this example.** If you want to group them with a third set of data, you would need to create two pie charts, or you would otherwise need to put the data into a clustered bar chart, column chart, line chart, or the like.

To create the **pie chart**, do the following:

Step 1: Select the data you want to reflect in the chart. In this example, you will want the *Last Name* and *Total Sales* columns to be graphed. In this case, the cells with the sales agents' *Last Name* are not adjacent to the cells with the *Total Sales*. Please recall therefore that you will need to press and hold the **Ctrl key** down as you select those cell ranges. **Also, please remember that you must NOT include totals and that you cannot stop and start again as you are selecting the SAME column of data** (Figure 7-11).

Step 2: Click the **Insert Ribbon** (Figure 7-12, Arrow A).

Step 3: Click the tiny arrow to the right of the small **Pie Chart Icon** on the **Insert Menu** (Figure 7-12, Arrow B). That will give you a list of different pie chart types from which to choose. In this example, click **3-D Pie** (Figure 7-12, Arrow C).

That will give you the chart you see in Figure 7-12.

Step 4: In this case, let's make sure that the chart is a sheet. Therefore, click **Move Chart** (Figure 7-13, Arrow).

Step 5: Name the chart *Territorial Sales Chart.*

Step 6: Click **OK**.

That will take you to the screen you see in Figure 7-14.

Step 7: Add labels for the chart by right-clicking anywhere on it, and that will bring up the **Shortcut Menu** (Figure 7-15).

	A	B	C	D
1	**Last Name**	**Sales Area**	**Total Sales**	
2	Herzog	Midwest	$360,487	
3	Jackson	Northwest	$547,135	
4	Preminger	Southeast	$658,214	
5	Markus	Southwest	$998,657	
6	Danielson	Northeast	$125,423	
7	Hamilton	Southeast	$654,871	
8	Kreidler	Midwest	$265,874	
9	McManus	Southwest	$124,115	
10	O'Brien	Midwest	$866,365	
11	Wallington	Southwest	$788,457	
12	Martinson	Southeast	$447,458	
13	Haid	Northwest	$233,478	
14	Remington	Southwest	$987,897	
15	Baldwin	Midwest	$916,767	
16	Haley	Southeast	$171,582	
17	Harrington	Northwest	$976,917	
18	Preston	Northeast	$891,987	
19	McAtee	Northeast	$752,391	
20	Kelly	Northwest	$852,456	
21	Mateo	Midwest	$654,951	
22		**Average**	$613,774	
23				

FIGURE 7-11

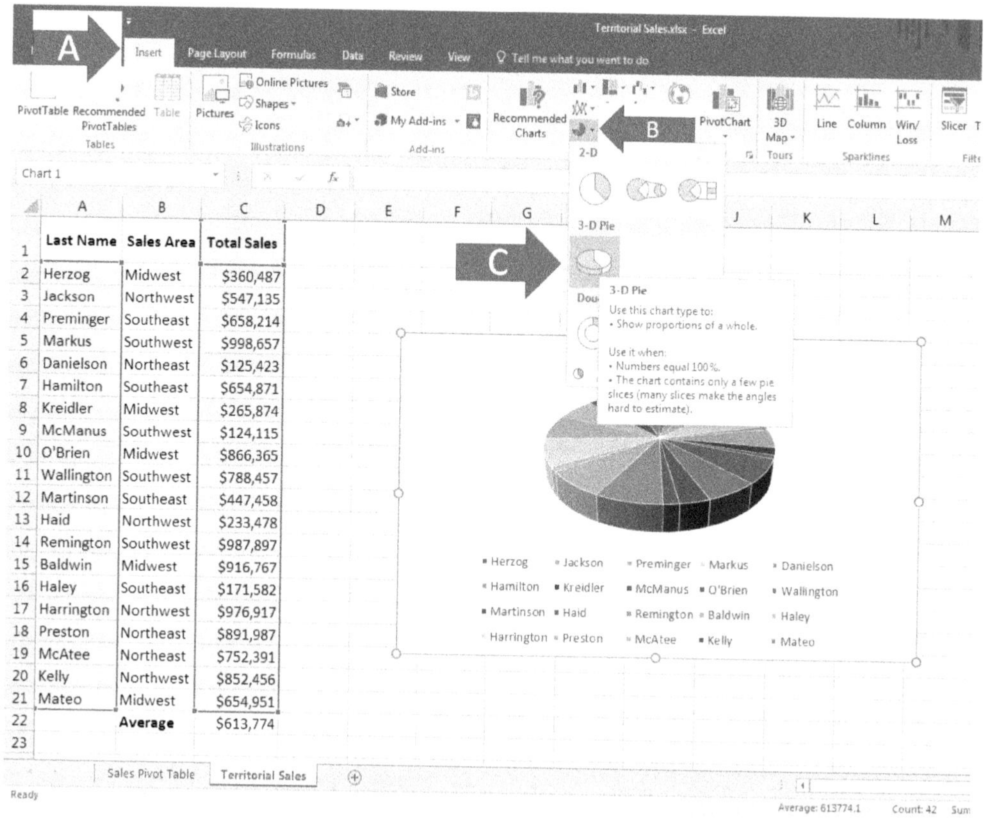

FIGURE 7-12

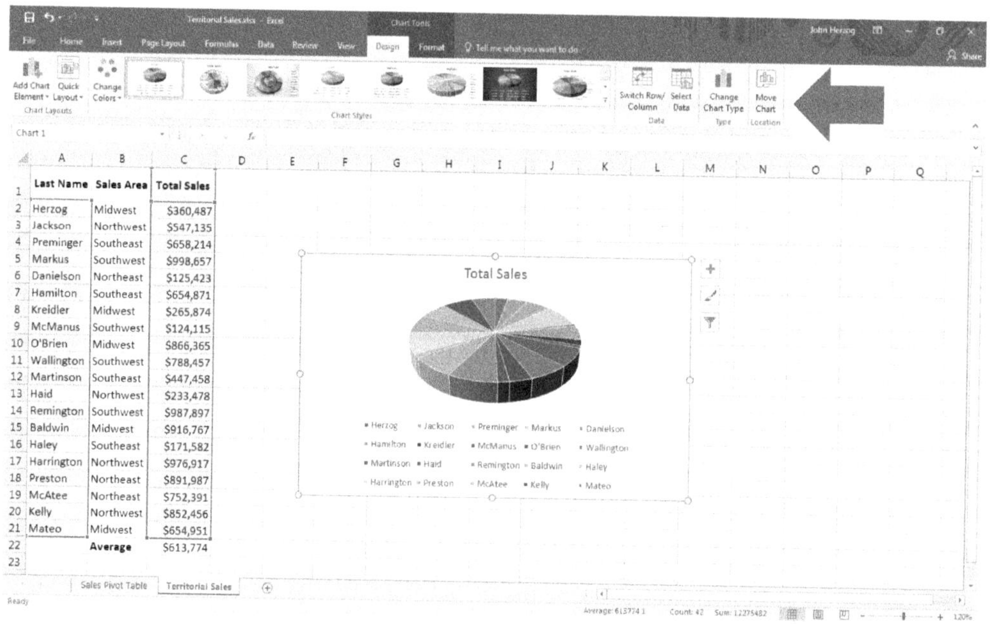

FIGURE 7-13

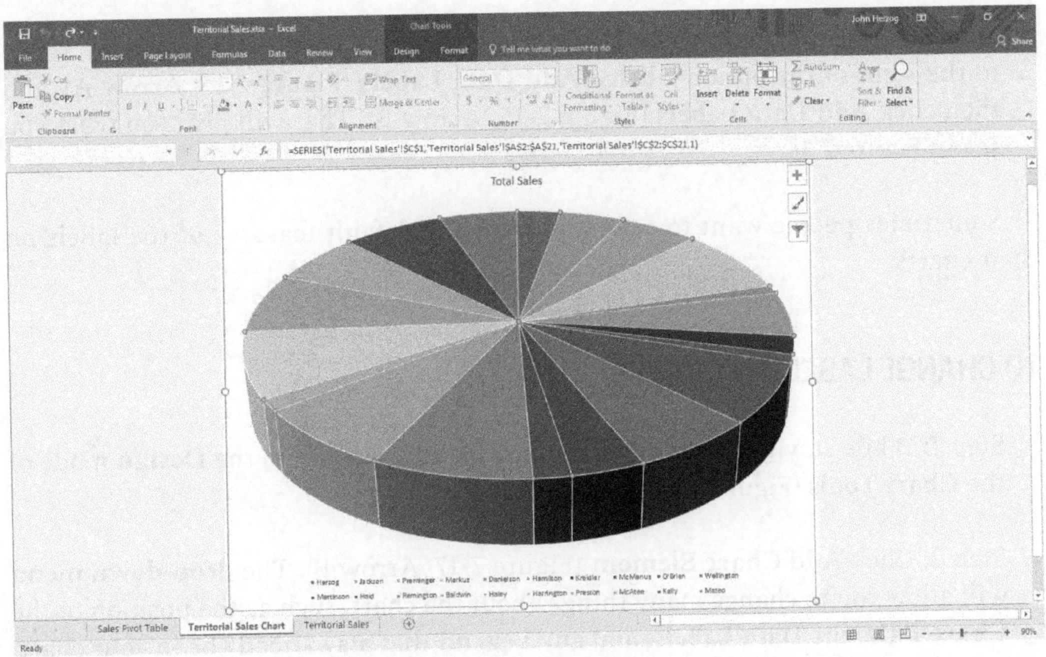

FIGURE 7-14

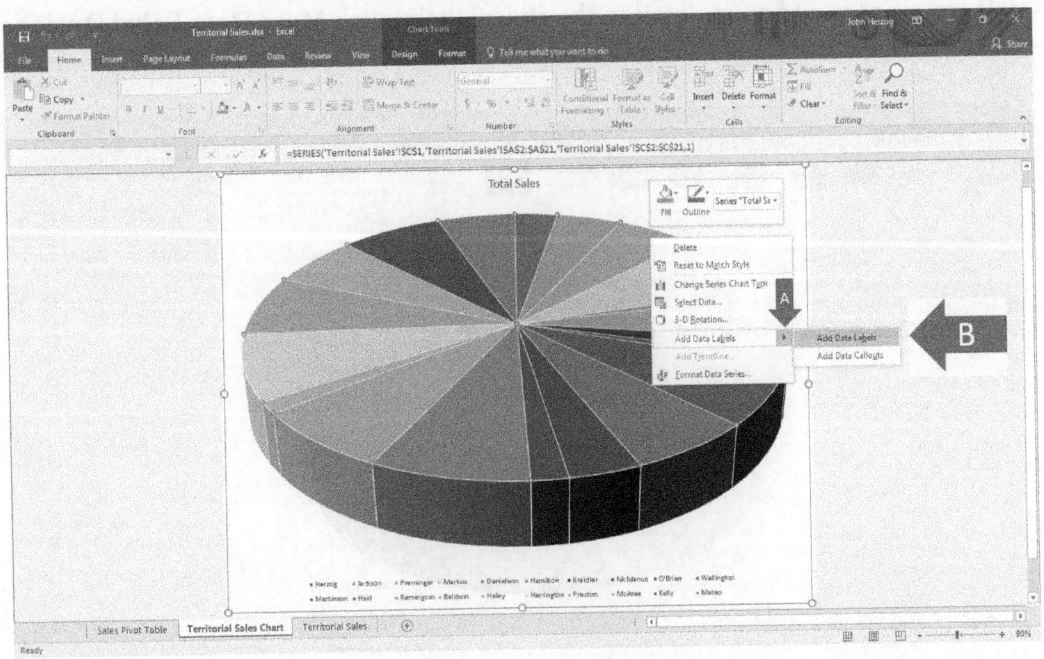

FIGURE 7-15

Step 8: In the **Shortcut Menu (or Context-Sensitive Menu)**, click the tiny arrow to the right of the option labeled **Add Data Labels** (Figure 7-15, Arrow A), and then click **Add Data Labels** (Figure 7-15, Arrow B). That will give you what you see in Figure 7-16.

Sometimes people want to see more than the default features of the labels on their charts.

TO CHANGE LABELS IN CHARTS

Step 1: While staying in the chart, make sure that you are in the **Design** mode of the **Chart Tools** (Figure 7-17, Arrow A).

Step 2: Click **Add Chart Element** (Figure 7-17, Arrow B). The **drop-down menu** will allow you to change many things about the chart, such as the position of the **Chart Title**, the **Data Labels**, and any **Legend** that may already be in your chart. For example, in this case, **Best Fit** (Figure 7-17, Arrow C) has been chosen regarding the labels, but if you want them to be outside and at the end of the chart, click **Outside End** of the chart (Figure 7-17, Arrow D).

If you want to add more detail to the chart labels, click **More Data Label Options** (Figure 7-17, Arrow E). That will take you to what you see in Figure 7-18. This menu

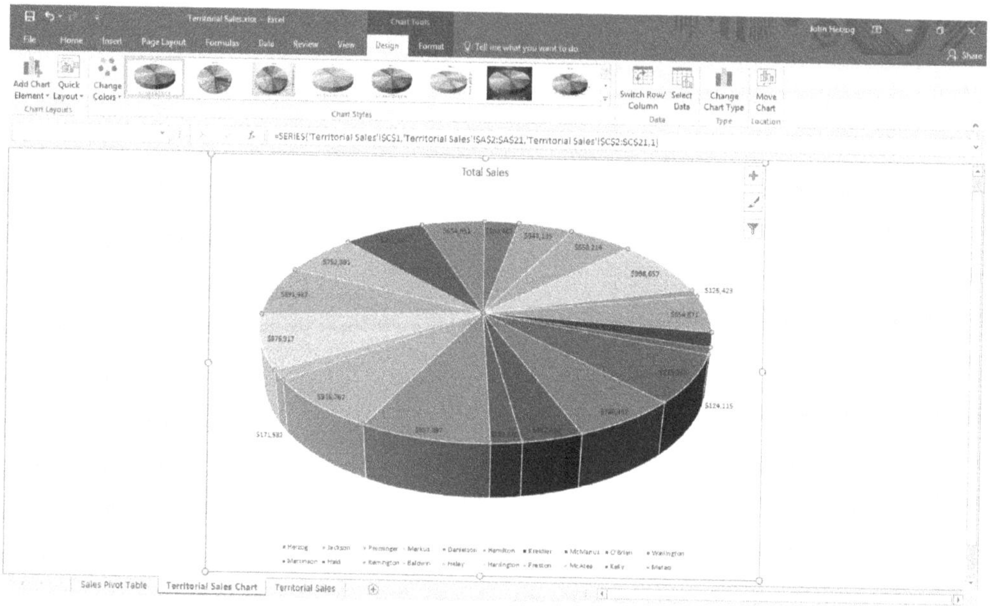

FIGURE 7-16

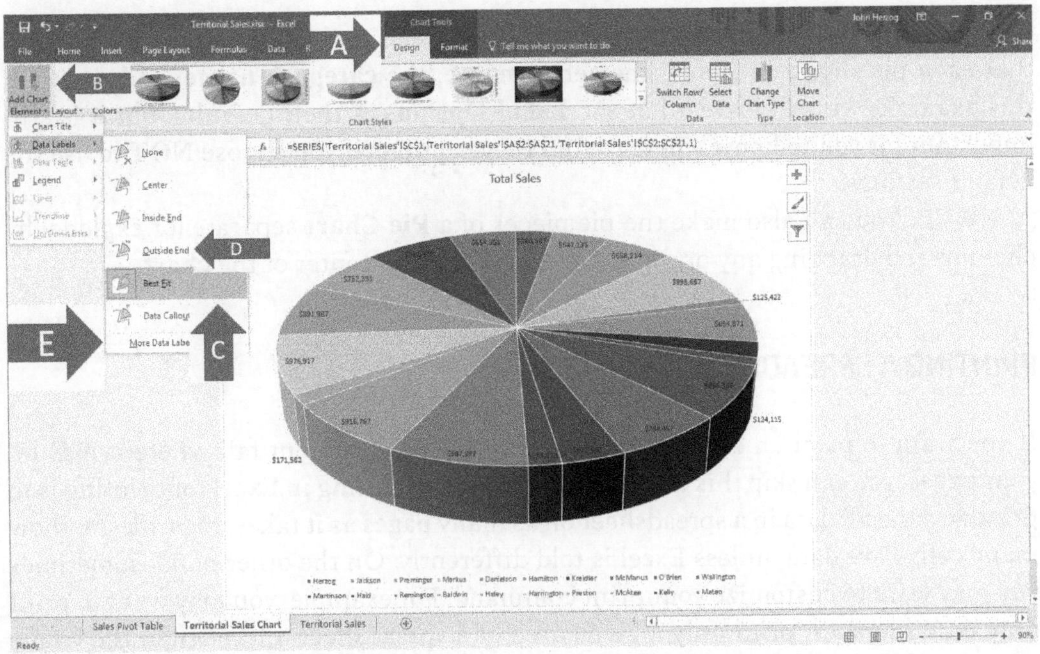

FIGURE 7-17

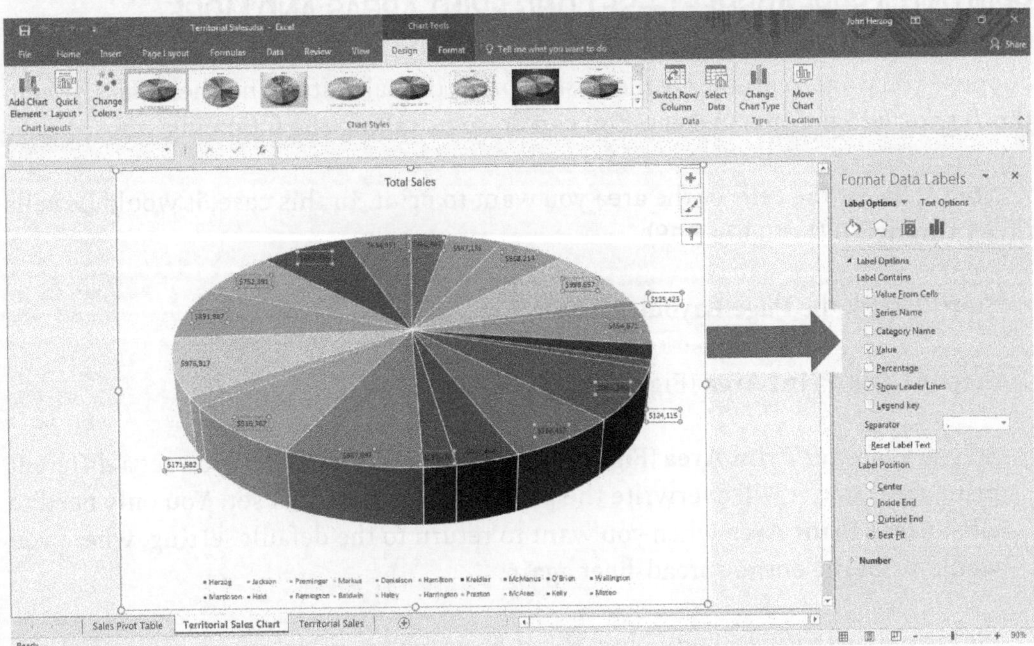

FIGURE 7-18

will allow you to show more available data here, such as the **percentage of the total** that each pie slice represents, the **series name**, the **category name**, and more. In this example, we will leave all of the **Label Options** at their defaults, but it is still important for you to know where to find these options if you choose **NOT** to use the default settings.

NOTE: You can also make the pie pieces of a **Pie Chart** separate (or explode) by clicking and dragging any one of them away from the center of the chart.

PRINTING A SPREADSHEET

If you want to print an entire spreadsheet and let the content fall where it may on your pages, you can skip this discussion. The default setting in Excel for printing is to print any and all data in a spreadsheet on as many pages as it takes, regardless of how many cells have data, unless Excel is told differently. On the other hand, sometimes you may want to customize your print command. For example, you may want to print only certain pages, print only a given area of a spreadsheet (rather than the whole spreadsheet), print your data vertically (portrait) rather than horizontally (landscape), or use other varying options.

PRINTING A SPREADSHEET—SETTING PRINT AREAS AND MORE

Suppose you want to print only the *Last Name* of the sales staff and their *Sales Area* in the *Territorial Sales* spreadsheet. You can do so by taking the following steps:

Step 1: Select the cells of the area you want to print. In this case, it would be cells A1 through B21 (Figure 7-19).

Step 2: Click the **Page Layout Ribbon** (Figure 7-20A, Arrow A).

Step 3: Click **Print Area** (Figure 7-20A, Arrow B).

Step 4: Click **Set Print Area** (Figure 7-20A, Arrow C). NOTE: If you set a different print area later, it will overwrite the print area you have just set. You only need to click **Clear Print Area** when you want to return to the default setting, where you would print the entire spreadsheet again.

Please notice that Excel will put a thin line around your print area to remind you where you set it (Figure 7-20B).

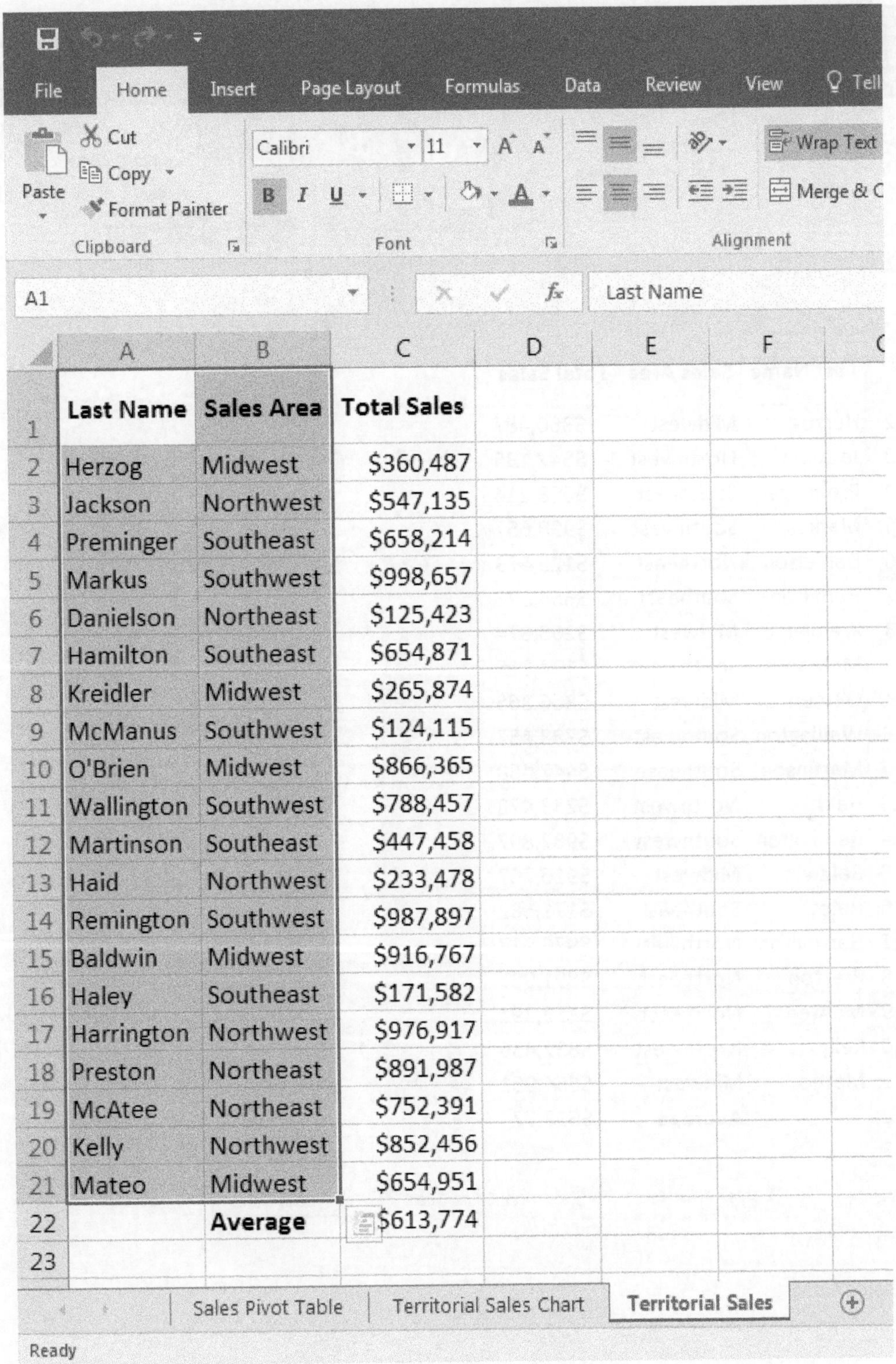

	A	B	C
1	Last Name	Sales Area	Total Sales
2	Herzog	Midwest	$360,487
3	Jackson	Northwest	$547,135
4	Preminger	Southeast	$658,214
5	Markus	Southwest	$998,657
6	Danielson	Northeast	$125,423
7	Hamilton	Southeast	$654,871
8	Kreidler	Midwest	$265,874
9	McManus	Southwest	$124,115
10	O'Brien	Midwest	$866,365
11	Wallington	Southwest	$788,457
12	Martinson	Southeast	$447,458
13	Haid	Northwest	$233,478
14	Remington	Southwest	$987,897
15	Baldwin	Midwest	$916,767
16	Haley	Southeast	$171,582
17	Harrington	Northwest	$976,917
18	Preston	Northeast	$891,987
19	McAtee	Northeast	$752,391
20	Kelly	Northwest	$852,456
21	Mateo	Midwest	$654,951
22		Average	$613,774
23			

FIGURE 7-19

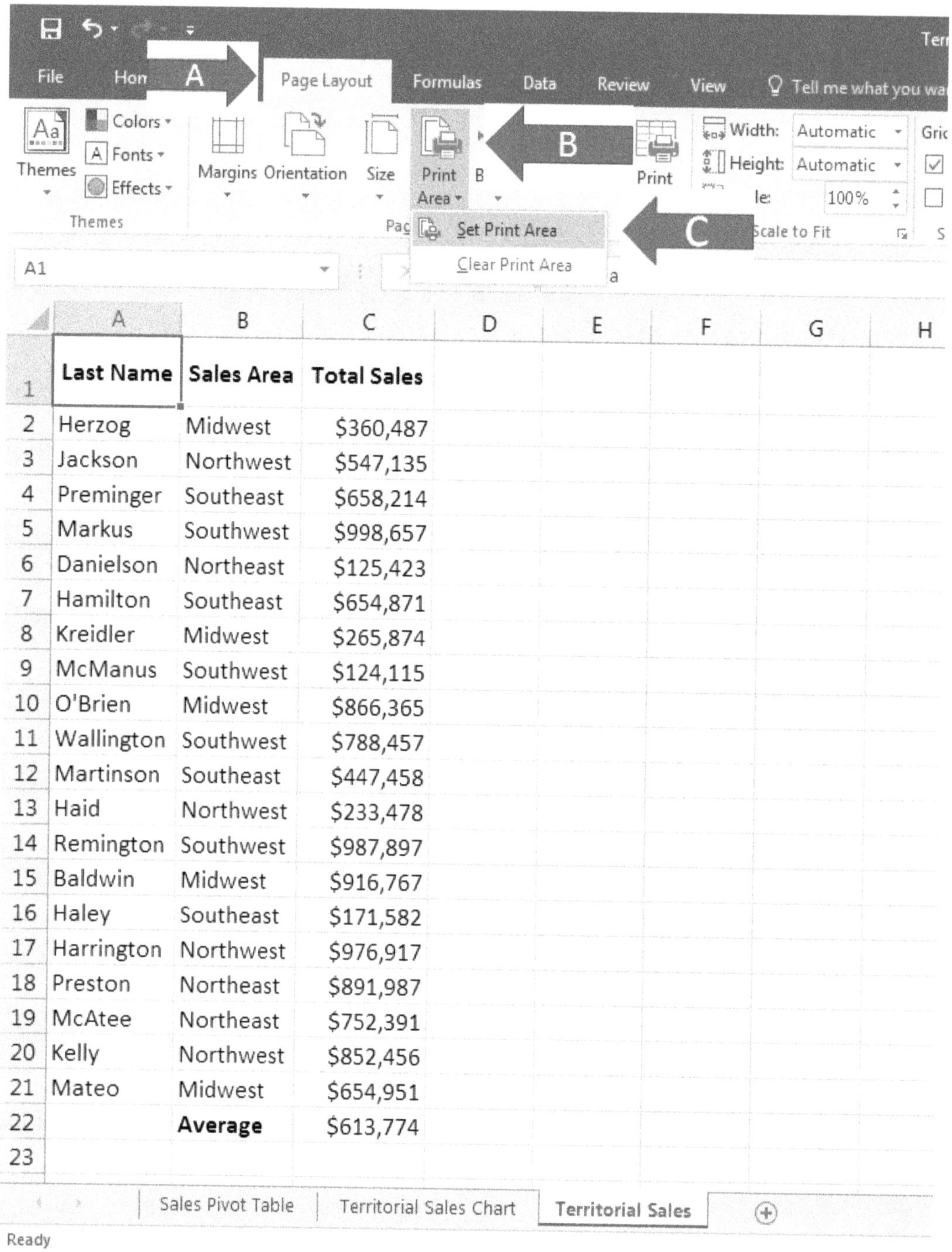
File
Hon
A
Page Layout
Formulas
Data
Review
View
Tell me what you wa
B
C
Themes
Colors
Fonts
Effects
Margins
Orientation
Size
Print Area
Print
Width: Automatic
Height: Automatic
100%
Scale to Fit
Set Print Area
Clear Print Area
A1
Last Name
Sales Area
Total Sales
Herzog
Midwest
$360,487
Jackson
Northwest
$547,135
Preminger
Southeast
$658,214
Markus
Southwest
$998,657
Danielson
Northeast
$125,423
Hamilton
Southeast
$654,871
Kreidler
Midwest
$265,874
McManus
Southwest
$124,115
O'Brien
Midwest
$866,365
Wallington
Southwest
$788,457
Martinson
Southeast
$447,458
Haid
Northwest
$233,478
Remington
Southwest
$987,897
Baldwin
Midwest
$916,767
Haley
Southeast
$171,582
Harrington
Northwest
$976,917
Preston
Northeast
$891,987
McAtee
Northeast
$752,391
Kelly
Northwest
$852,456
Mateo
Midwest
$654,951
Average
$613,774
Sales Pivot Table
Territorial Sales Chart
Territorial Sales
Ready

FIGURE 7-20A

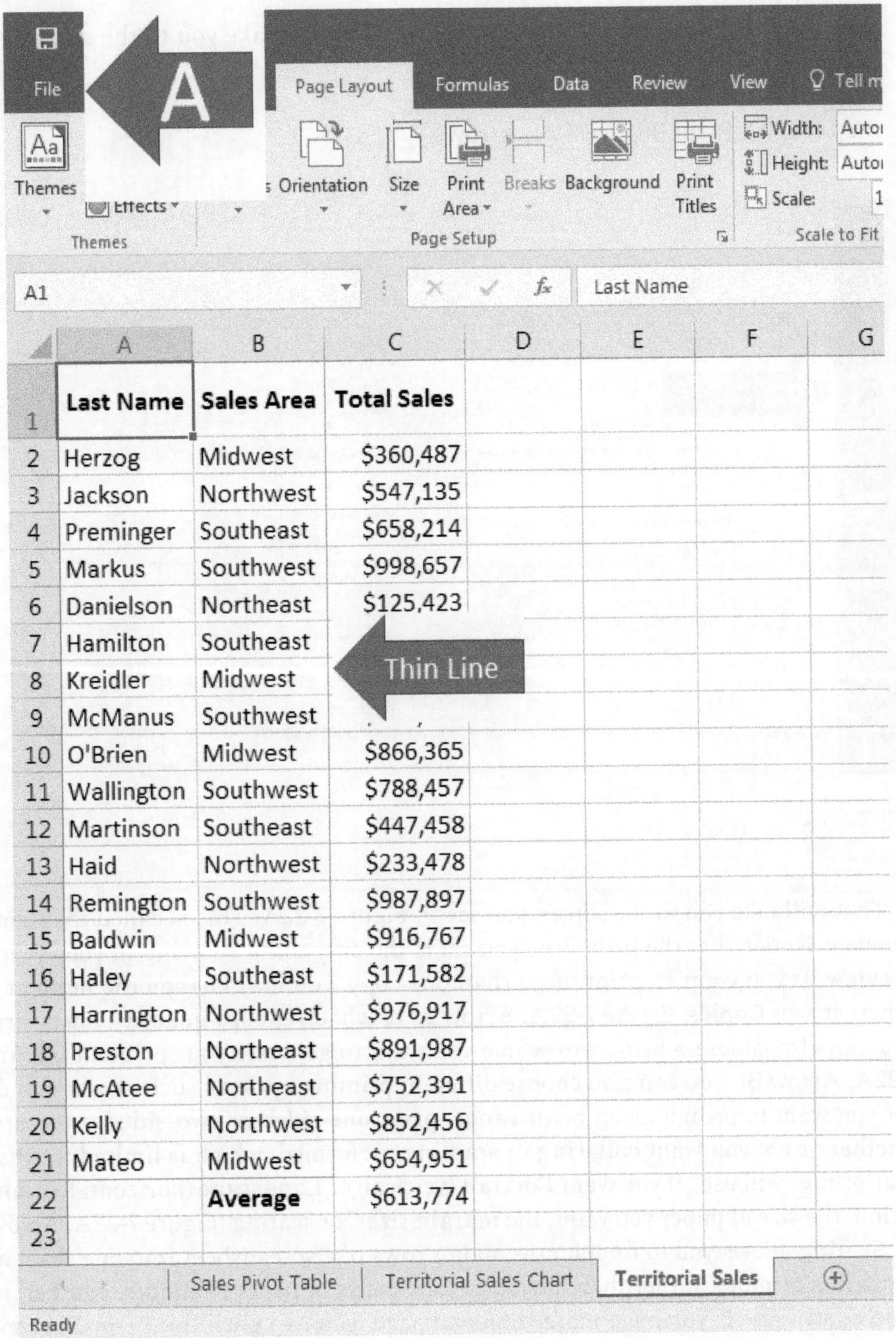
File
A
Page Layout
Formulas
Data
Review
View
Themes
Effects
Orientation
Size
Print Area
Breaks
Background
Print Titles
Width:
Height:
Scale:
Page Setup
Scale to Fit
A1
Last Name
Last Name
Sales Area
Total Sales
Herzog
Midwest
$360,487
Jackson
Northwest
$547,135
Preminger
Southeast
$658,214
Markus
Southwest
$998,657
Danielson
Northeast
$125,423
Hamilton
Southeast
Kreidler
Midwest
McManus
Southwest
Thin Line
O'Brien
Midwest
$866,365
Wallington
Southwest
$788,457
Martinson
Southeast
$447,458
Haid
Northwest
$233,478
Remington
Southwest
$987,897
Baldwin
Midwest
$916,767
Haley
Southeast
$171,582
Harrington
Northwest
$976,917
Preston
Northeast
$891,987
McAtee
Northeast
$752,391
Kelly
Northwest
$852,456
Mateo
Midwest
$654,951
Average
$613,774
Sales Pivot Table
Territorial Sales Chart
Territorial Sales
Ready

FIGURE 7-20B

Step 5: Click **File** (Figure 7-20B, Arrow A). That will take you to the screen you see in Figure 7-21.

Step 6: Click **Print** (Figure 7-21, Arrow).

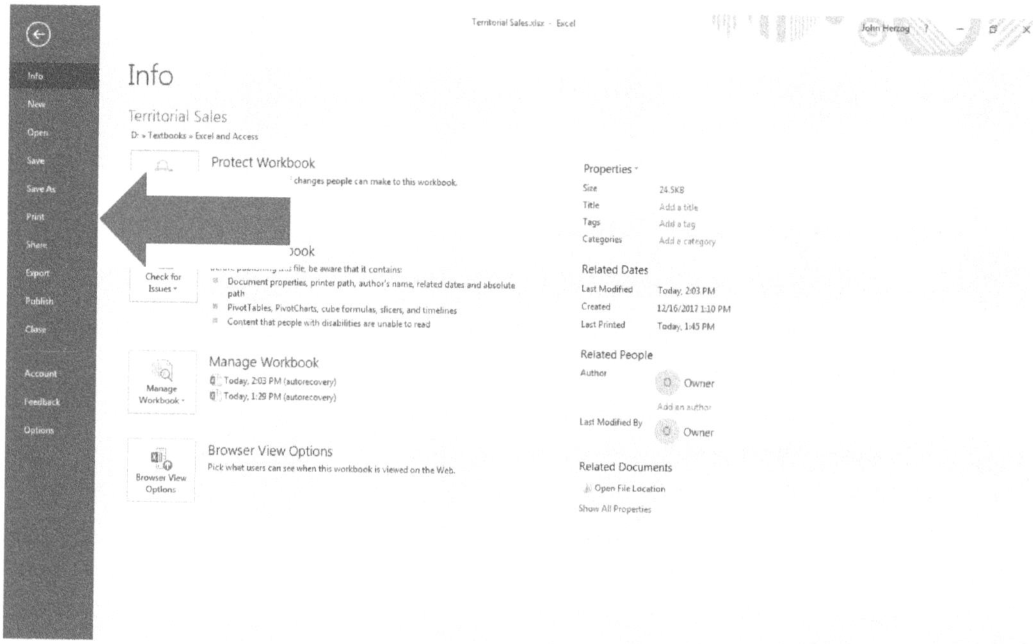

FIGURE 7-21

That will take you to the screen you see in Figure 7-22A. This is called the **Print Preview**. Notice that the print area you set is the only thing showing in your **Print Preview**. If you want to print more than one copy, you may change the number 1 (where it says **Copies**, Figure 7-22A, Arrow A) to whatever copy quantity you desire. You can also choose a printer to which you want to generate your printout (Figure 7-22A, Arrow B). You can also choose different printout options, such as whether or not you want to print a given **print range**, **print one-sided** or **two-sided** printouts, whether or not you want **collating** or **stapling** performed (which is limited to what your printer will use), if you want **Portrait** (vertical) or **Landscape** (horizontal) **orientation**, the **size of paper** you want, the **margin sizes**, or **scaling** (Figure 7-22A, Arrow C). **Scaling** allows you to fit columns and/or rows of a spreadsheet into one sheet of paper in a printout (rather than having to tape pages of paper printouts together) to keep continuity. If you have more than one page, you can click the **Print Preview Page Navigator** to see more pages (Figure 7-22A, Arrow D).

Step 6: If you are satisfied with the settings you have chosen, click **Print** (Figure 7-22A, Arrow E).

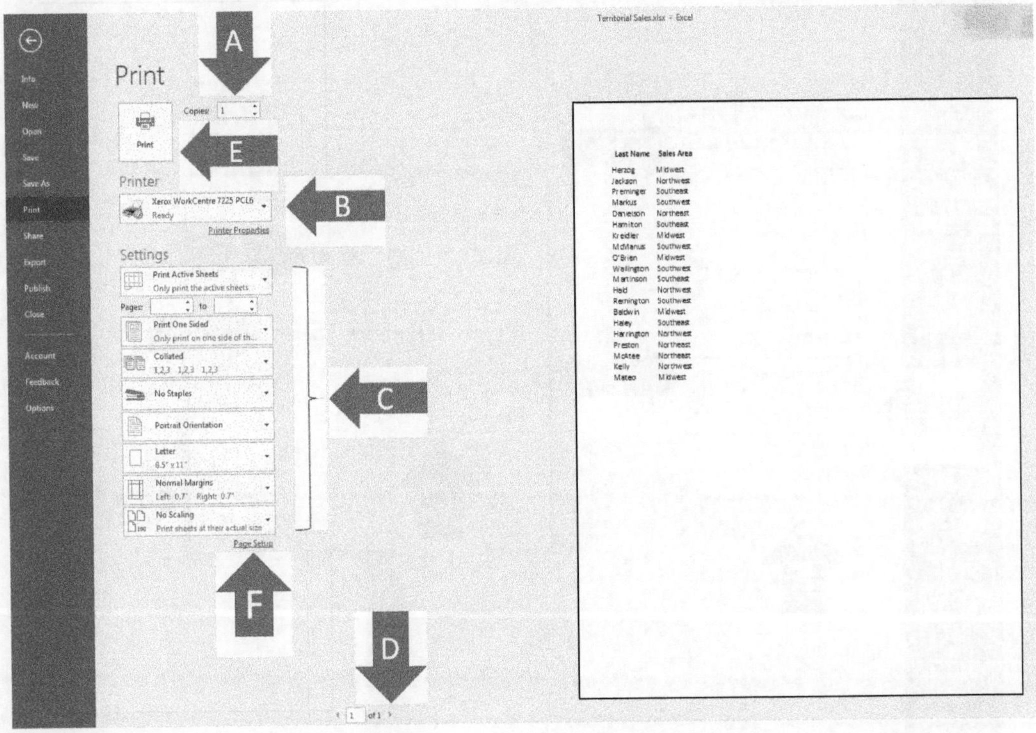

FIGURE 7-22A

PRINTING OPTIONS IN THE PAGE SETUP MENU

Sometimes there are other options you will need to use when you print a spreadsheet. Such options can be found in the **Page Setup** menu in the **Print Menu** shown in Figure 7-22A, Arrow F. If you click it, you will see several tabs that can be of good use. Many of the options in those tabs make it possible to accomplish things in a different way from what we have already discussed. Let's look at them now in Figure 7-22B.

Arrow A shows that you can change the orientation of the printout from **Portrait** (vertical) to **Landscape** (horizontal), just as you can in the **Portrait Orientation** area under **Settings** in the **Print Menu** shown in Figure 7-22B, Arrow E (mentioned briefly earlier). Clicking on the tiny arrow to the right of the **Portrait Orientation** command (Figure 7-22B, Arrow E) will allow you to change it, just as by clicking the tiny arrow on **any one** of the **Settings** commands will allow you to do.

Arrow B shows you that you can also do **Scaling** of a printout as can also be done under **Settings** in the **Print Menu** (Figure 7-22B, Arrow F). If the data from your

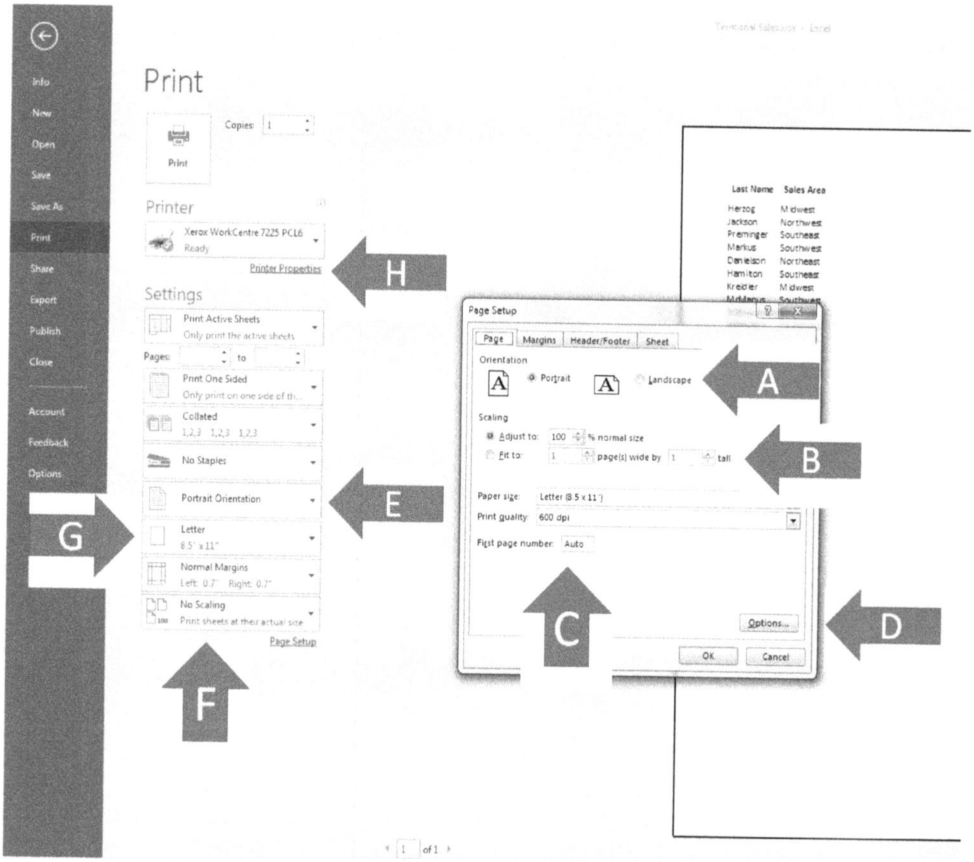

FIGURE 7-22B

spreadsheet is too small to read after putting all of it into one page, you can still make sure that columns will be on one page (1 page wide), while the rows will then print on many pages (perhaps 10 pages tall). In that case, you would set the scaling at **Fit to 1 page(s) wide by 10 tall** (Figure 7-22B, Arrow B). The default for scaling is always set for **No Scaling**.

Arrow C shows you options where you can set things such as the paper size you want (also available under the **Setting** area of the **Print Menu**, Figure 7-22B, Arrow G); the quality of printout you need—a higher **DPI** (**Dots Per Inch**) will give you a greater print quality; and what page you want to print first upon printing. If you leave the **First page number** setting at **Auto**, Excel will print your first page of data as page one rather than skipping it.

Arrow D shows the **Options** button. It is the same as the **Printer Properties** under the **Printer** area in the **Print Menu** (Figure 7-22B, Arrow H). That button will display a screen that will allow you to send commands to the printer. These options include whether or not you want to print in color, with staples, with more than one page on a piece of paper, and much more. Such commands and availability of them will vary dramatically depending upon what printer to which you are sending the print command. Therefore, we will not attempt to cover all of those possibilities, as there are hundreds of different types of printers available.

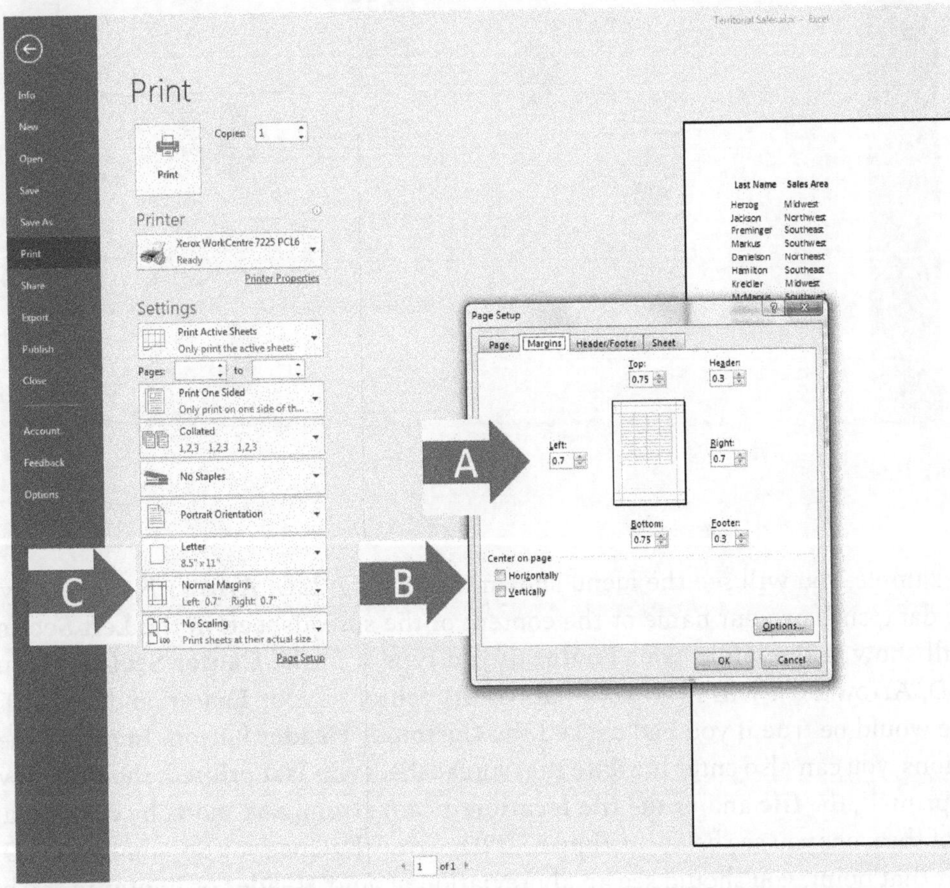

FIGURE 7-22C

If you click the **Margins Tab** in the **Page Setup Menu**, you will see that it gives you the ability to set margins (Figure 7-22C, Arrow A) as you can also do in the **Settings** area (Figure 7-22C, Arrow C). Spreadsheets can also be printed with your spreadsheet centered horizontally and/or vertically on a page by clicking the **Center on page** settings shown in Figure 7-22C, Arrow B.

If you click the **Header/Footer** tab in the **Page Setup** menu, you will see the screen shown in Figure 7-22D, Arrows A and C. If you want to add a **Header** or **Footer** in the printout of a spreadsheet, you can do so by clicking either **Customer Header** or **Customer Footer** (Figure 7-22D, Arrow A). If you click the **Customer Footer** button,

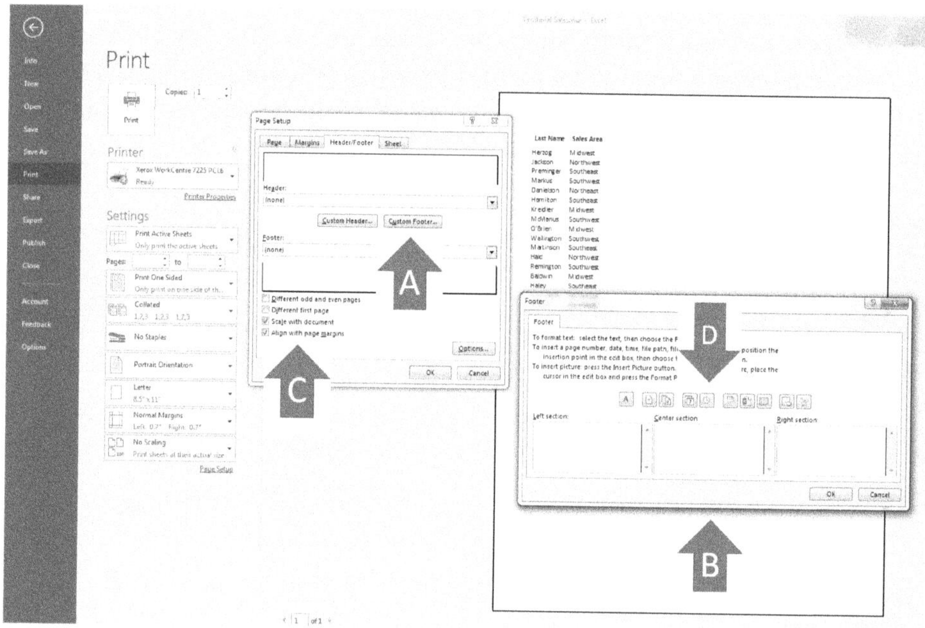

FIGURE 7-22D

for example, you will see the menu shown in Figure 7-22D, Arrows B and D. If you type data, such as your name or the content of the spreadsheet, in the **Left Section**, it will show at the left of your **Footer**. If you type it in the **Center Section** (Figure 7-22D, Arrow B), it will show your data at the center of your **Footer**, and so on. The same would be true if you had clicked the **Customer Header** button. In any of these sections, you can also enter the **date** that spreadsheet was last printed, the **time** it was last printed, the **file** and/or the **file location** it came from, and more, by clicking any one of the icons you see below in Figure 7-22D, Arrow D. Once you do, simply click **OK.**

At that point, you should see an abbreviation of your **Header** or **Footer** in the box above the word **Header:** and/or in the box below the word **Footer**, depending upon whether you entered a **header or footer**. If nothing is entered in either the **Header** or the **Footer**, it will read (none) beneath the word **Header** and (none) beneath the word **Footer**.

Also notice the options above the area shown by Figure 7-22D, Arrow C. You can have your spreadsheet print different odd or even pages, use a different first page, and you can also control scaling and margins here, too. After changing any one of these settings, simply click **OK**.

On the **Sheet tab** of the menu, once again after having clicked **Page Setup** (Figure 7-22E, Arrow A) as discussed in Chapter 6, you can also instruct Excel to print **comments** (Figure 7-22E, Arrow B). Below that box you can tell Excel to print any errors in cells or to make them blank in your spreadsheet. Errors in cells will occur, for example, if you enter formulas that attempt to calculate cells that only hold text or if functions have not been entered correctly.

Please notice many of the other options you have in the **Sheet tab** of the **Page Setup** menu, including a means to show **gridlines** when you print and to print a specific row on each page of your printout if it has more than one page. For example, you may want the top row of your spreadsheet to repeat at the top of each page of your printout because it will usually hold your column labels or description of the data in them. This is helpful if your spreadsheet has many pages. If the reader of your printouts is on Page 5, for example, without descriptions of the data in the columns on each page, they may easily forget the data they are reading. In that case, you would want row 1 to be repeated at the top of each page; thus you would enter the number 1 in the **Rows to repeat at top:** box. You would enter rows 1:2 in that box if there are descriptions in more than one row.

You can also change the print area in this menu as well, which was discussed earlier.

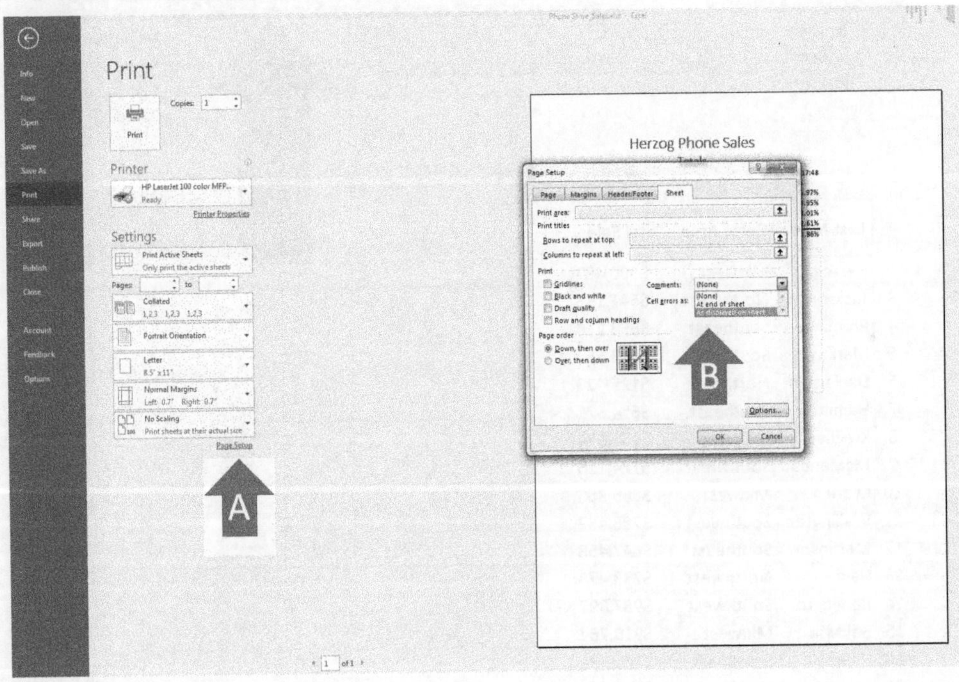

FIGURE 7-22E

CREATING MACROS WITH BUTTONS

Sometimes you will find yourself doing things in spreadsheets that are repetitive. For instance, in the setting of the **print area** just discussed, many times people want to set different print areas of a given spreadsheet each day. Rather than having to repeat the steps we just did, you can create a button with a **Macro** attached to it that will do it for you. Dozens of things can be automated, far more than discussed in this example. Another example would be that you may want to apply a certain format to a cell that you use often.

Any tasks you can do in a spreadsheet can be recorded, and when you tell the system to **Run** the **Macro** (or recording) that you create, Excel will perform those same steps in the same order that you did them, as many times as you run the **Macro**.

Let's create a **Macro** first that will perform the task of setting the **Print Area** of the spreadsheet by doing the following:

Step 1: Make sure you have the **Developer Ribbon** on your computer version of Excel (Figure 7-23, Arrow). If you don't have the **Developer Ribbon**, right-click on

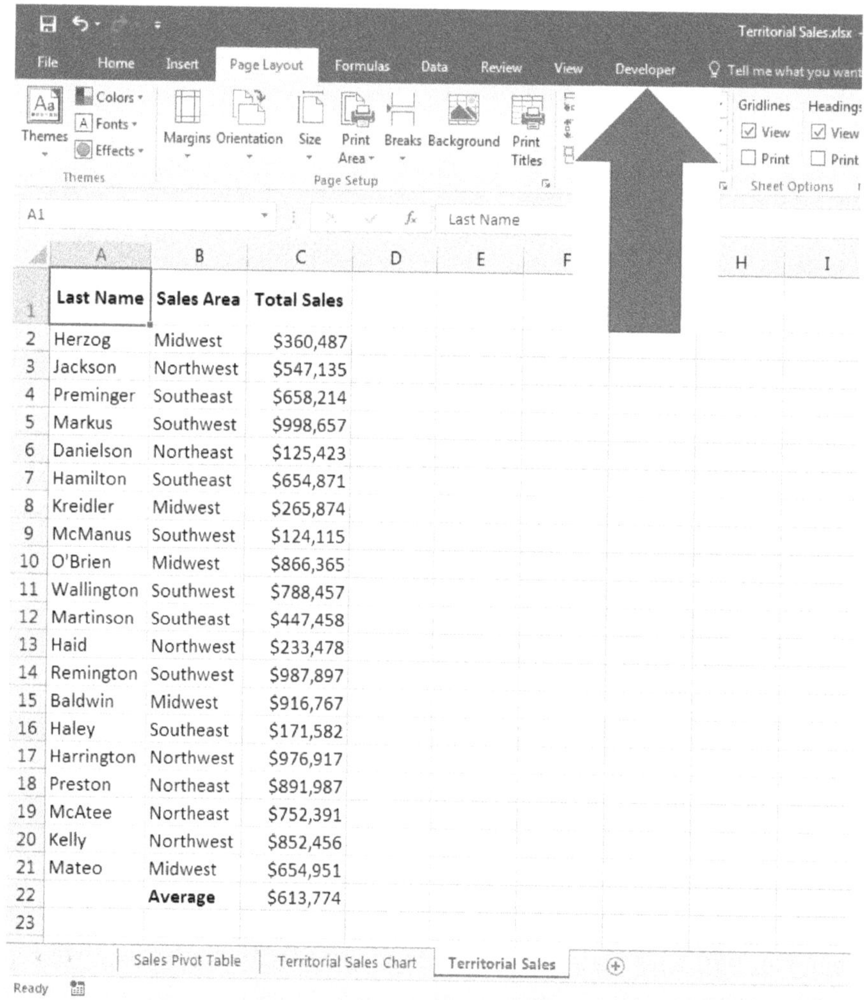

FIGURE 7-23

any one of the **Ribbon** names at the top of your menu, and then click **Customize the Ribbon ...** (Figure 7-24, Arrow).

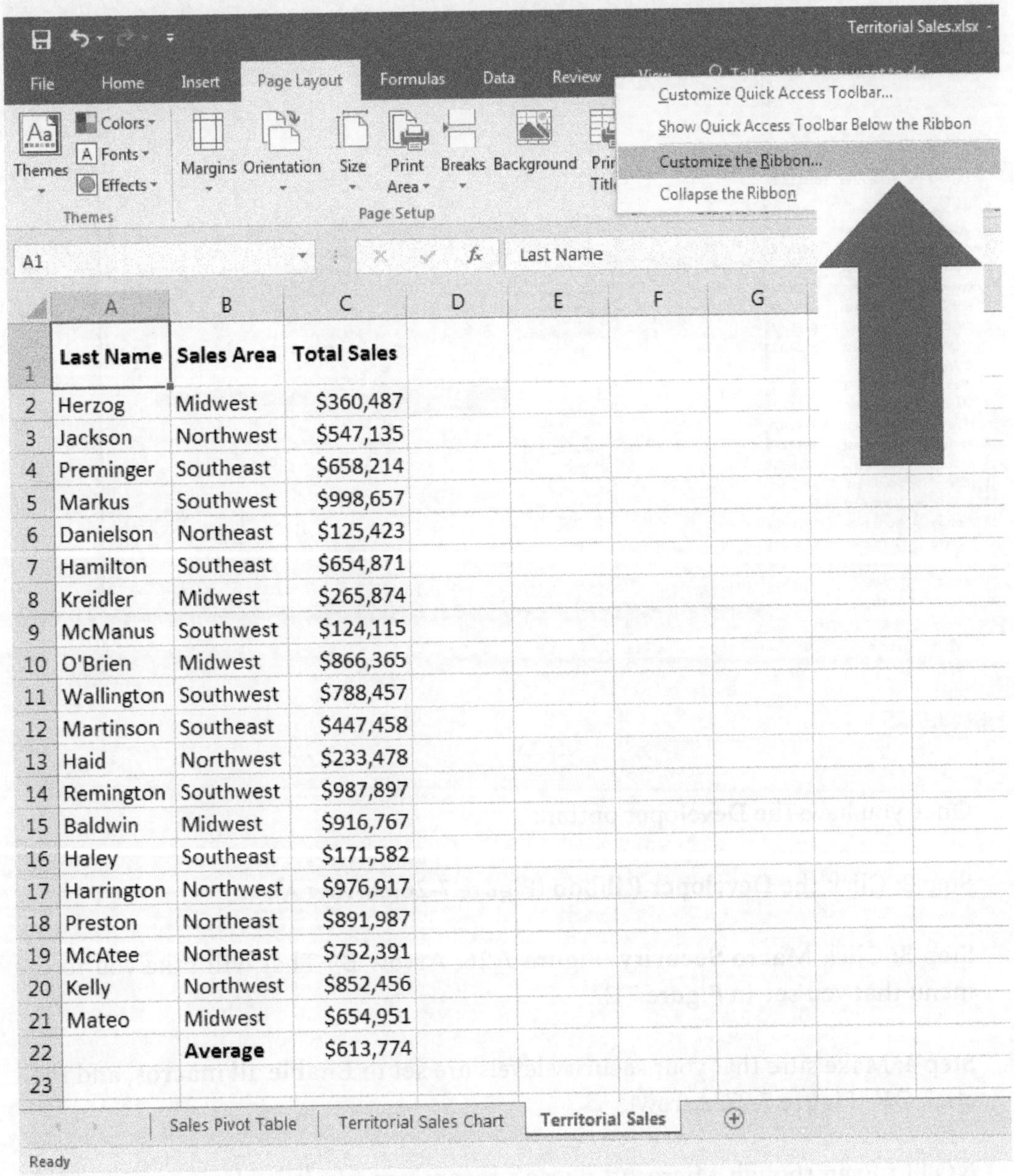

FIGURE 7-24

That will take you to the menu shown in Figure 7-24. Make sure that the **Developer** option is checked (Figure 7-25, Arrow), and then click **OK**.

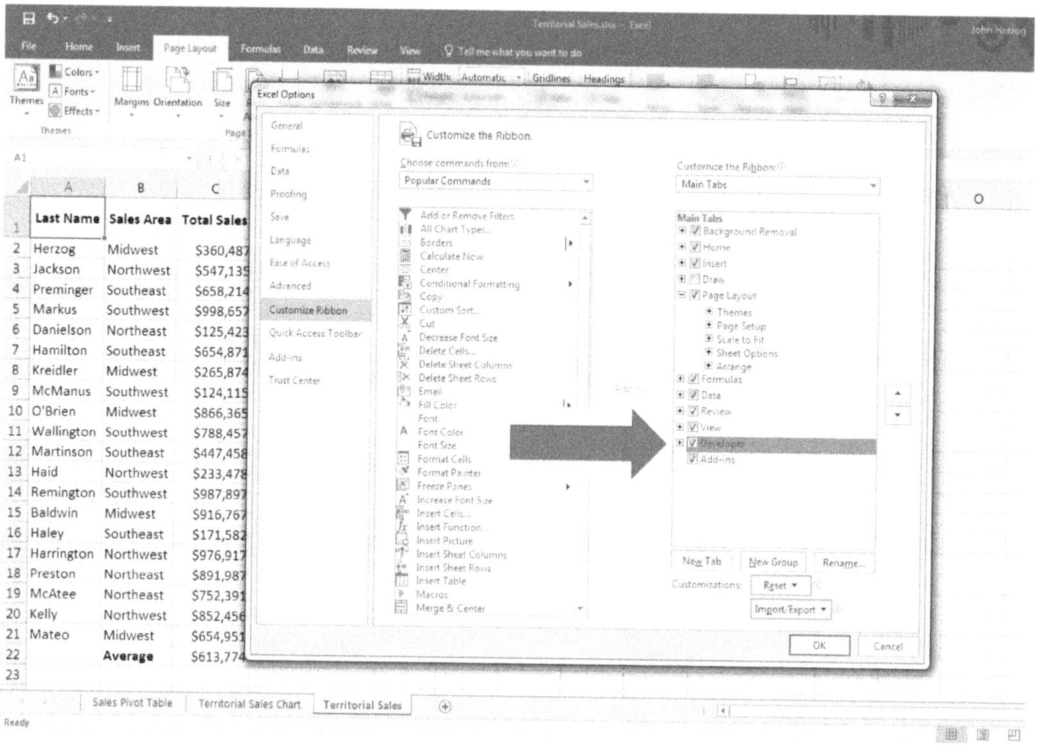

FIGURE 7-25

Once you have the **Developer** option:

Step 2: Click the **Developer Ribbon** (Figure 7-26, Arrow A).

Step 3: Click **Macro Security** (Figure 7-26, Arrow B). That will take you to the menu that you see in Figure 7-27.

Step 4: Make sure that your security levels are set to **Enable all macros,** and then click **OK** (Figure 7-27, Arrow).

NOTE: Even though Microsoft doesn't recommend it, it is okay to select it. By enabling things such as macros to run on your computer, you do so with the understanding that anything hazardous to your computer you may have inadvertently downloaded from the web, with coding that is similar to macros, can now be run on your computer.

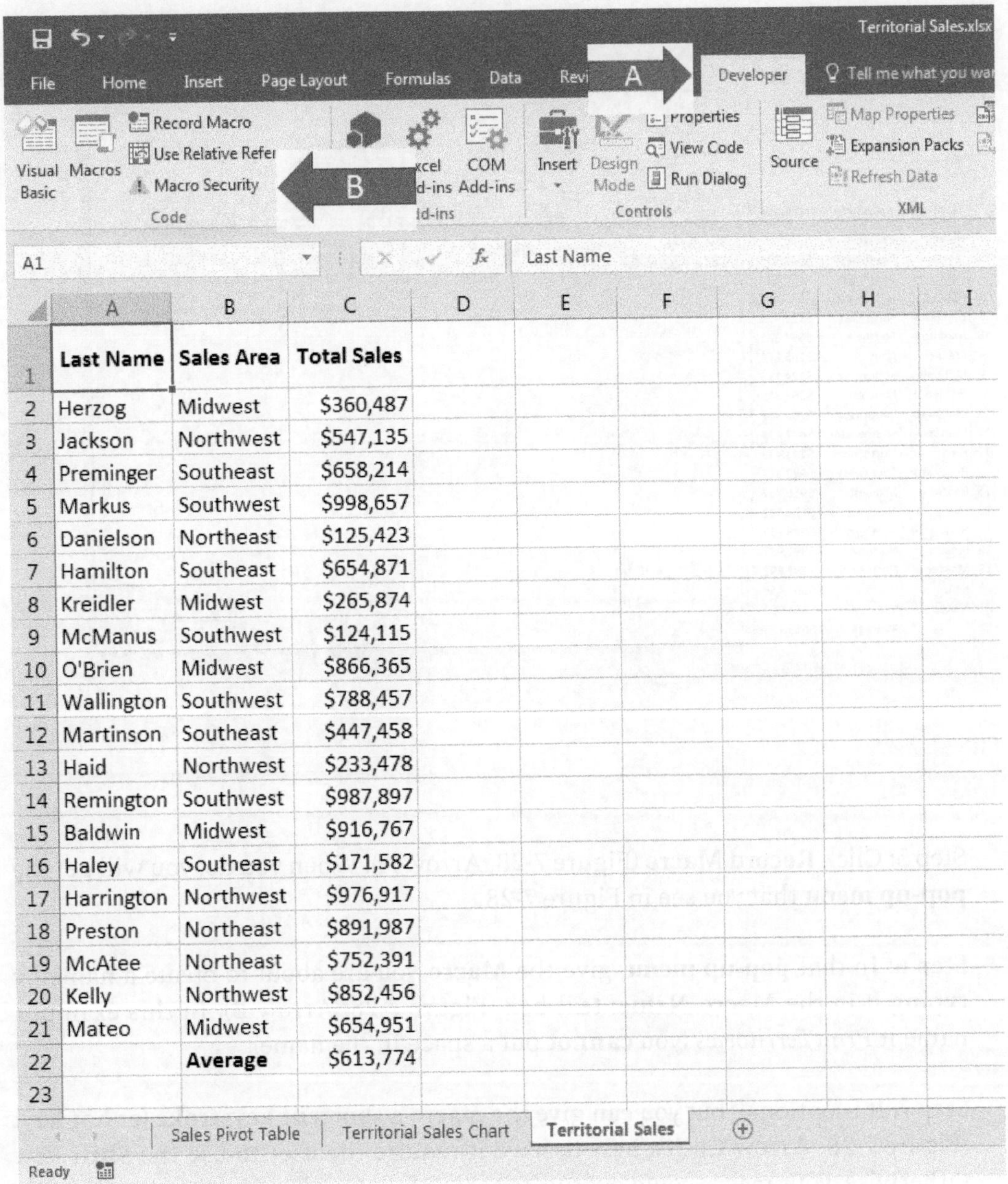
Territorial Sales.xlsx
File
Home
Insert
Page Layout
Formulas
Data
A
Developer
Record Macro
Use Relative Refer
Macro Security
Visual Basic
Macros
Code
B
COM Add-ins
Insert
Design Mode
Properties
View Code
Run Dialog
Controls
Source
Map Properties
Expansion Packs
Refresh Data
XML
A1
Last Name
Last Name | Sales Area | Total Sales
Herzog | Midwest | $360,487
Jackson | Northwest | $547,135
Preminger | Southeast | $658,214
Markus | Southwest | $998,657
Danielson | Northeast | $125,423
Hamilton | Southeast | $654,871
Kreidler | Midwest | $265,874
McManus | Southwest | $124,115
O'Brien | Midwest | $866,365
Wallington | Southwest | $788,457
Martinson | Southeast | $447,458
Haid | Northwest | $233,478
Remington | Southwest | $987,897
Baldwin | Midwest | $916,767
Haley | Southeast | $171,582
Harrington | Northwest | $976,917
Preston | Northeast | $891,987
McAtee | Northeast | $752,391
Kelly | Northwest | $852,456
Mateo | Midwest | $654,951
Average | $613,774
Sales Pivot Table
Territorial Sales Chart
Territorial Sales
Ready

FIGURE 7-26

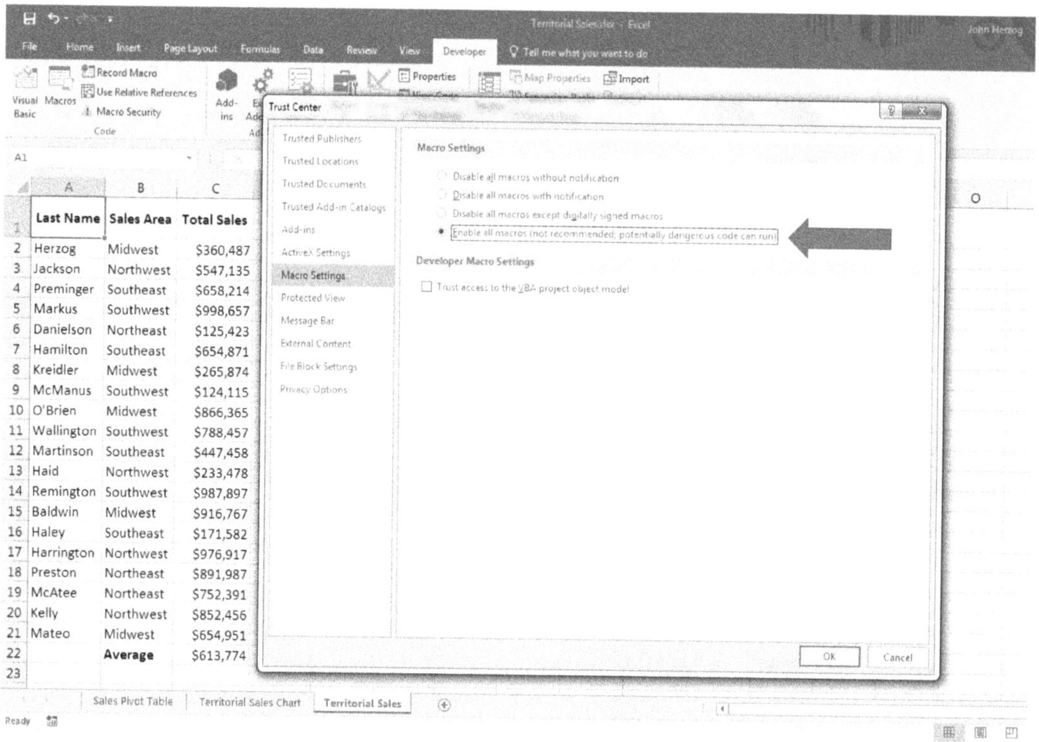

FIGURE 7-27

Step 5: Click **Record Macro** (Figure 7-28, Arrow A). When you do, you will get the **pop-up menu** that you see in Figure 7-28.

Step 6: In that **pop-up menu**, give the **Macro** you are about to create a name by typing it in the **Macro Name:** text box (Figure 7-28, Arrow B). In this example, name it *PrintTerritories* (you **cannot** put a space in the name).

Step 7: It is optional, but you can give the macro a **shortcut keystroke** (or **hot key**) (Figure 7-28, Arrow C). It's best to have it require the user to use the **Shift** key. Otherwise, your macro could overwrite some of the existing **shortcut commands** that are standard with Excel (such as **Ctrl P** to Print; **Ctrl C** to Copy; **Ctrl X** to Cut; **Ctrl V** to Paste; and so on). Those shortcut keys will no longer work if you use them in this or any macro you create. In this example, press and hold the **Shift** key, and then type *T* in the box. That would make the **Shortcut key** in this example **Ctrl Shift** *T*.

Step 8: Click **OK**. Notice that when you do, the **Developer Ribbon** will indicate that you are recording because now the **Record Macro** command has been replaced

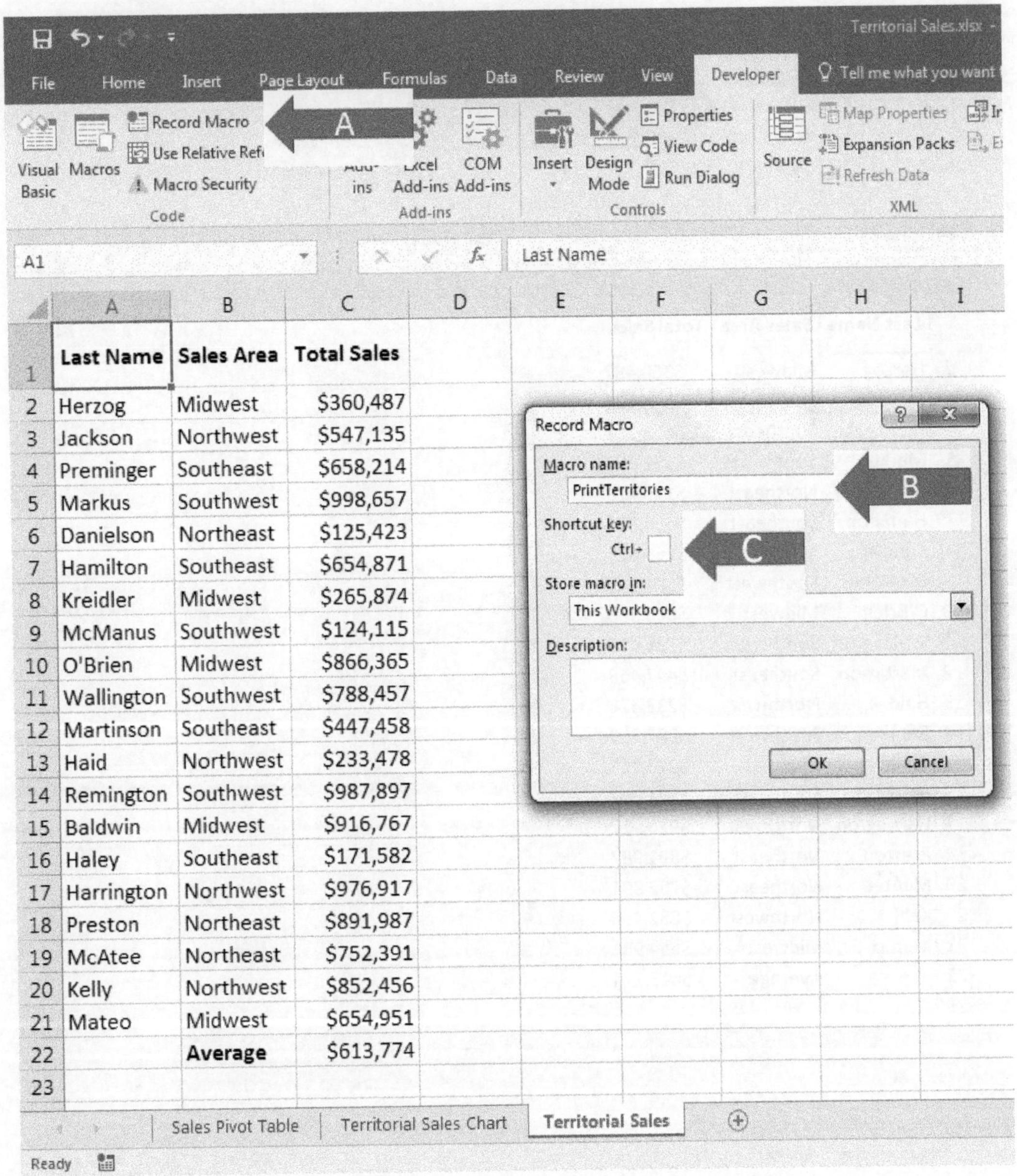

FIGURE 7-28

by the **Stop Recording** command (Figure 7-29). **Anything you do in the next few steps will be recorded in this macro.**

Step 9: Now you can begin to do the task that you want to record in this macro. Consequently, for this example, select cells A1 through B21 and set them as the **Print Area** as was shown earlier in this chapter.

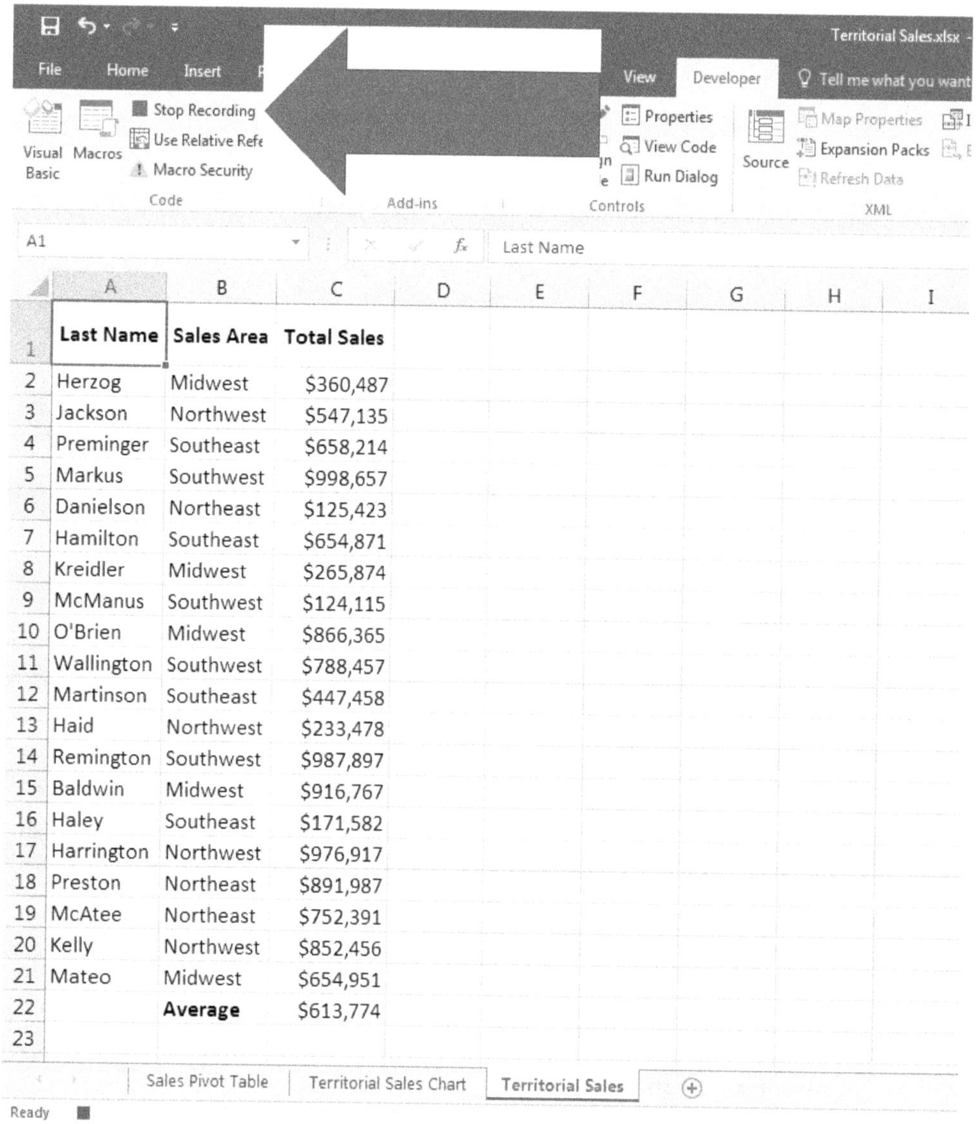

FIGURE 7-29

Step 10: Click **Stop Recording** (Figure 7-29, Arrow).

Step 11: Make another macro, this time setting cells A1 through C22 as a **Print Area**. To do so, select those cells, and then repeat steps 5 through 10. Be sure to name this new macro as *PrintAll* and give it a Shortcut key of **Ctrl Shift** *A*.

Once you have created these macros, you should see them work by using the shortcut key combination you gave them (**Ctrl Shift** *T* and **Ctrl Shift** *A*).

TO ADD A BUTTON TO RUN MACROS

If you forget the shortcut keys for your macros, or if others use your spreadsheet and don't know the shortcut keys you assigned to them, you will have a difficult time running the macros you just created. Therefore, to solve this problem, let's put buttons on the spreadsheet and then attach the macros to them that we want them to run.

To do so, use the following steps:

Step 1: In the same **Developer Ribbon** (Figure 7-30, Arrow A), click on the tiny arrow below the **Insert** command (often referred to as the **Toolbox**) (Figure 7-30, Arrow B).

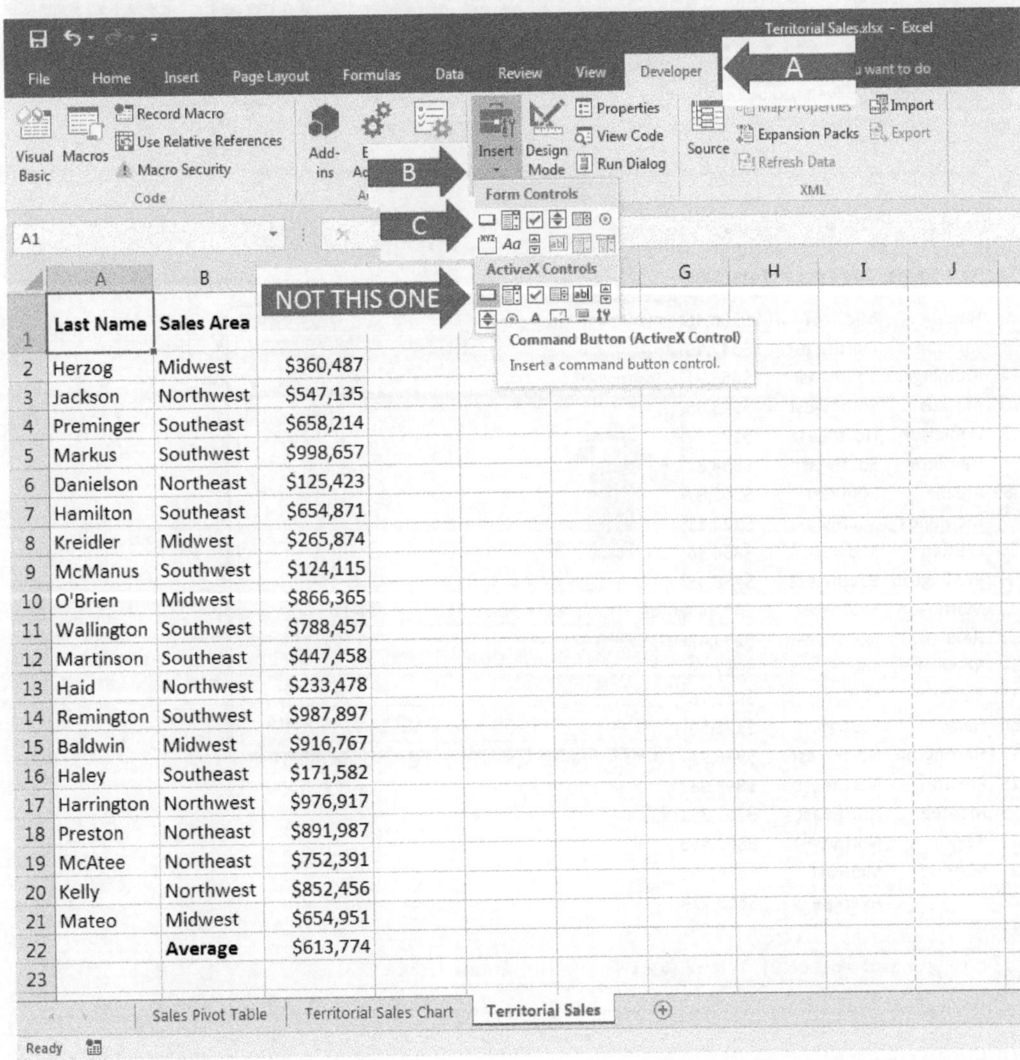

FIGURE 7-30

Step 2: Click once on the first button (**Form Control**) (Figure 7-30, Arrow C). **Do not double-click. Do not get this one mixed up with the lower button** (Figure 7-30, Arrow labeled as *Not This One*).

Step 3: Click where you want the button. **Do not click and drag; just click once.** In this case, we want it to be in the upper left of cell E2. NOTE: Initially, the pop-up menu will cover the button. At this point, you will see the menu in Figure 7-31.

Step 4: Choose the macro in the box labeled as **Macro name:** that you want to apply to this button. Here it is the one named *PrintTerritories* (Figure 7-31, Arrow).

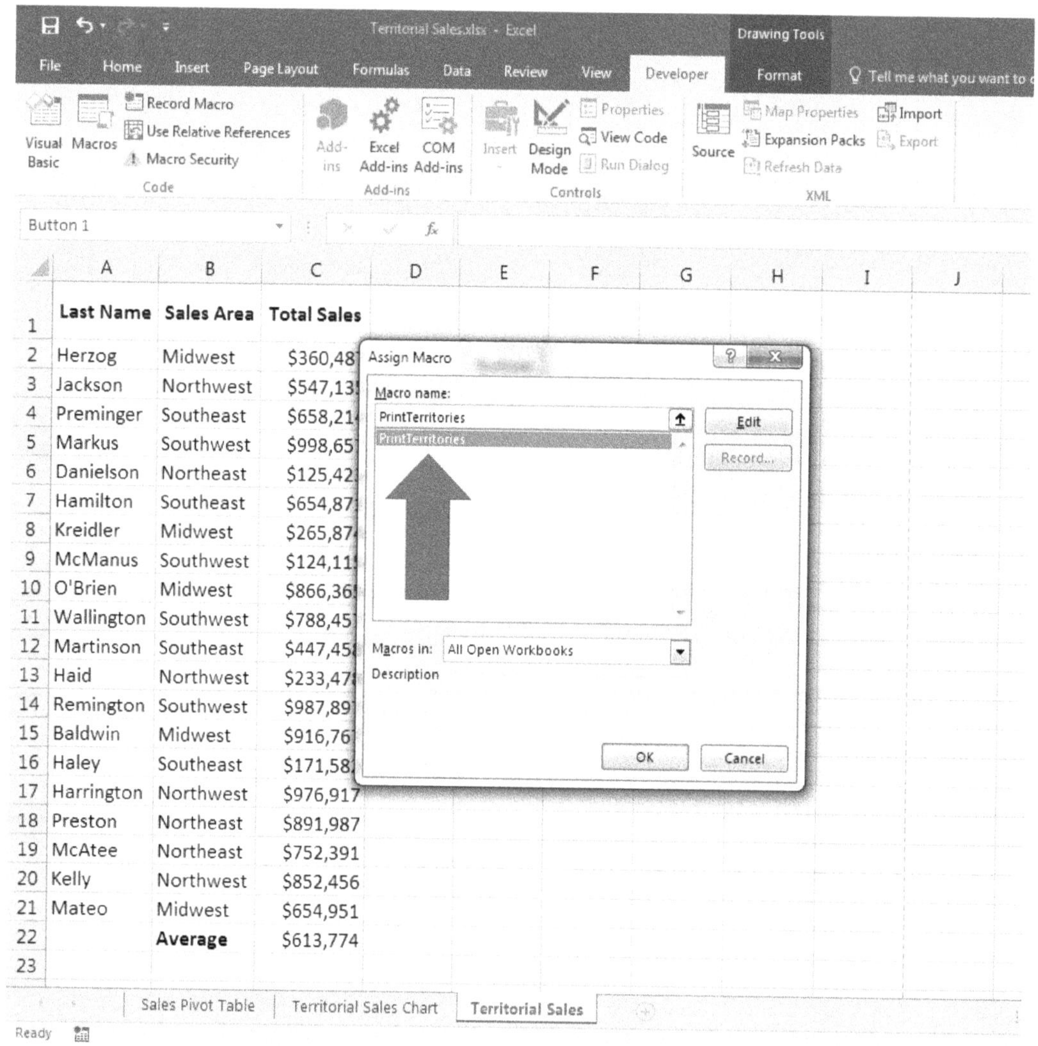

FIGURE 7-31

Step 5: Click **OK**.

Step 6: That will take you to the screen in Figure 7-32.

Step 7: Resize the button just as you would resize a **Clipart** in Word or PowerPoint, until it fills cells E2 through F5 as shown in Figure 7-33.

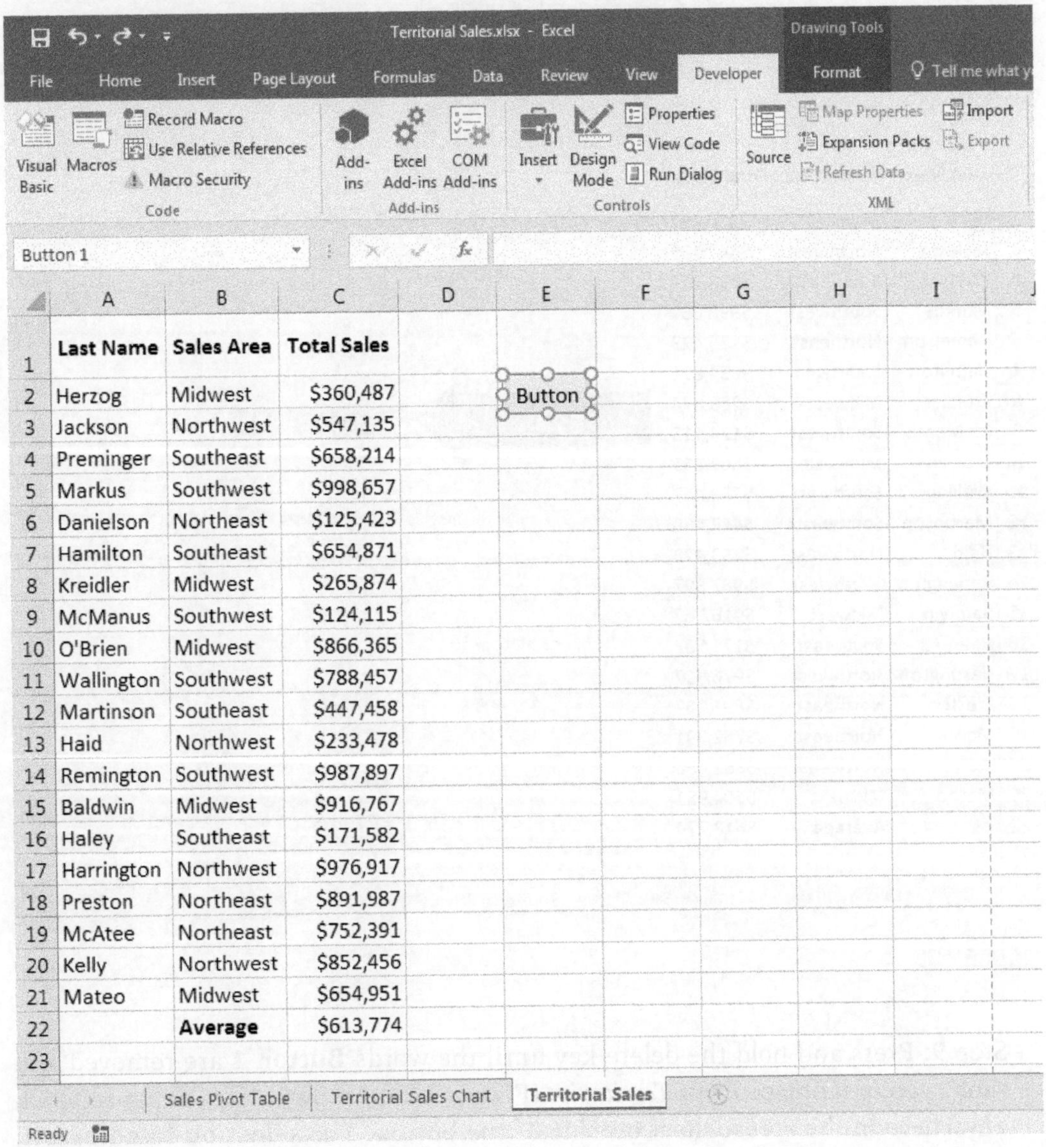

FIGURE 7-32

Step 8: Right-click on the button and in the **shortcut menu**, click **Edit Text** (Figure 7-33, Arrow).

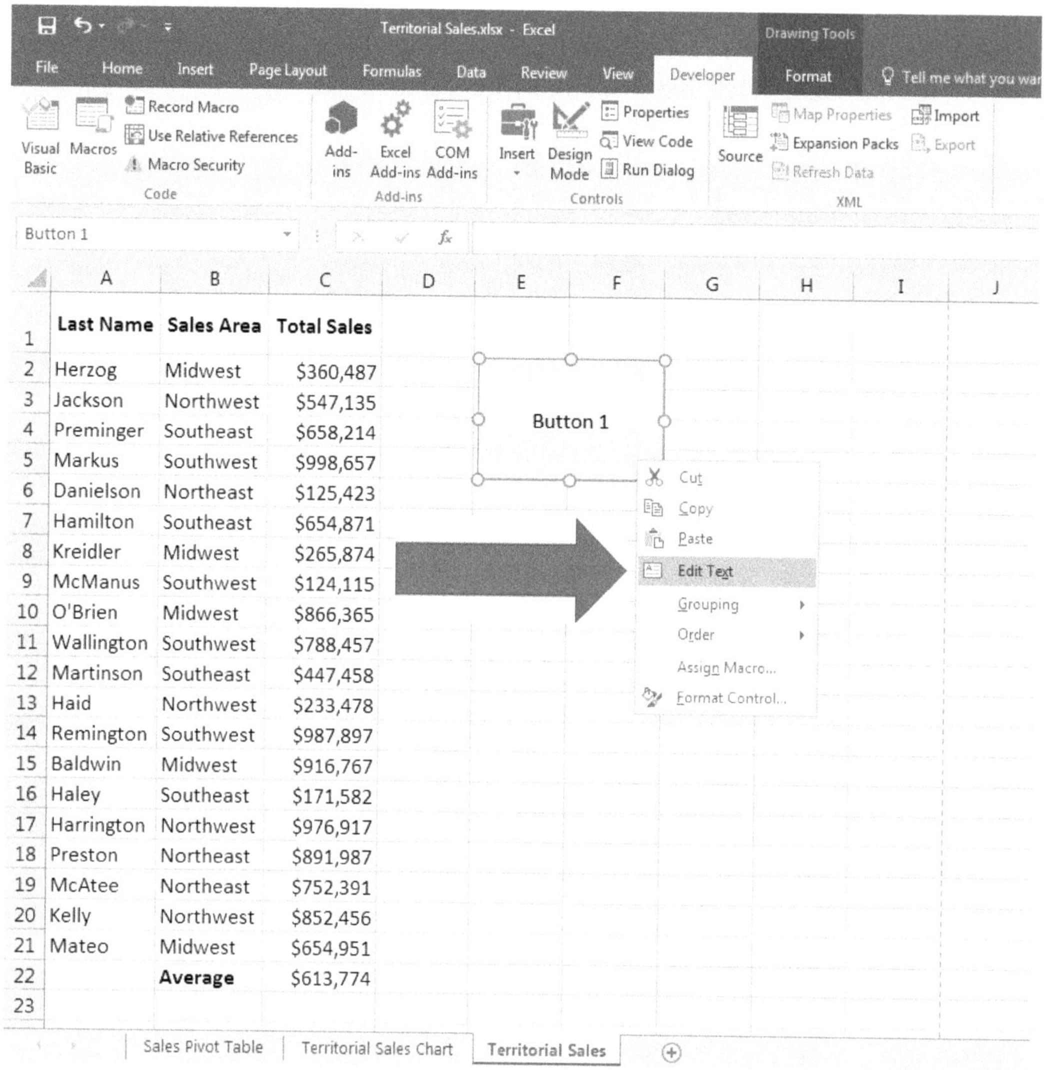

FIGURE 7-33

Step 9: Press and hold the delete key until the words **Button 1** are removed and then type in its place *Print Territories* (Figure 7-34, Arrow). After doing so, click anywhere in the spreadsheet outside of the button. The **selection handles** will then disappear.

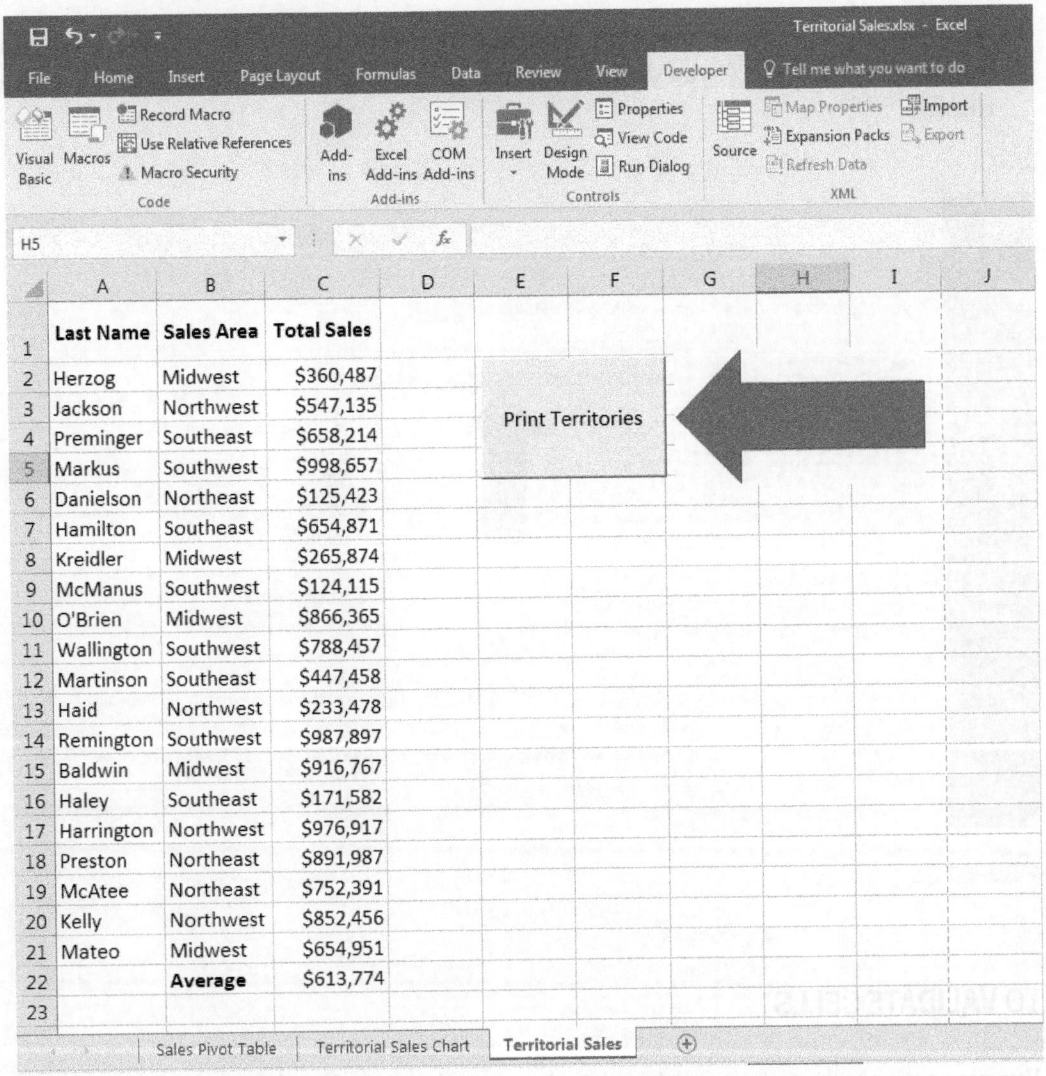

	A	B	C
1	**Last Name**	**Sales Area**	**Total Sales**
2	Herzog	Midwest	$360,487
3	Jackson	Northwest	$547,135
4	Preminger	Southeast	$658,214
5	Markus	Southwest	$998,657
6	Danielson	Northeast	$125,423
7	Hamilton	Southeast	$654,871
8	Kreidler	Midwest	$265,874
9	McManus	Southwest	$124,115
10	O'Brien	Midwest	$866,365
11	Wallington	Southwest	$788,457
12	Martinson	Southeast	$447,458
13	Haid	Northwest	$233,478
14	Remington	Southwest	$987,897
15	Baldwin	Midwest	$916,767
16	Haley	Southeast	$171,582
17	Harrington	Northwest	$976,917
18	Preston	Northeast	$891,987
19	McAtee	Northeast	$752,391
20	Kelly	Northwest	$852,456
21	Mateo	Midwest	$654,951
22		**Average**	$613,774
23			

FIGURE 7-34

Step 10: Click the button and try it out. When you point to the button from this point forward, your mouse will appear as an index finger as you would see when you choose things on the Internet.

Step 11: Repeat steps 1 through 10 to add the button for the *PrintAll* macro.

Step 12: Save this workbook, using the procedure discussed earlier, but be sure to save it as an **Excel Macro-Enabled Workbook**. NOTE: You must save workbooks with **Macros** as an **Excel Macro-Enabled Workbook** with the **.xlsm extension** (Figure 7-35, Arrow). **If you don't, your macros will be erased as soon as you close the file!**

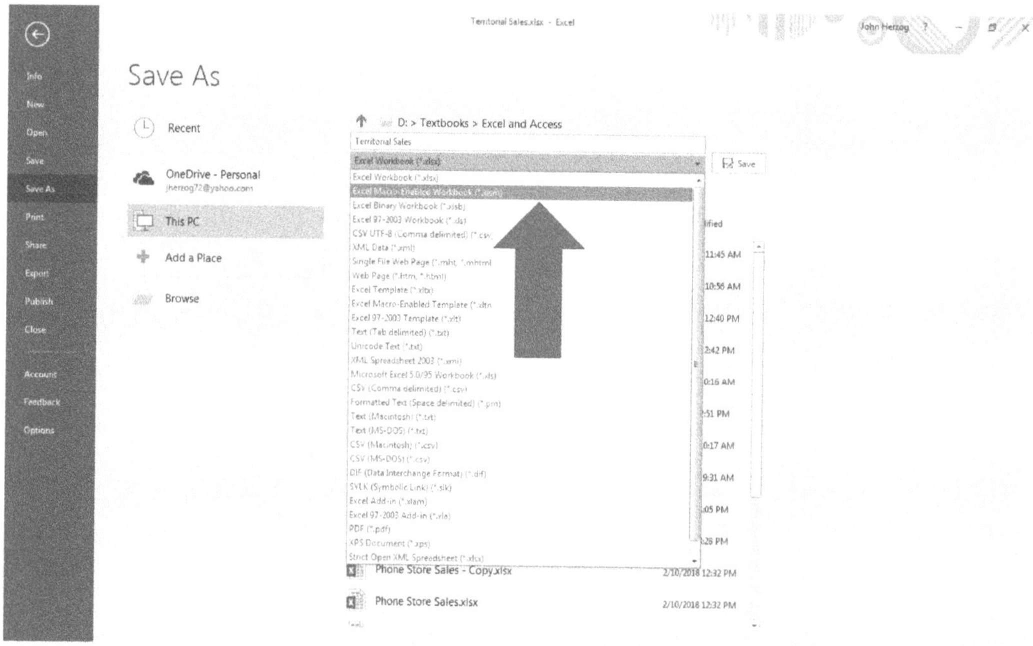

FIGURE 7-35

TO VALIDATE CELLS

Everyone wants to make sure that data is entered correctly. The problem is that the default setting for Excel is to allow any kind of data to be entered into any cell. However, cells can be set up to **validate** all entered data. For example, if you have cells where dates are entered, you can set the spreadsheet up to never allow invalid dates, such as 7/43/2018. Suppose, in this case, that you don't want anyone to enter negative numbers by accident, nor do you want them to put in some astronomical sales number. Also, suppose that you want to make sure that the *Sales Area* entries are not mistyped by someone misspelling *Northwest* or *Northeast* or the like. Otherwise, if a person did mistype it, filters would not be accurate.

Let's validate the *Total Sales* numbers first.

Step 1: Select the column where the *Total Sales* will be entered. In this example, it will be cells C2 through C21 as shown in Figure 7-36, Arrow A.

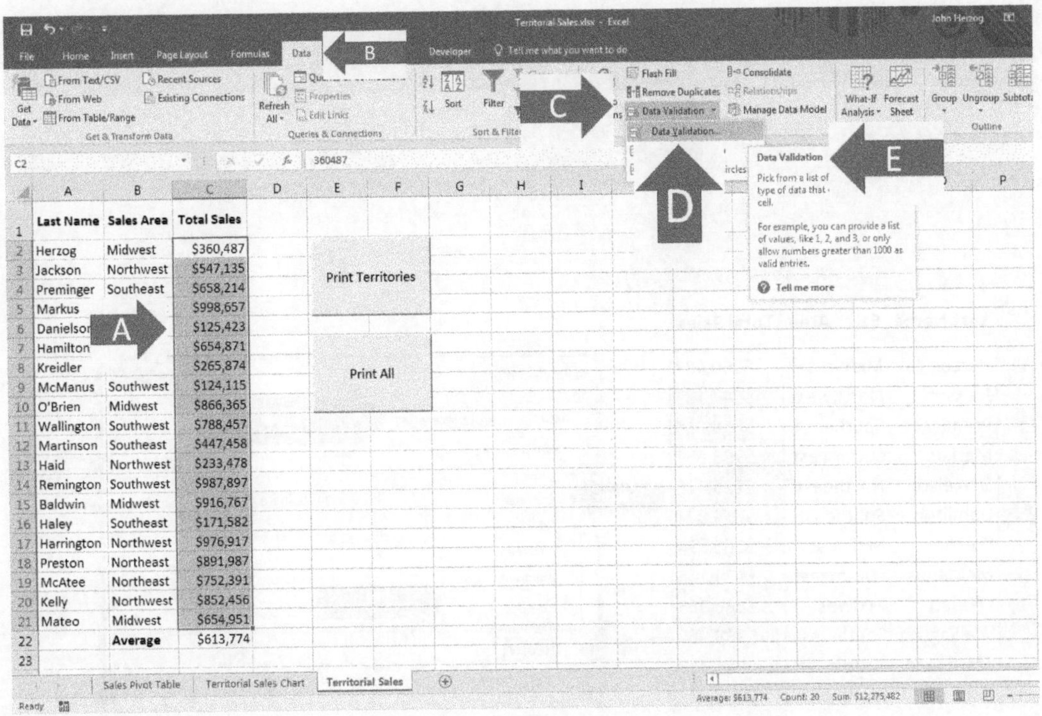

FIGURE 7-36

Step 2: Click the **Data Ribbon** (Figure 7-36, Arrow B).

Step 3: Click the **Data Validation** option (Figure 7-36, Arrow C).

Step 4: Click the **Data Validation** option in the **drop-down list** (Figure 7-36, Arrow D).

NOTE: There is often an input message that will appear when you point to a menu option that will give you a description of what the menu option will do when chosen (Figure 7-36, Arrow E).

Step 5: In the menu that displays as shown in Figure 7-37, choose **Decimal** (Figure 7-37, Arrow A). NOTE: For numbers other than dollars that need integers, choose **Whole Numbers**. Also notice that the default of this menu is always allowing **Any Value.**

Step 6: Choose a **Minimum** value (in this example, it would be zero) (Figure 7-37, Arrow B) and a **Maximum** value that would be acceptable (in this case, it will be 10000000 with no commas) (Figure 7-37, Arrow C). After you do so, **ONLY**

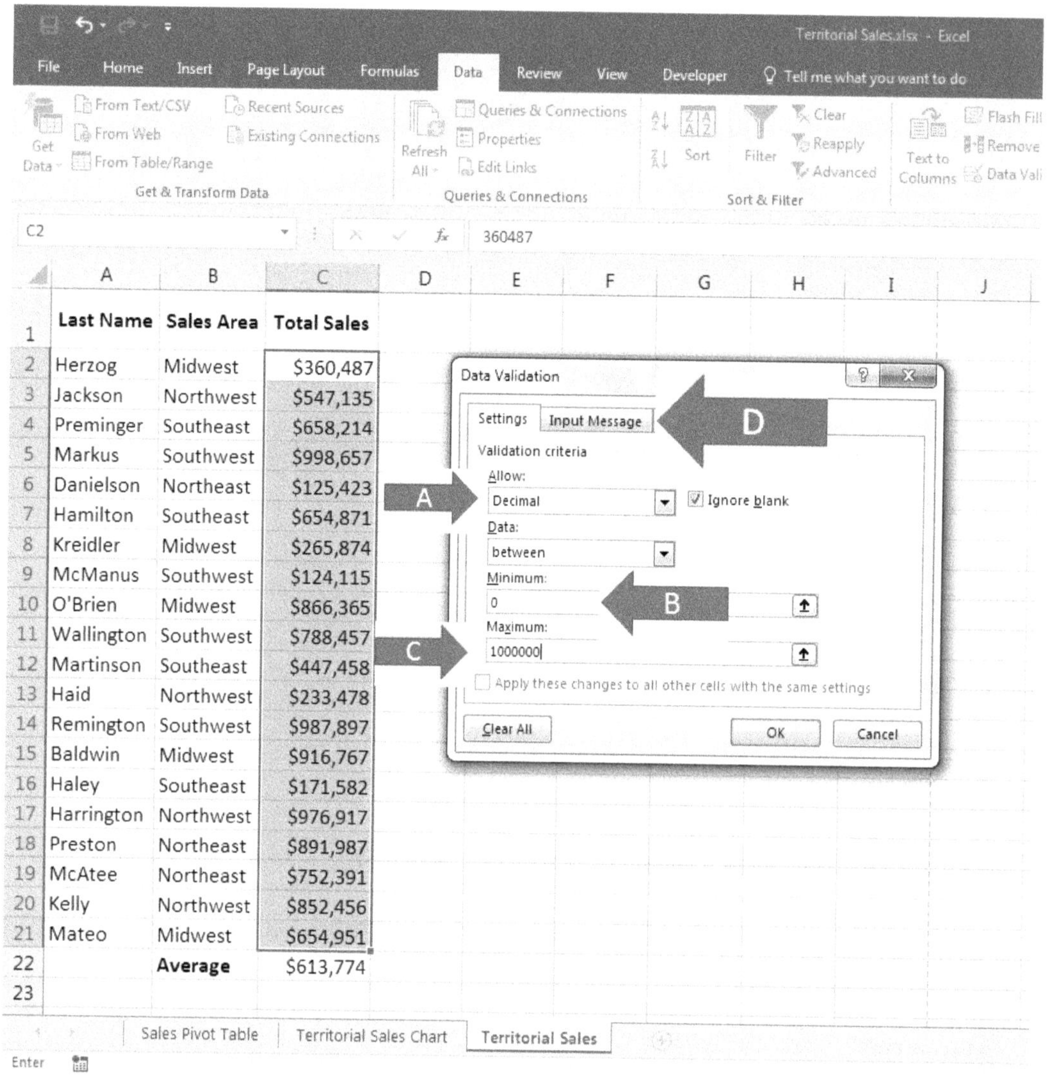

FIGURE 7-37

numbers in that range from 0 to $10,000,000 will be permitted to be entered in these cells. This will prevent entry of a negative number or a number that is beyond their scope of possible sales.

When people click or select cells that have restrictions such as these, you will want to let them know about those restrictions that you set as they enter data, by creating an **Input Message**. You will also want to have the cell display an **Error Message** if users enter invalid data. You can do these things with the following steps.

Step 1: Click **Input Message** (Figure 7-37, Arrow D). That will take you to the menu shown in Figure 7-38.

Step 2: Inside the **Title** box (Figure 7-38, Arrow A), type *Be Careful.* Inside the **Input message** box (Figure 7-38, Arrow B), type *Enter only numbers that are greater than or equal to zero and less than $10,000,000.* NOTE: It is a good idea to select that text and invoke the copy command by pressing **Ctrl C** at this point because in a moment you will need to enter those same words again when you create the **Error Message**.

Step 3: Click **Error Alert** (Figure 7-38, Arrow C).

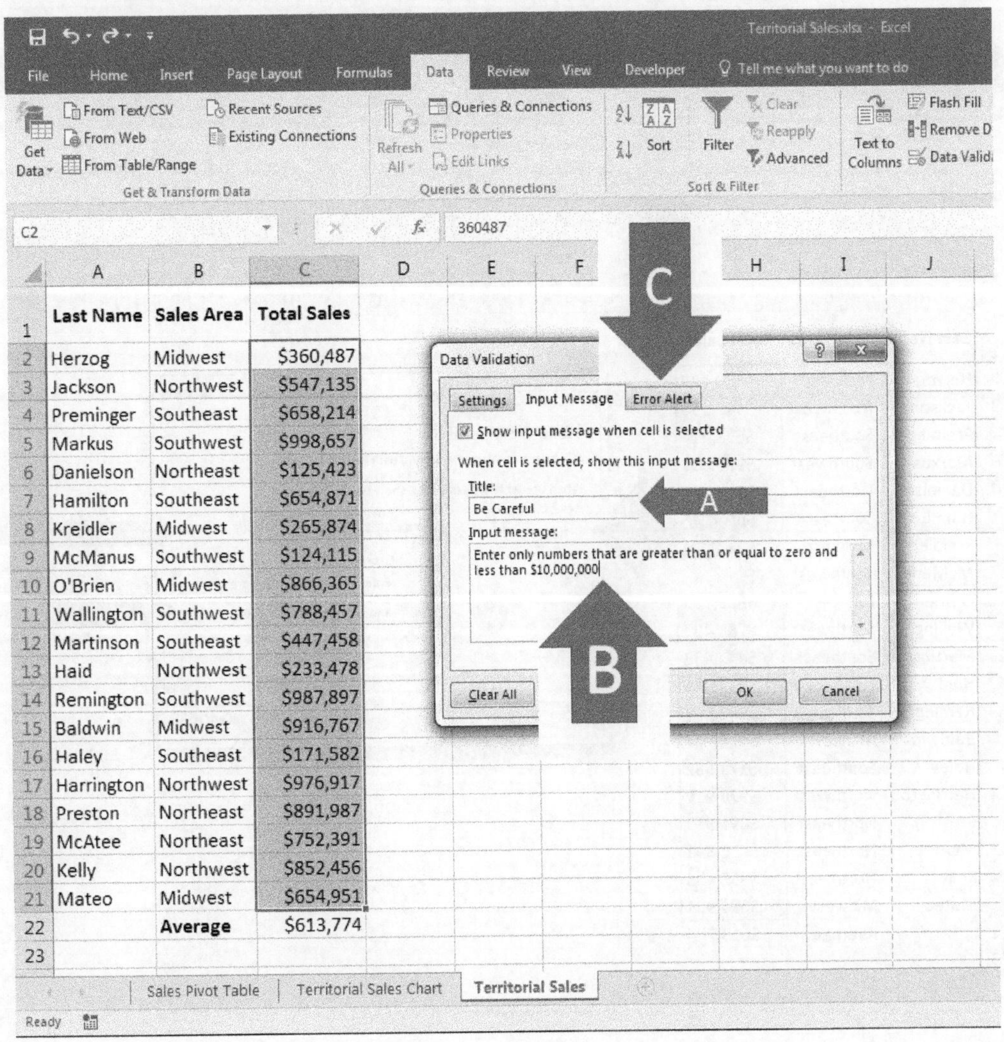

FIGURE 7-38

Step 4: That will take you to the menu you see in Figure 7-39. Put the same words in the **Title** box that will read *Be Careful* (Figure 7-39, Arrow A), and in the **Error message** box enter the words *Enter only numbers that are greater than or equal to zero and less than $10,000,000* (Figure 7-39, Arrow B) that you used in the **Input Message**. NOTE: If you copied the **Input Message** from the previous screen as suggested, at this point all that you need to do is press **Ctrl V** (or Paste). In this example, keep the **Error Message** style as the **Stop Style**, as it is appropriate for this type of message (Figure 7-39, Arrow C). By doing so, the X below the **Stop Style** box will appear in the **Error Message**.

Step 5: Click **OK**.

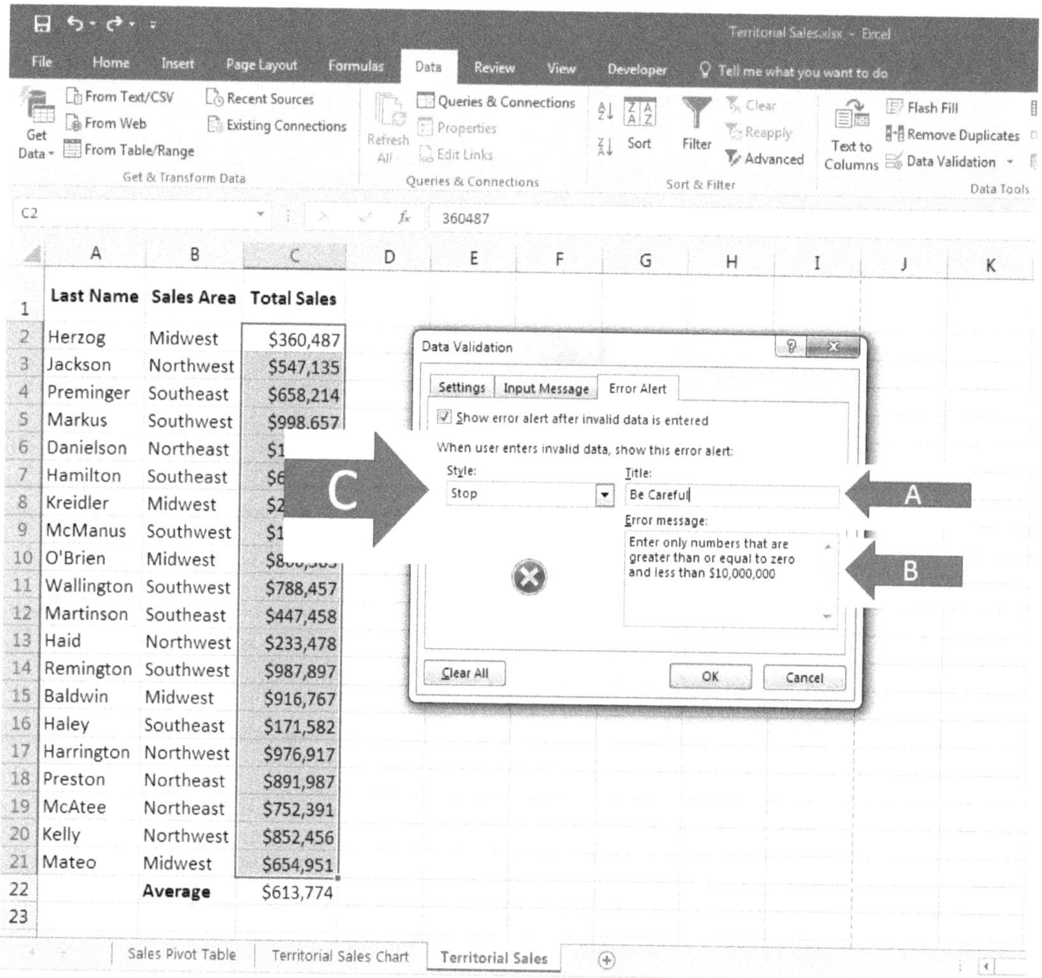

FIGURE 7-39

Now try to enter some data. First, when you click on any one of those cells, the **Input Message** appears, explaining what data is and is not appropriate for the cells. If you enter a negative number in any one of the cells, for example, the **Error Message** you created will pop up (Figure 7-40). Please notice that the **Stop Style** (or the red *X*) is shown on the **Error Message** as well as your error message and the error message **Title**, which reads *Be Careful.*

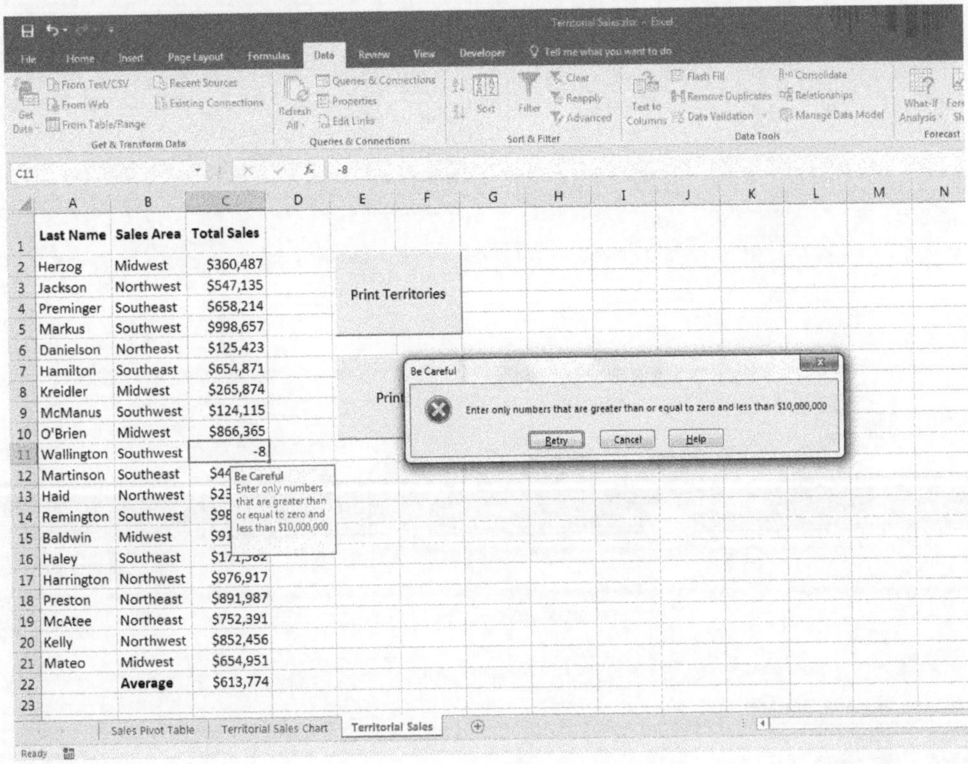

FIGURE 7-40

Now let's create validation in column B so that only *Southeast*, *Southwest*, *Northeast*, *Northwest*, and *Midwest* can be entered in the **Sales Area** column. First, we will create a list of just those words. You will want to do it far away from your main spreadsheet so that no one will inadvertently erase it or type over it. You can do so by using the following steps:

Step 1: Go to a cell far away from the list—perhaps cell AZ1—and type *Southeast.* Then in cell AZ2, type *Southwest.* In cell AZ3, type *Northeast.* In cell AZ4, type *Northwest*, and in cell AZ5, type *Midwest* (Figure 7-41).

Step 2: Select cells B2 through B21.

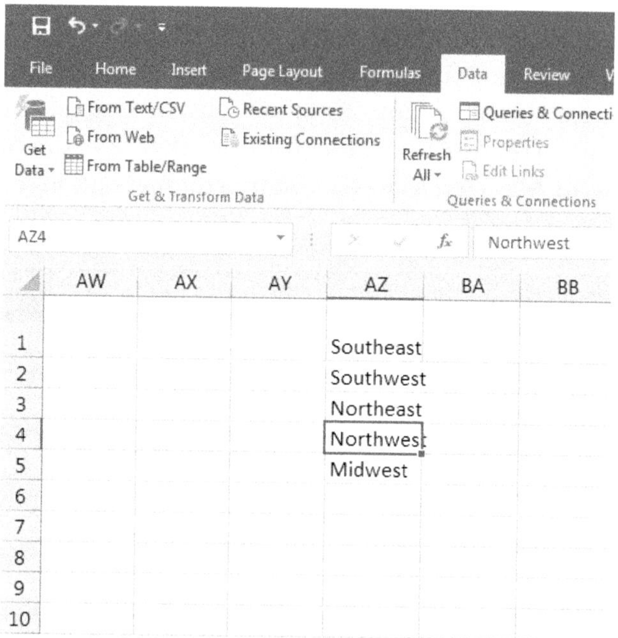

FIGURE 7-41

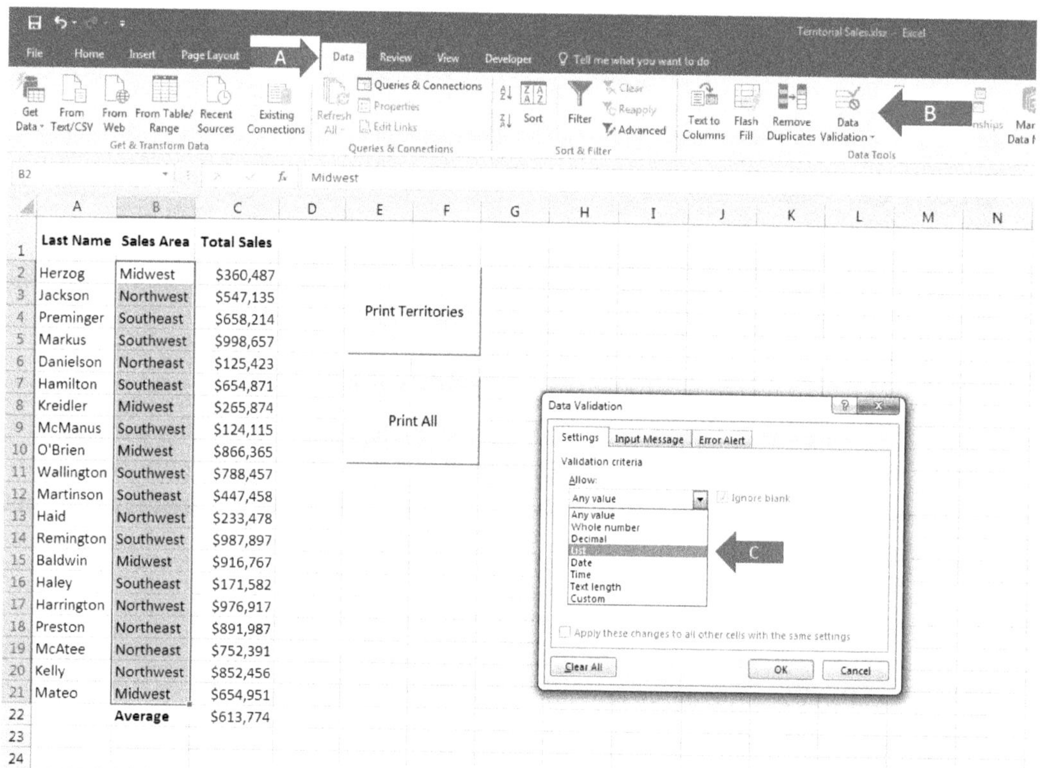

FIGURE 7-42

Step 3: Click the **Data Ribbon** (Figure 7-42, Arrow A).

Step 4: Click **Data Validation** (Figure 7-42, Arrow B).

Step 5: When the **Data Validation** menu pops up, select the **List** option (Figure 7-42, Arrow C) in the **Settings** tab **drop-down list**. That will take you to the menu seen in Figure 7-43.

Step 6: Click inside the **Source box** (Figure 7-43, Arrow A).

Step 7: Scroll to the right of the spreadsheet over to cells AZ1 and select the cells of the list you created a moment ago from cells AZ1 through AZ5. Notice that it will automatically place that cell range into the **Source box** (Figure 7-43, Arrow A) of the **Data Validation pop-up menu**.

Step 8: Click **OK**.

Step 9: Add any **Input message** and **Error Alert** that you feel is appropriate.

Step 10: Click **OK**.

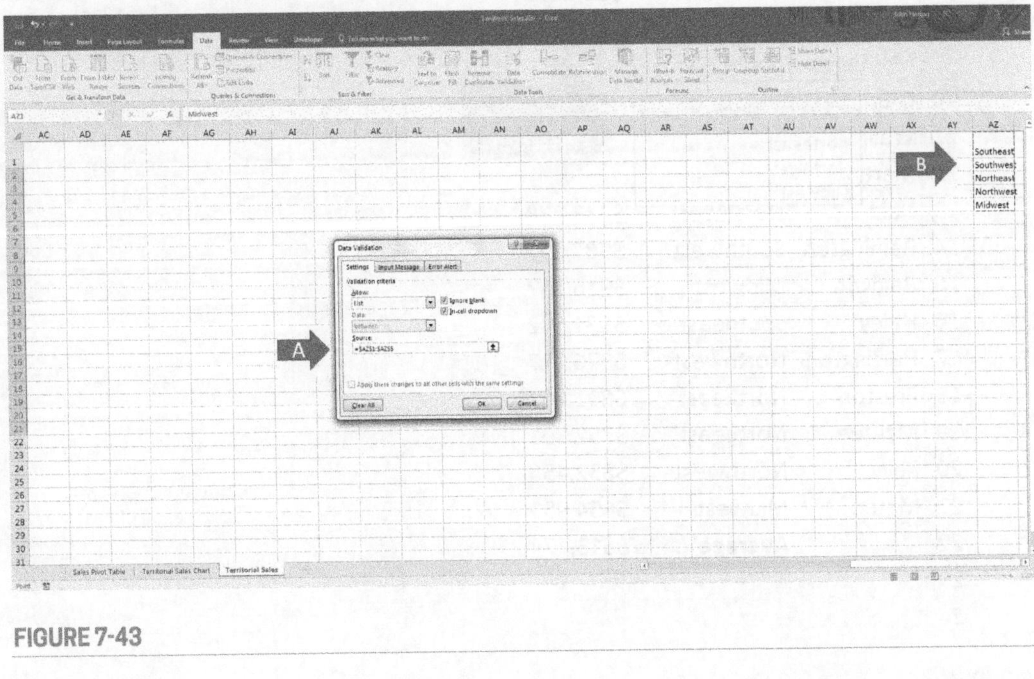

FIGURE 7-43

Notice now that when you click on any one of the cells in the cell range of B2 through B21, a **drop-down list box** will appear. When you click it, the program gives only the items from your list and will not permit any other data to be entered (Figure 7-44, Arrow).

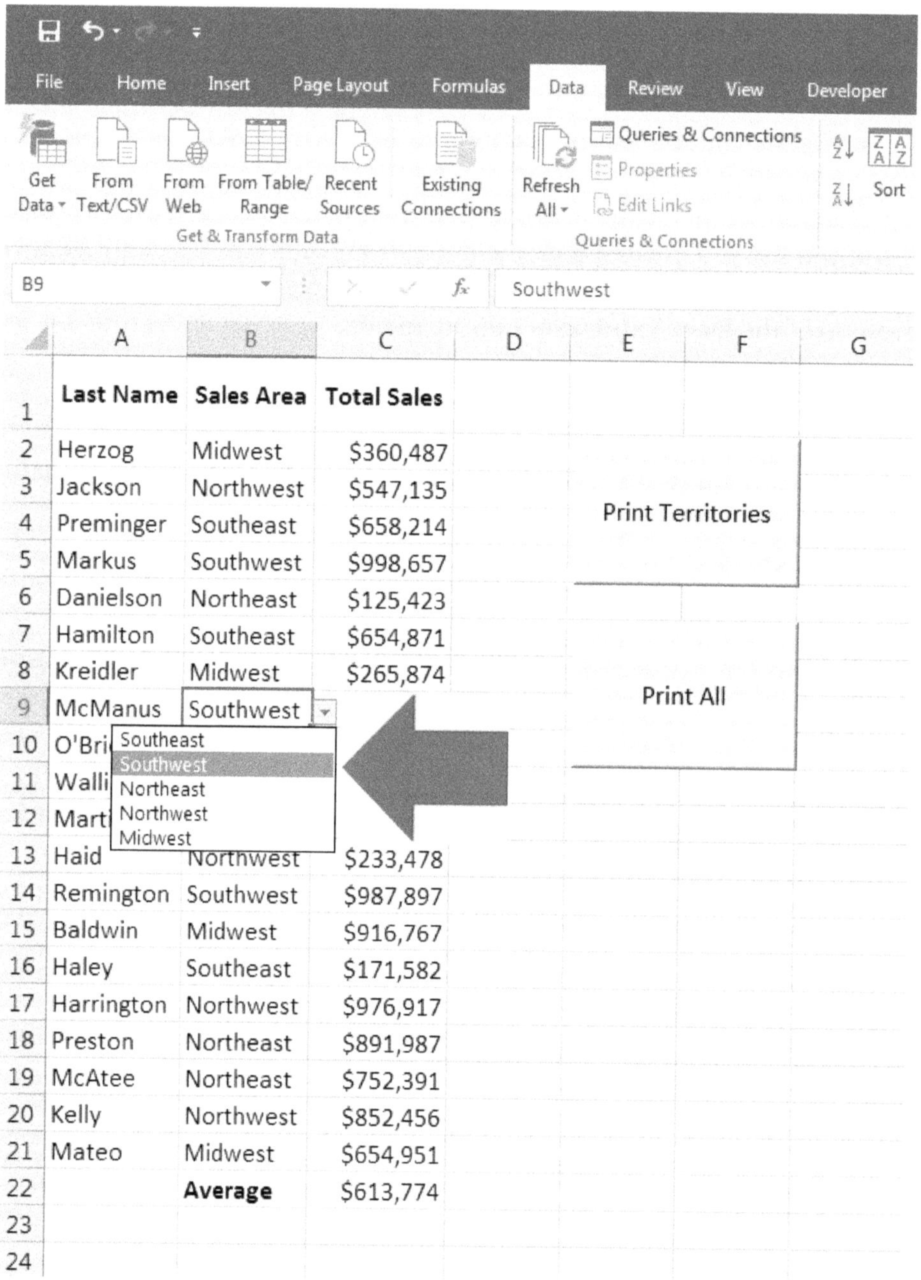

FIGURE 7-44

CHAPTER 7—ASSIGNMENT (SUMMARY OF TASKS)

1. Open the workbook entitled *Territorial Sales* associated with this textbook, shown in Figure 7-45.

	A	B	C	D
1	Last Name	Sales Area	Total Sales	
2	Herzog	Midwest	$360,487	
3	Jackson	Northwest	$547,135	
4	Preminger	Southeast	$658,214	
5	Markus	Southwest	$998,657	
6	Danielson	Northeast	$125,423	
7	Hamilton	Southeast	$654,871	
8	Kreidler	Midwest	$265,874	
9	McManus	Southwest	$124,115	
10	O'Brien	Midwest	$866,365	
11	Wallington	Southwest	$788,457	
12	Martinson	Southeast	$447,458	
13	Haid	Northwest	$233,478	
14	Remington	Southwest	$987,897	
15	Baldwin	Midwest	$916,767	
16	Haley	Southeast	$171,582	
17	Harrington	Northwest	$976,917	
18	Preston	Northeast	$891,987	
19	McAtee	Northeast	$752,391	
20	Kelly	Northwest	$852,456	
21	Mateo	Midwest	$654,951	
22				
23				

FIGURE 7-45

2. Validate cells with the following specifications:
    a. Make sure that the *Total Sales* numbers in cells C2 through C21 are validated so that they permit only entries that are greater than or equal to zero (0) and less than $10,000,000.
    b. When the cell is selected, make sure that a help message (**Input Message**) appears that says *Enter only numbers that are greater than or equal to zero and less than $10,000,000*, and make sure the **Title** reads *Be Careful*.
    c. Make sure that an **Error Message** pops up that says *Enter only numbers that are greater than or equal to zero and less than $10,000,000*, and make sure the **Title** reads *Be Careful*.
    d. Make a list of the different sales areas. In cell AZ1 type *Southeast*. Then in cell AZ2, type *Southwest*. In cell AZ3 type *Northeast*. In cell AZ4, type *Northwest*. In cell AZ5, type *Midwest*.
    e. Make sure that when you click any cell in cells B2 through B21, a drop-down list will become available that will only allow you to enter those *Sales Areas*.
    f. Add an **Input Message** with a **Title** and an **Error Message** with a **Title** that would be appropriate for that cell range.
3. Create a **Pivot Table** using cells A1 through C21 that will show the sum of the *Total Sales* by *Last Name* (as the **row headings**) and by *Sales Area* (as the **column headings**) with the following specifications:
    a. Make sure the **Pivot Table** is sent to a new spreadsheet.
    b. Change the widths of all of its columns to *16* characters.
    c. Change the style of the **Pivot Table** to Medium 2.
    d. Change the format of the **Pivot Table** dollars to **Currency**. At this point, the **Pivot Table** should look like what you see in Figure 7-46.
    e. Filter the **Pivot Table** so that it will **NOT** show sales from *Northwest* and *Southeast*.
    f. Name the sheet *Pivot Table*.
4. Return to the original *Territorial Sales* spreadsheet.
    a. Enter the **Average** function in cell C22 that averages all of the *Total Sales*. Type the word *Average* in cell B22 and make it **bold**.
    b. Create a **Macro** that will set the **Print Area** of the sheet to cells A1 through B21 and name the macro *PrintTerritories*.
    c. Make a button that will run the **Macro** and make sure the button caption reads *Print Territories*. Place it in cells E2 through F5.
    d. Create a second **Macro** that will set the **Print Area** to cells A1 through C21 and name it *PrintAll*.
    e. Make a button that will run the **Macro**; make sure the caption reads *Print All* (Figure 7-47) and place it in cells E7 through F10.

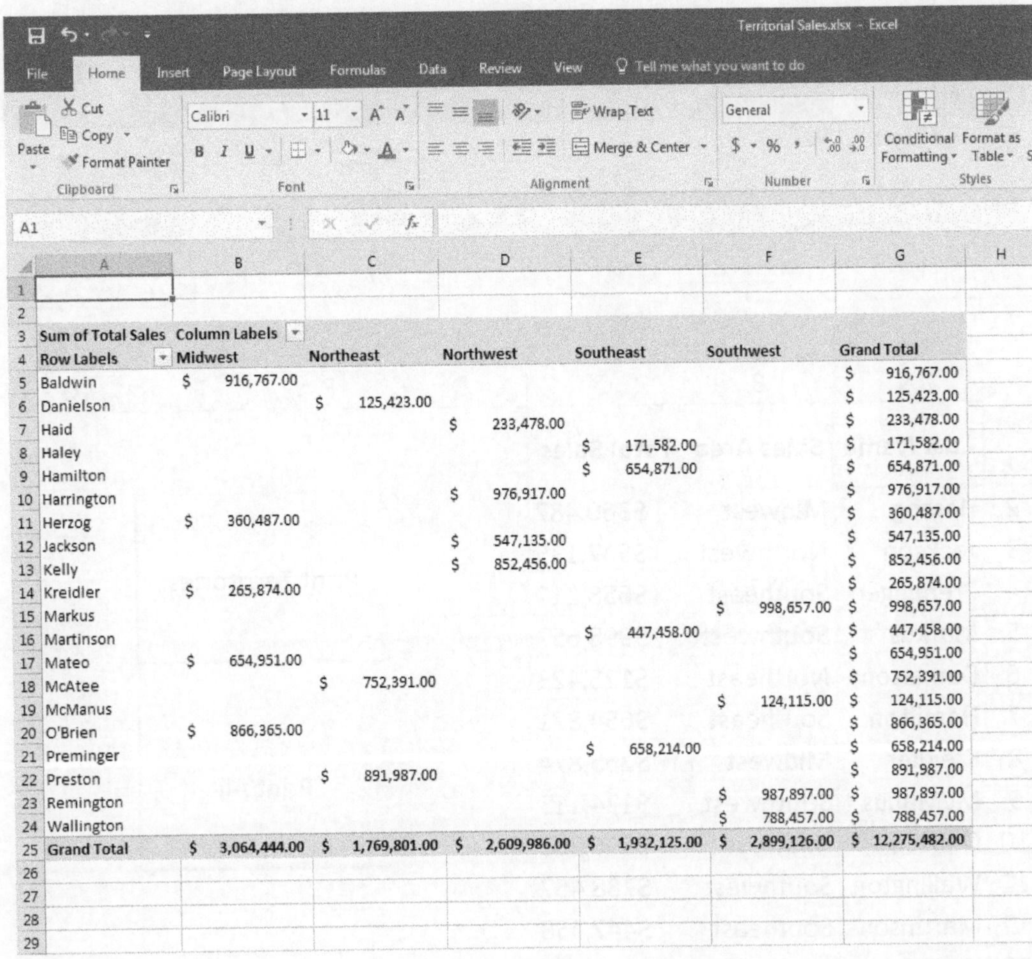

Sum of Total Sales	Column Labels					
Row Labels	**Midwest**	**Northeast**	**Northwest**	**Southeast**	**Southwest**	**Grand Total**
Baldwin	$ 916,767.00					$ 916,767.00
Danielson		$ 125,423.00				$ 125,423.00
Haid			$ 233,478.00			$ 233,478.00
Haley				$ 171,582.00		$ 171,582.00
Hamilton				$ 654,871.00		$ 654,871.00
Harrington			$ 976,917.00			$ 976,917.00
Herzog	$ 360,487.00					$ 360,487.00
Jackson			$ 547,135.00			$ 547,135.00
Kelly			$ 852,456.00			$ 852,456.00
Kreidler	$ 265,874.00					$ 265,874.00
Markus					$ 998,657.00	$ 998,657.00
Martinson				$ 447,458.00		$ 447,458.00
Mateo	$ 654,951.00					$ 654,951.00
McAtee		$ 752,391.00				$ 752,391.00
McManus					$ 124,115.00	$ 124,115.00
O'Brien	$ 866,365.00					$ 866,365.00
Preminger				$ 658,214.00		$ 658,214.00
Preston		$ 891,987.00				$ 891,987.00
Remington					$ 987,897.00	$ 987,897.00
Wallington					$ 788,457.00	$ 788,457.00
Grand Total	**$ 3,064,444.00**	**$ 1,769,801.00**	**$ 2,609,986.00**	**$ 1,932,125.00**	**$ 2,899,126.00**	**$ 12,275,482.00**

FIGURE 7-46

5. Make a **Pie Chart** similar to the one you see in Figure 7-48, using the *Last Names* of the sales staff and the *Total Sales* only (not the average). Make the chart its own sheet.
6. Save the workbook. NOTE: You must save this assignment to a separate workbook as an **Excel Macro-Enabled Workbook** in order for the macros to work. That means it should have the extension of .xlsm.

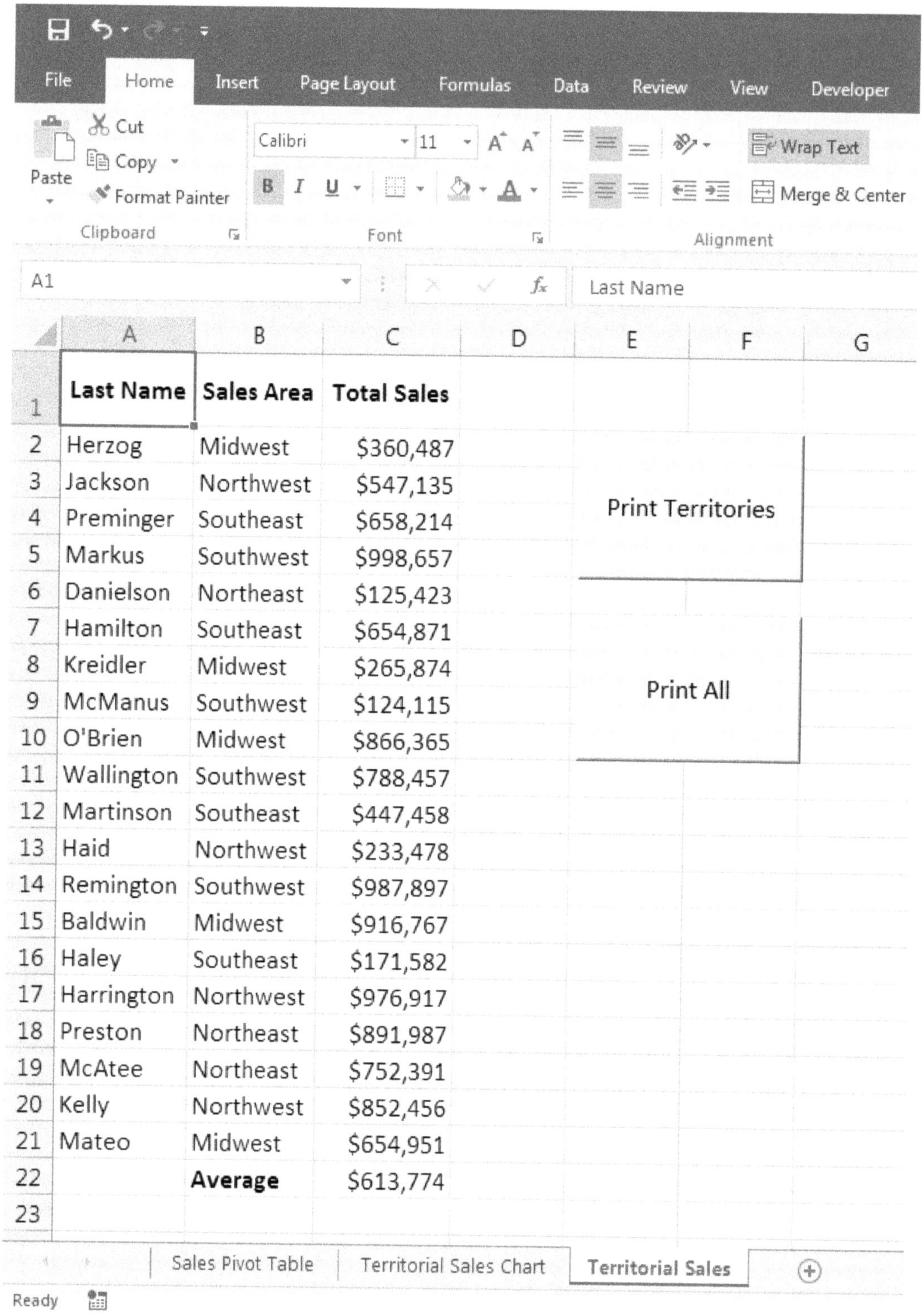
File
Home
Insert
Page Layout
Formulas
Data
Review
View
Developer
Cut
Copy
Paste
Format Painter
Clipboard
Calibri
11
Font
Wrap Text
Merge & Center
Alignment
A1
Last Name
Last Name
Sales Area
Total Sales
Herzog Midwest $360,487
Jackson Northwest $547,135
Preminger Southeast $658,214
Markus Southwest $998,657
Danielson Northeast $125,423
Hamilton Southeast $654,871
Kreidler Midwest $265,874
McManus Southwest $124,115
O'Brien Midwest $866,365
Wallington Southwest $788,457
Martinson Southeast $447,458
Haid Northwest $233,478
Remington Southwest $987,897
Baldwin Midwest $916,767
Haley Southeast $171,582
Harrington Northwest $976,917
Preston Northeast $891,987
McAtee Northeast $752,391
Kelly Northwest $852,456
Mateo Midwest $654,951
Average $613,774
Print Territories
Print All
Sales Pivot Table
Territorial Sales Chart
Territorial Sales
Ready

FIGURE 7-47

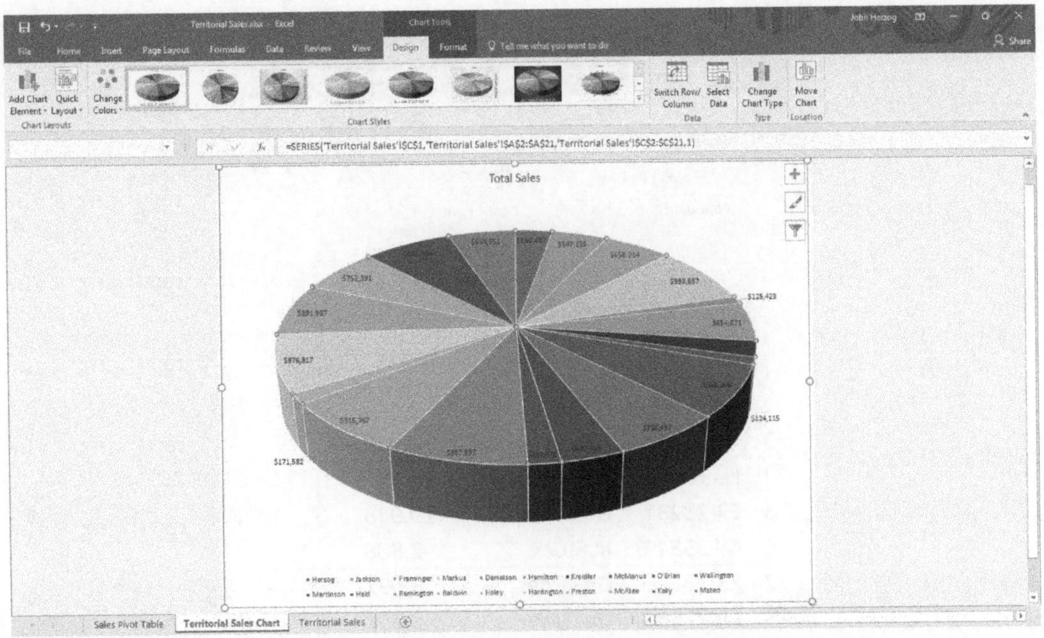

FIGURE 7-48

CHAPTER 7—ALTERNATE ASSIGNMENT (SUMMARY OF TASKS)

1. Open the workbook entitled *Inventory* associated with this book, shown in Figure 7-49.
2. Validate cells with the following specifications:
   a. Make sure that the *Quantity* numbers in cells C2 through C21 are validated so that they permit only entries that are greater than or equal to zero and less than 100,000.
   b. Make sure that a help message (**Input Message**) appears when the cell is selected that says *Enter only numbers that are greater than or equal to zero and less than 100,000*, and make sure the **Title** reads *Be Careful*.
   c. Make sure that an **Error Message** pops up that says *Enter only numbers that are greater than or equal to zero and less than 100,000*. Make sure the title reads *Be Careful*.
   d. Make a list that will show the different types of inventory *Status*. In cell AZ1, type *In Stock*. Then in cell AZ2, type *On Order*.
   e. Make sure that when you click any cell in cells B2 through B21 a drop-down list will become available that will only allow you to enter those types of *Status*.
   f. Add an **Input Message** with a **Title** and an **Error Message** with a **Title** that would be appropriate for that cell range.

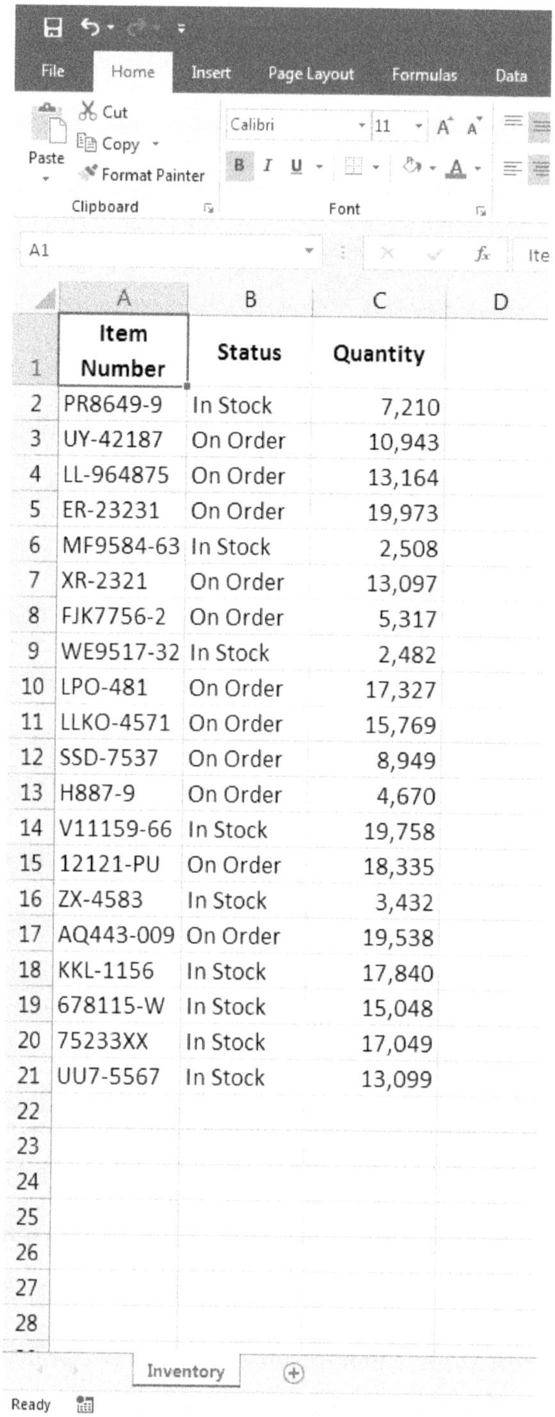
File
Home
Insert
Page Layout
Formulas
Data
Cut
Copy
Paste
Format Painter
Clipboard
Calibri
11
Font
A1
Ite
A
B
C
D
Item Number
Status
Quantity
PR8649-9
In Stock
7,210
UY-42187
On Order
10,943
LL-964875
On Order
13,164
ER-23231
On Order
19,973
MF9584-63
In Stock
2,508
XR-2321
On Order
13,097
FJK7756-2
On Order
5,317
WE9517-32
In Stock
2,482
LPO-481
On Order
17,327
LLKO-4571
On Order
15,769
SSD-7537
On Order
8,949
H887-9
On Order
4,670
V11159-66
In Stock
19,758
12121-PU
On Order
18,335
ZX-4583
In Stock
3,432
AQ443-009
On Order
19,538
KKL-1156
In Stock
17,840
678115-W
In Stock
15,048
75233XX
In Stock
17,049
UU7-5567
In Stock
13,099
Inventory
Ready

FIGURE 7-49

3. Create a **Pivot Table** using cells A1 through C21 that will show the sum of the *Quantity* by *Item Number* (as the **row headings**) and by *Status* (as the **column headings**) and with the following specification:
    a. Make sure the **Pivot Table** is sent to a new spreadsheet.
    b. Change the style of the **Pivot Table** to Medium 2.
    c. Change the format of the **Pivot Table** numbers to **General** with a comma separator and no decimals.
    d. Make the columns a width of *16* characters. At this point, the **Pivot Table** should look like what you see in Figure 7-50.
    e. Filter the **Pivot Table** so that it will **NOT** show the column with items that are *On Order.*
    f. Name the sheet *Pivot Table.*
4. Return to the original *Inventory* spreadsheet.
    a. Enter the **Average** function in cell C22 that averages all of the numbers in the *Quantity* column. Type the word *Average* in cell B22 and make it **bold**.
    b. Create a **Macro** that will format any cell with a *yellow* background, *red* letters, and with *bold* and *italic* font styles. Name the macro *Format.*
    c. Make a button that will run the macro and make sure the button caption reads *Change Font.* Place it in cells E2 through F5.
    d. Create a second **Macro** that will set the **Print Area** to cells A1 through C21; name it *PrintAll.*
    e. Make a button that will run the **Macro** and make sure the caption reads *Print All* (Figure 7-51). Place it in cells E7 through F10.
5. Make a **Pie Chart** similar to the one you see in Figure 7-52, using the *Item Numbers* and the *Quantities* only, and make the chart its own sheet.
6. Make sure that the labels on the pie slices appear as you see them in Figure 7-52.
7. Change the **X Rotation** of the chart to *50* degrees, and explode the pie slice with the value of *13,164* as shown.
8. Save the workbook. NOTE: You must save this assignment in a separate workbook as an **Excel Macro-Enabled Workbook** in order for the macros to work. That means it should have the extension of **.xlsm**.

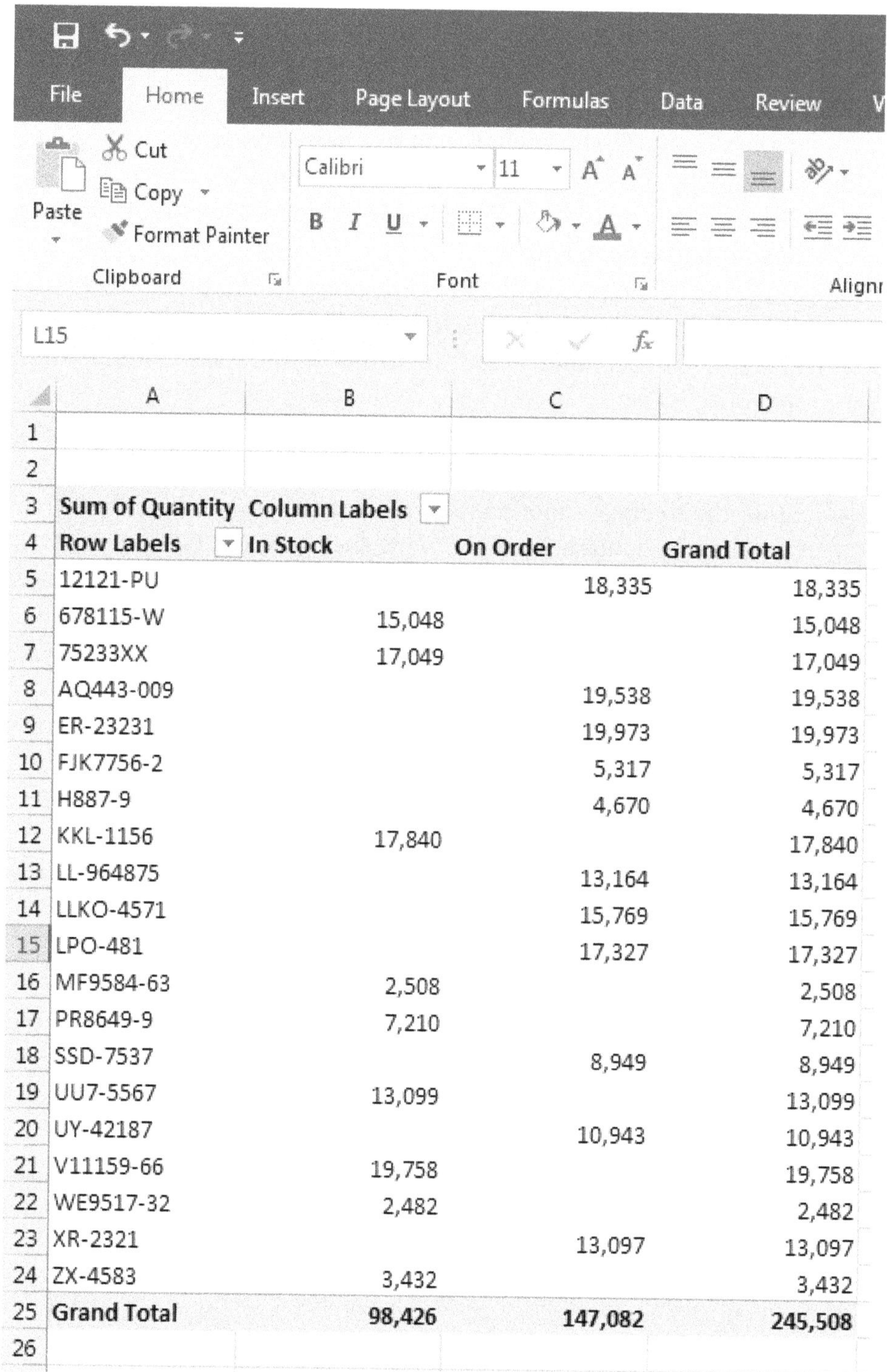

	A	B	C	D
1				
2				
3	**Sum of Quantity**	**Column Labels**		
4	**Row Labels**	**In Stock**	**On Order**	**Grand Total**
5	12121-PU		18,335	18,335
6	678115-W	15,048		15,048
7	75233XX	17,049		17,049
8	AQ443-009		19,538	19,538
9	ER-23231		19,973	19,973
10	FJK7756-2		5,317	5,317
11	H887-9		4,670	4,670
12	KKL-1156	17,840		17,840
13	LL-964875		13,164	13,164
14	LLKO-4571		15,769	15,769
15	LPO-481		17,327	17,327
16	MF9584-63	2,508		2,508
17	PR8649-9	7,210		7,210
18	SSD-7537		8,949	8,949
19	UU7-5567	13,099		13,099
20	UY-42187		10,943	10,943
21	V11159-66	19,758		19,758
22	WE9517-32	2,482		2,482
23	XR-2321		13,097	13,097
24	ZX-4583	3,432		3,432
25	**Grand Total**	**98,426**	**147,082**	**245,508**
26				

FIGURE 7-50

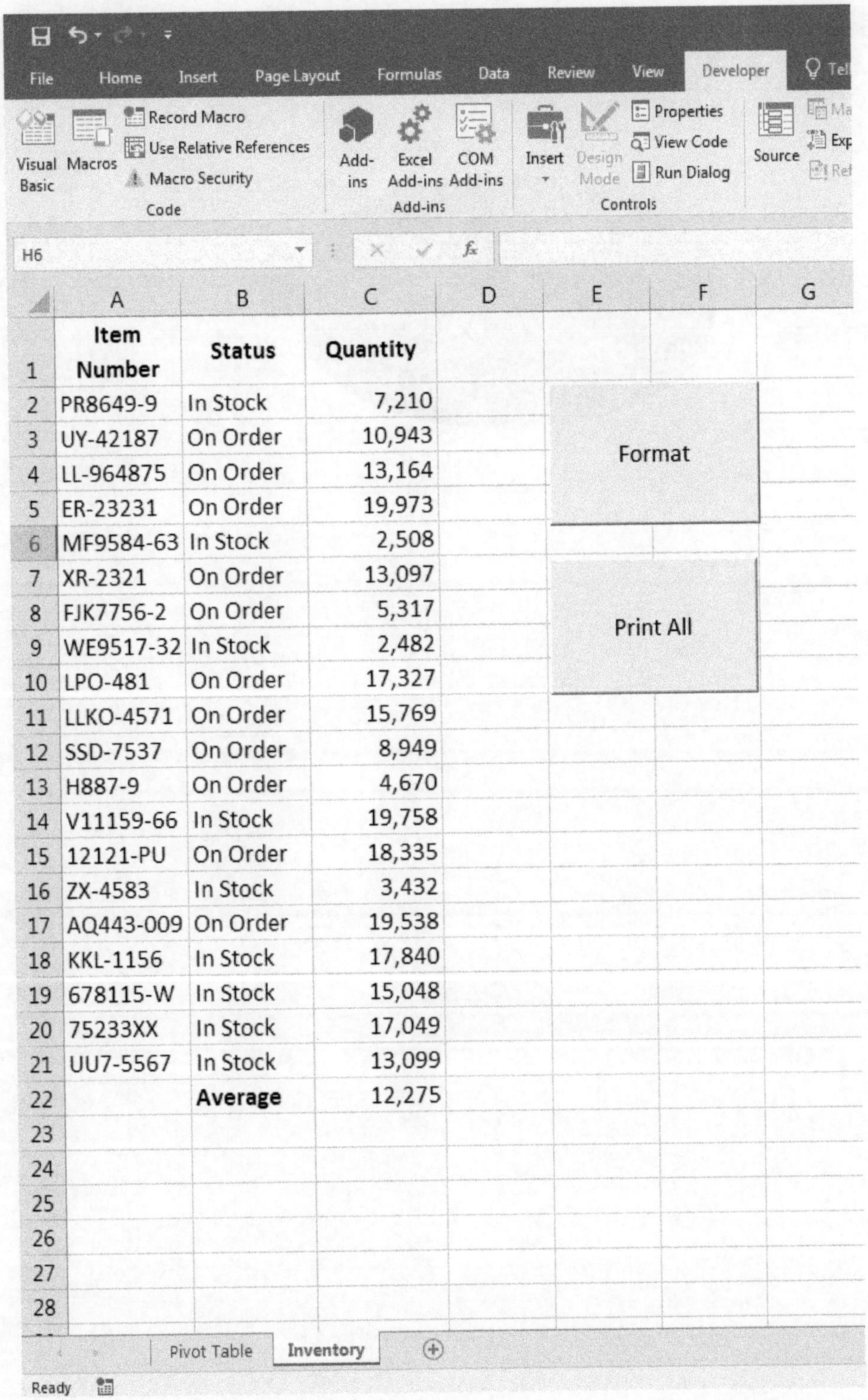

	A	B	C
1	**Item Number**	**Status**	**Quantity**
2	PR8649-9	In Stock	7,210
3	UY-42187	On Order	10,943
4	LL-964875	On Order	13,164
5	ER-23231	On Order	19,973
6	MF9584-63	In Stock	2,508
7	XR-2321	On Order	13,097
8	FJK7756-2	On Order	5,317
9	WE9517-32	In Stock	2,482
10	LPO-481	On Order	17,327
11	LLKO-4571	On Order	15,769
12	SSD-7537	On Order	8,949
13	H887-9	On Order	4,670
14	V11159-66	In Stock	19,758
15	12121-PU	On Order	18,335
16	ZX-4583	In Stock	3,432
17	AQ443-009	On Order	19,538
18	KKL-1156	In Stock	17,840
19	678115-W	In Stock	15,048
20	75233XX	In Stock	17,049
21	UU7-5567	In Stock	13,099
22		**Average**	12,275

FIGURE 7-51

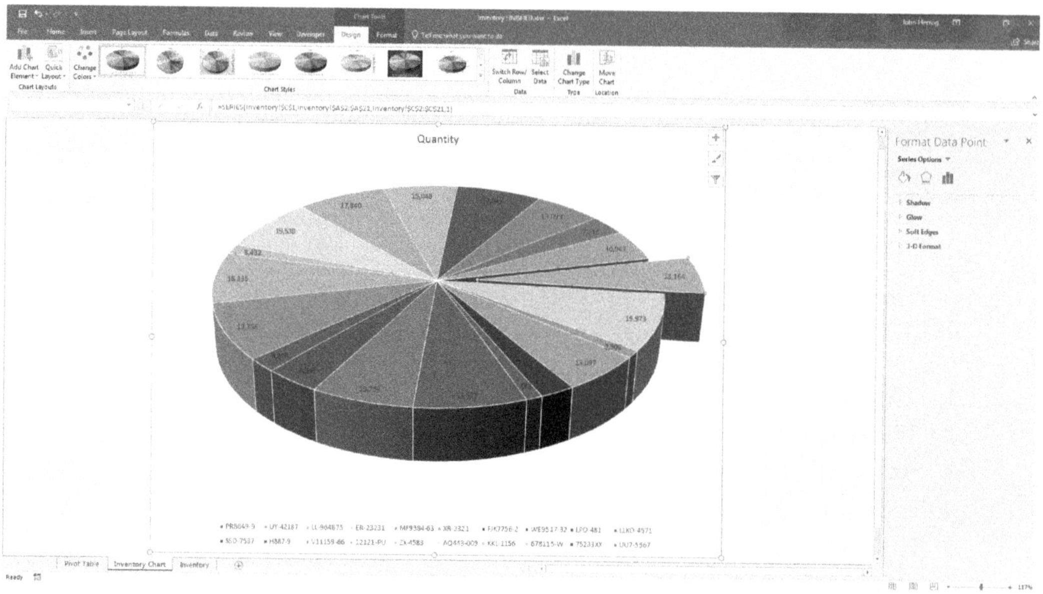

FIGURE 7-52

CHAPTER EIGHT

Using Visual Basic Within Microsoft Excel

BY THE END of this chapter, the user should have a better understanding of how to create **Pop-Ups** (also known as **Input Boxes**) in Excel by the use of **Visual Basic Programming**.

CHAPTER 8—TASK: CREATE A PROFIT AND LOSS STATEMENT USING POP-UP MENUS (ALSO KNOWN AS INPUT BOXES)

There will be times when you need features in a spreadsheet that the usual menus will not provide. Such features may be needed to make a spreadsheet more user-friendly. For example, suppose you are creating a *Profit and Loss Statement* for an organization and you want to prompt the user who is entering the data with **pop-up menus** as you see in Figure 8-1. Before doing so, let us first set up the initial spreadsheet.

1. Start a new spreadsheet.
2. Enter the words *Profit and Loss* in cell A1, making the words **bold** at a font size of *16* points high.
3. Leave cell A2 blank.
4. In cell A3, type the word *Revenue.*
5. Leave cell A4 blank.
6. In cell A5, type the word *Rent.*
7. In cell A6, type the words *Costs of Goods Sold.*
8. In cell A7, type the word *Utilities.*
9. In cell A8, type the word *Advertising.*
10. In cell A9, type the words *Telephone Expenses.*
11. In cell A10, type the words *Salaries and Wages.*
12. In cell A11, type the word *Maintenance.*
13. In cell A12, type the word *Supplies.*

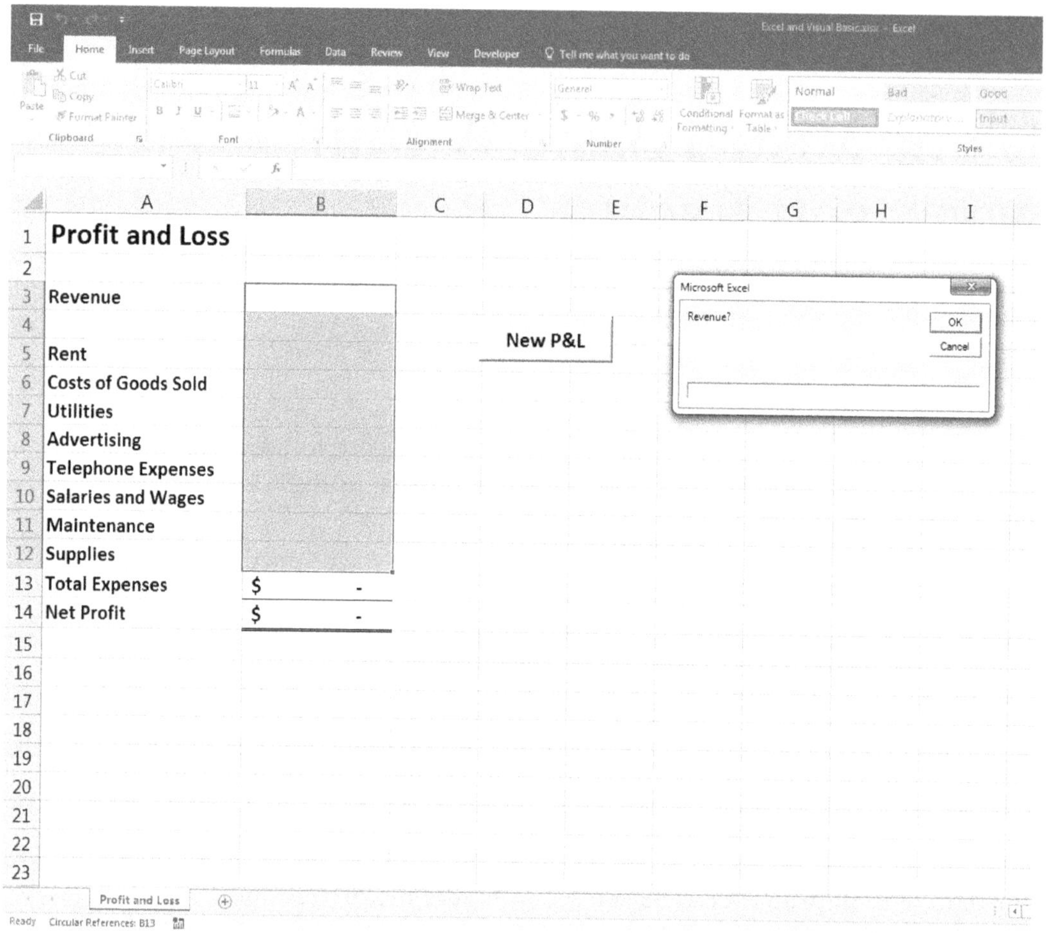

FIGURE 8-1

14. In cell A13, type the words *Total Expenses.*
15. In cell A14, type the words *Net Profit.*
16. Select cells B3 through B14 and make the format of them to be **Accounting** (by clicking the dollar sign [$] in the **Home Ribbon** in the **Number Format Group**).
17. Apply the **Thick Bottom Border** to cell B13.
18. Apply the **Top and Double Bottom Border** to cell B14.
19. Name the sheet tab *Profit and Loss.* NOTE: It is important that you do this before you attempt to write the **Visual Basic programming code** because you will be asking **Visual Basic** to go to (or select) the sheet with the name *Profit and Loss.*

One way to create **Input Boxes** in a spreadsheet is to write a **Visual Basic** programming sequence. Unlike the macros you created in Chapter 7, you must create the button for it first. The programming is then embedded inside that button. The button you use is also different from the one used in Chapter 7 with the recorded macro.

TO PLACE A NEW BUTTON IN A SPREADSHEET AND BEGIN WRITING VISUAL BASIC CODE

To place a new button in the spreadsheet to create a **Visual Basic** sequence, do the following:

Step 1: Click on the **Developer Ribbon** (Figure 8-2, Arrow A). NOTE: If you don't have the **Developer Ribbon** available, refer to the procedure in Chapter 7 that shows how to make it visible.

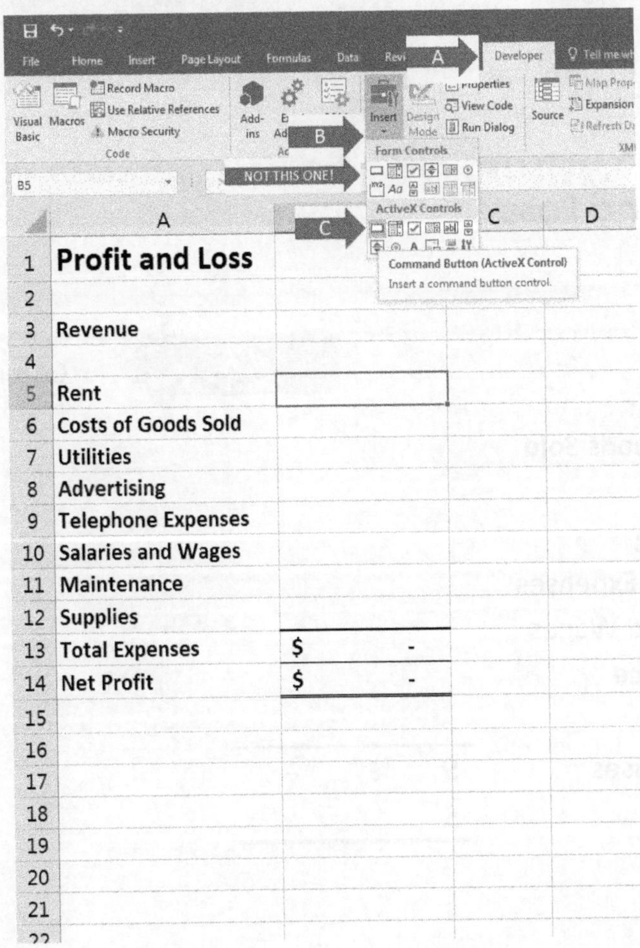

FIGURE 8-2

Step 2: Click the tiny arrow below the option labeled **Insert** (Figure 8-2, Arrow B). The **Insert** option is also called the **Toolbox**.

Step 3: Click the first **command button** in the **ActiveXControls** area in the lower portion of the drop-down menu (Figure 8-2, Arrow C).

Step 4: Click once on the upper-left corner of cell D4 (**do not double-click and do not click and drag**). Your screen will look like Figure 8-3 at the arrow.

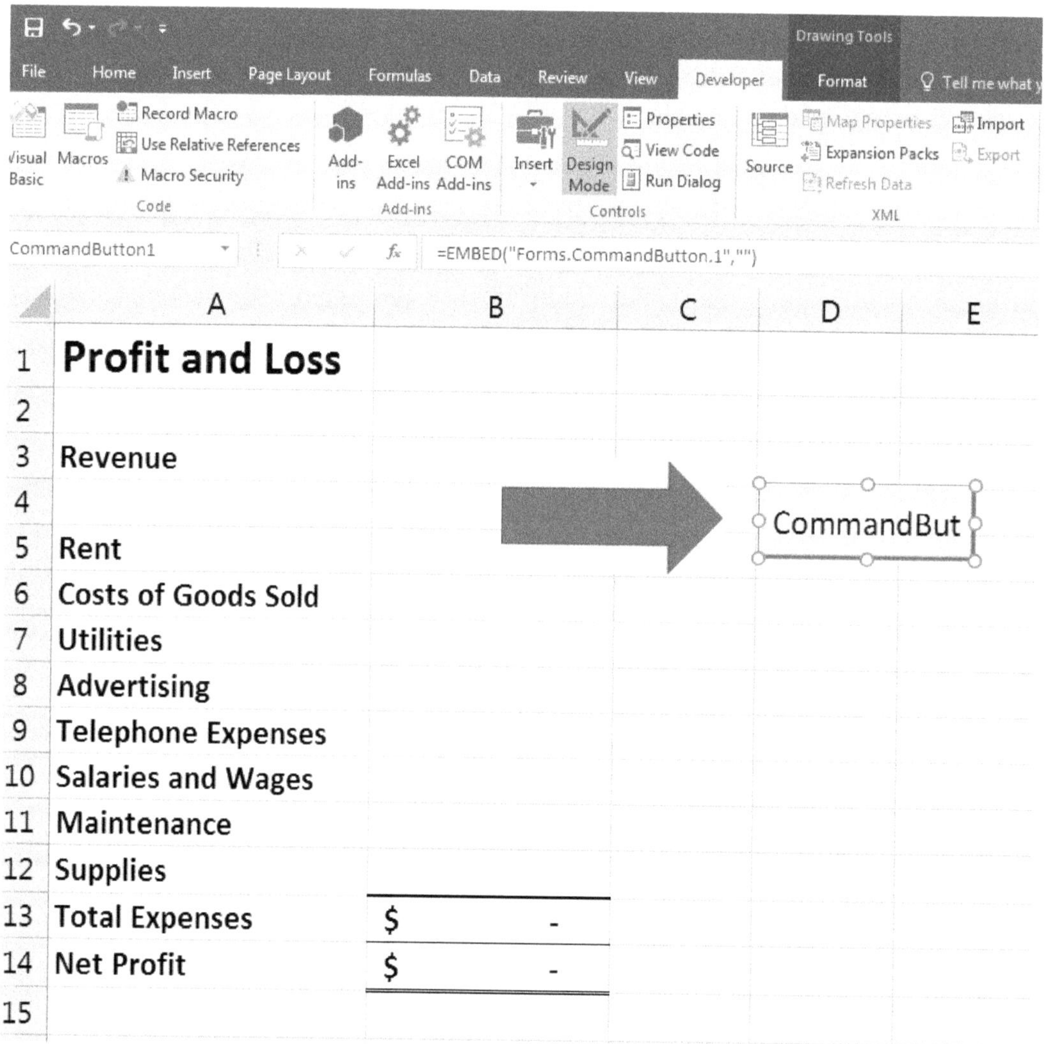

FIGURE 8-3

Step 5: Double-click the **command button** that you have just placed in the spreadsheet, and it will launch you into **Visual Basic**, which will appear as you see in Figure 8-4.

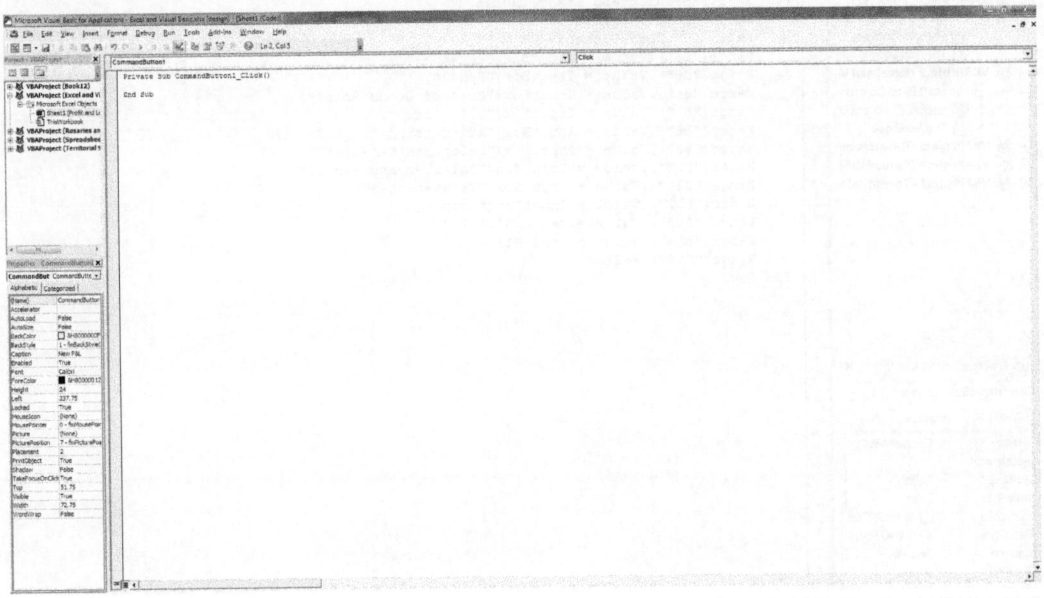

FIGURE 8-4

Between the lines that read **Private Sub CommandButton1_Click**() and **End Sub**, you need to type the following items, **VERBATIM**, including the placement of all punctuation:

```
Sheets("Profit and Loss").Select
Range("B3:B14").Select
Selection.ClearContents
Range("B3").Value = InputBox("Revenue?")
Range("B5").Value = InputBox("Rent?")
Range("B6").Value = InputBox("Costs of Goods Sold")
Range("B7").Value = InputBox("Utilities?")
Range("B8").Value = InputBox("Advertising?")
Range("B9").Value = InputBox("Telephone Expenses?")
Range("B10").Value = InputBox("Salaries and Wages?")
Range("B11").Value = InputBox("Maintenance?")
Range("B12").Value = InputBox("Supplies?")
Range("B13").Value = "=SUM(B5:B12)"
Range("B14").Value = "=B3-B13"
Range("A16").Select
```

FIGURE 8-5

When you have finished typing in the code, your screen should look like Figure 8-5.

It's important to note some very important concepts with this sequence of **Visual Basic Programming Code**.

First, the lines can be—but don't necessarily have to be—indented. The indentation is used only to make it easier to read the code if you want to troubleshoot later should an error occur. Second, you do not need to put in spaces between the coding.

When it is needed, **Visual Basic** will put them in for you. The only times that you will want to put spaces in the sequence is in the captions of the input boxes, such as *Costs of Goods Sold?* or *Telephone Expenses?* Third, the coding is not case-sensitive.

The following is an explanation of what **Visual Basic** is commanding **Excel** to do in each line.

Code Line 1 (that reads *Sheets("Profit and Loss").Select*): This line, below the **Private Sub** line, is telling Excel to go to (and therefore to select) the sheet named **Profit and Loss**. That's why it is very important that you name the sheet **Profit and Loss** as stated earlier.

Code Line 2 (that reads *Range("B3:B14").Select*): This line is telling Excel to select the entire cell range of cells B3 through B14 in the active spreadsheet.

Code Line 3 (that reads *Selection.ClearContents*): This line is telling Excel to clear or delete whatever is currently selected. In this example, the cell range of B3 through B14 was selected by the command in the previous line (Code Line 2).

Code Lines 4 through 11 (the first of which reads *Range("B3").Value = InputBox ("Revenue?")*): These lines are telling Excel to ask for input by the use of the **Input Boxes**. Whatever data is entered into the **Input Boxes** is to be placed in the cell range that begins the line. For example, Code Line 4 is saying that whatever data is entered in the **Input Box** (which asks the question *Revenue?)* is to be placed into cell B3, and whatever data is entered in the **Input Box** that asks the question *Rent?* is to be placed into cell B5, and so on.

Code Line 12 (that reads *Range("B13").Value = "=SUM(B5:B12)"*): This line is telling Excel to put a function into cell B13 that will sum the cells holding the *expense* data, cells B5 through B12.

Code Line 13 (that reads *Range("B14").Value = "=B3-B13)*: This line is telling Excel to put a formula into cell B14 that will subtract the *Total Expenses* in cell B13 from *Revenue* in cell B3. Notice that when you tell **Visual Basic** to enter a function or a formula, whatever you enter between the quotation marks that follow **.Value**=, will literally be entered verbatim into the cell designated in the **Range.**

Code Line 14 (that reads *Range("A16").Select*): This line is telling Excel to go to a cell away from the newly entered data in order that it can be more easily read.

After you have entered all of the code, make sure that all quotation marks and punctuation is in their proper places.

Click on the **close button** at the upper right of your screen (Figure 8-6, Arrow), and that will take you back to your spreadsheet. NOTE: When you close the **Visual Basic** screen, the coding will automatically be retained in the button. It is not necessary to save the workbook at this point, but it will be very important to do so later as a **macro-enabled workbook**. Saving the workbook saves the content of your spreadsheet, the button you have created, and the **Visual Basic Code** that the button contains.

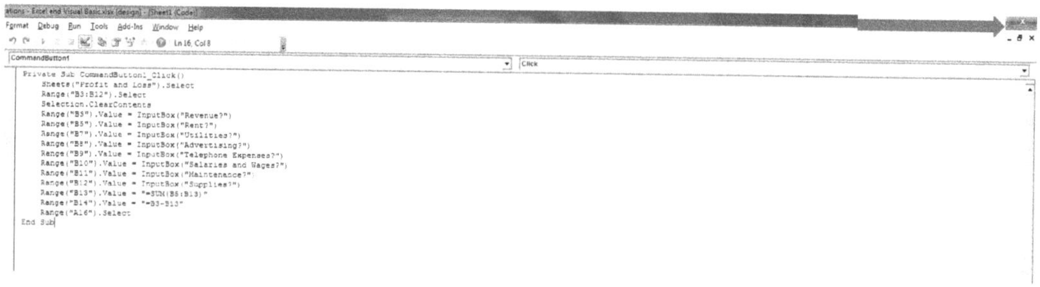

FIGURE 8-6

TO CHANGE THE BUTTON CAPTION AND THEN ACTIVATE THE BUTTON TO RUN VISUAL BASIC

To change the caption on the button:

Step 1: Right-click on the newly added command button (Figure 8-7, Arrow A). That will give you a **shortcut menu**.

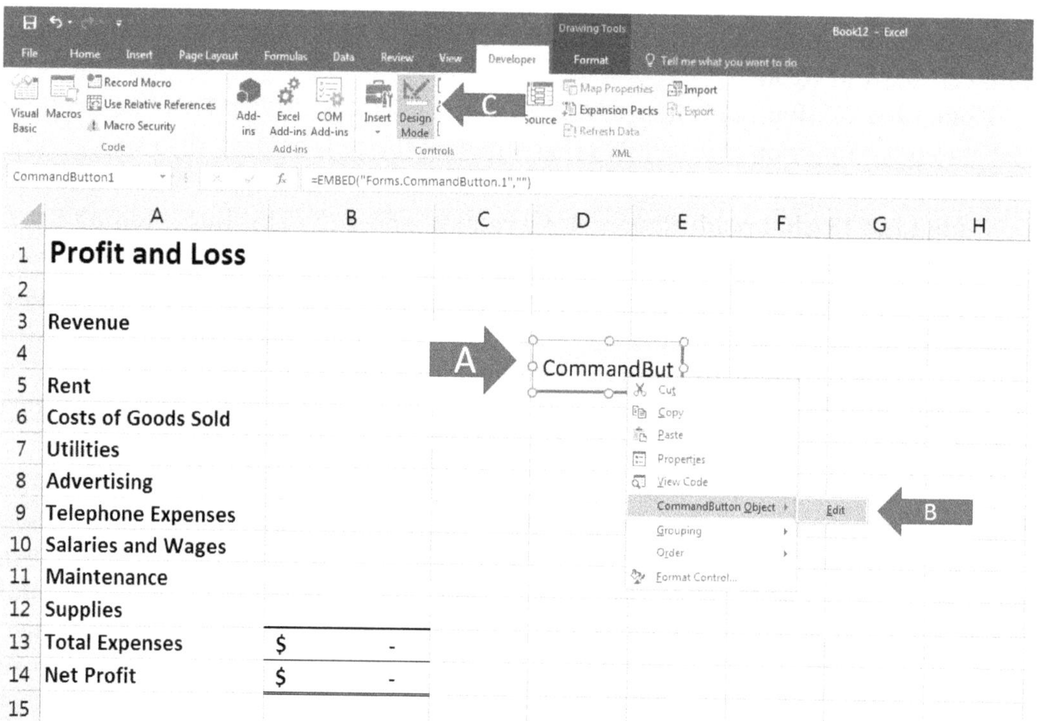

FIGURE 8-7

Step 2: On the **shortcut menu**, click **CommandButton Object**; a small **submenu** that reads **Edit** will appear.

Step 3: Click **Edit** (Figure 8-7, Arrow B).

Step 4: After you click **Edit**, the cursor will move inside the button. Press and hold the backspace key to remove the words **CommandButton1**.

Step 5: In place of the words **CommandButton1**, type *New P&L*.

Step 6: To activate the button, click the **Design Mode** icon (Figure 8-7, Arrow C) in the **Developer Ribbon**. So far through this process, the **Design Mode** icon has been recessed, which means that the **Design Mode** had been running. When you click on it, you are turning off the **Design Mode** and the **Design Mode** button will no longer be recessed. By doing so, the button is now ready to use.

Step 7: Now point to the button with your mouse. An **index finger** will appear. When you click the button now, the **Input Boxes** will appear. At that point, suppose we want to enter the following numbers in our example:

1. When asked for *Revenue*, enter *1000000*.
2. When asked for *Rent*, enter *20000*.
3. When asked for *Costs of Goods Sold*, enter *10000*.
4. When asked for *Utilities*, enter *2000*.
5. When asked for *Advertising*, enter *4000*.
6. When asked for *Telephone Expenses*, enter *3000*.
7. When asked for *Salaries and Wages*, enter *30000*.
8. When asked for *Maintenance*, enter *500*.
9. When asked for *Supplies*, enter *3000*.

Your screen should look like Figure 8-8.

Step 8: Save the workbook as **Visual Basic Homework.xlsm** before you close the workbook. NOTE: Please remember that you must put the **.xlsm** extension on the file or your **Visual Basic** coding will be deleted when you close the file, just as it would be if you had recorded macros.

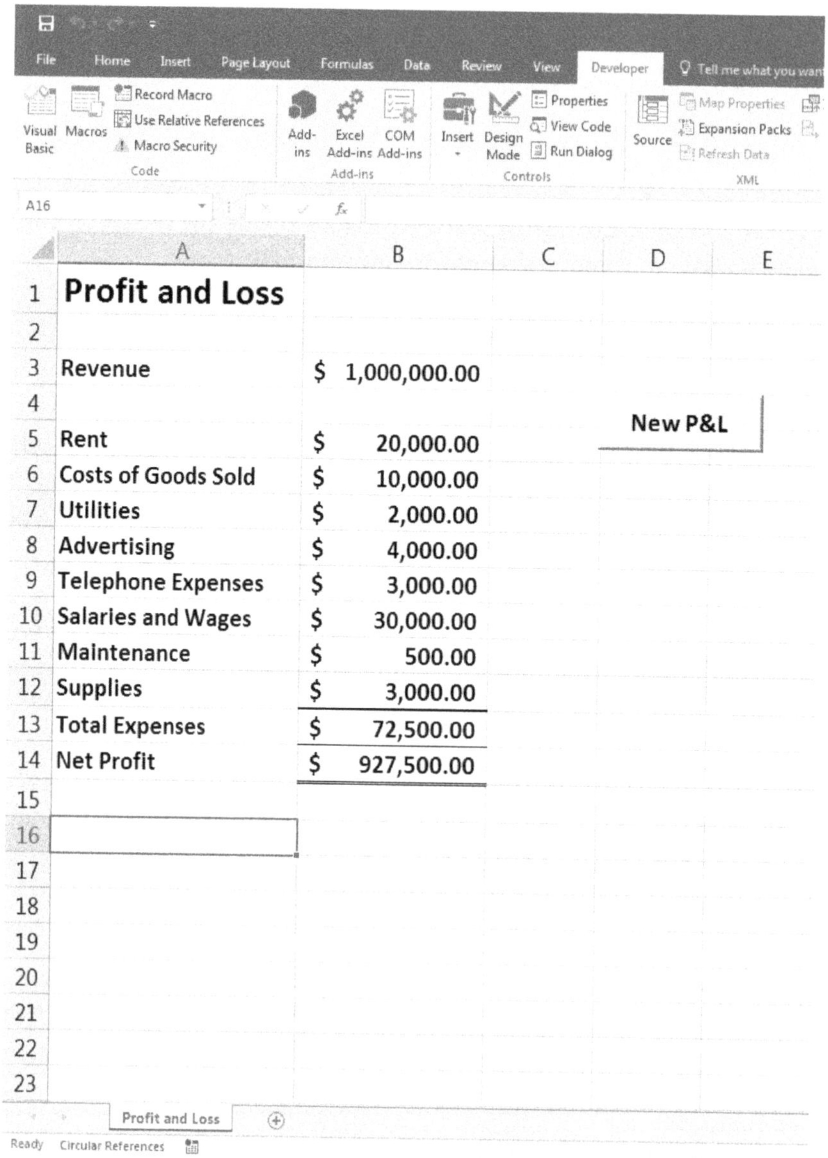

FIGURE 8-8

CHAPTER 8—ASSIGNMENT (SUMMARY OF TASKS)

1. Start a new spreadsheet, and then change the name of the sheet tab to **Profit and Loss**.
2. Enter the words *Profit and Loss* in cell A1, making the words bold and at a font size of *16* points high.

3. Leave cell A2 blank.
4. In cell A3, type the word *Revenue.*
5. Leave cell A4 blank.
6. In cell A5, type the word *Rent.*
7. In cell A6, type the words *Costs of Goods Sold.*
8. In cell A7, type the word *Utilities.*
9. In cell A8, type the word *Advertising.*
10. In cell A9, type the word *Telephone Expenses.*
11. In cell A10, type the word *Salaries and Wages.*
12. In cell A11, type the word *Maintenance.*
13. In cell A12, type the word *Supplies.*
14. In cell A13, type the words *Total Expenses.*
15. In cell A14, type the word *Net Profit.*
16. Select cells B3 through B14 and make the format of them to be **Accounting** (by clicking on the dollar sign [$] in the **Home Ribbon** in the **Number Format Group**).
17. Apply the **Thick Bottom** border to cell B13.
18. Apply the **Top and Double Bottom** border to cell B14.
19. Place a new button on the upper-left corner of cell D4 (do not click and drag).
20. Place the **Visual Basic Code** in the **Visual Basic window** between the lines that read Private Sub CommandButton1_Click() and End Sub as follows:

```
Sheets("Profit and Loss").Select
Range("B3:B12").Select
Selection.ClearContents
Range("B3").Value = InputBox("Revenue?")
Range("B5").Value = InputBox("Rent?")
Range("B6").Value = InputBox("Costs of Goods Sold")
Range("B7").Value = InputBox("Utilities?")
Range("B8").Value = InputBox("Advertising?")
Range("B9").Value = InputBox("Telephone Expenses?")
Range("B10").Value = InputBox("Salaries and Wages?")
Range("B11").Value = InputBox("Maintenance?")
Range("B12").Value = InputBox("Supplies?")
Range("B13").Value = "=SUM(B5:B13)"
Range("B14").Value = "=B3-B13"
Range("A16").Select
```

21. After entering the code sequence, in place of the words **CommandButton1**, type *New P&L*.
22. Turn off the **Design** mode of the button and then click the button.
23. Enter the following data when prompted.
    a. When asked for **Revenue**, enter *1000000*.
    b. When asked for **Rent**, enter *20000*.
    c. When asked for **Costs of Goods Sold**, enter *10000*.
    d. When asked for **Utilities**, enter *2000*.
    e. When asked for **Advertising**, enter *4000*.

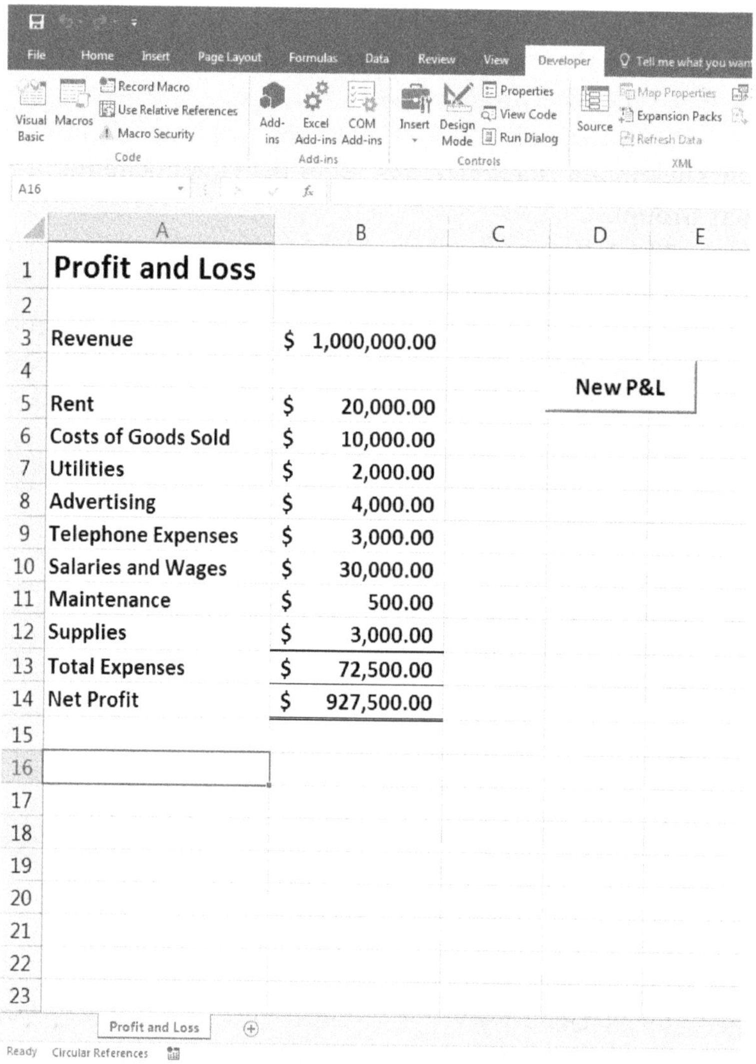

FIGURE 8-9

f. When asked for **Telephone Expenses**, enter *3000*.
g. When asked for **Salaries and Wages**, enter *30000*.
h. When asked for **Maintenance**, enter *500*.
i. When asked for **Supplies**, enter *3000*.

Your screen should look like Figure 8-9.

24. Save the workbook as **Visual Basic Homework.xlsm** on your desktop and submit it.

CHAPTER 8—ALTERNATE ASSIGNMENT (SUMMARY OF TASKS)

1. Start a new spreadsheet, and then change the name of the sheet tab to *Product on Hand Calculation*.
2. Enter the words *Product on Hand Calculation* in cell A1, making the words **bold** and at a font size of *16* points high.
3. Change the width of column A to *37* characters.
4. Leave cell A2 blank.
5. In cell A3, type the word *Product*.
6. In cell A4, type the words *Beginning Inventory*.
7. In cell A5, type the word *Ordered*.
8. In cell A6, type the word *Received*.
9. In cell A7, type the word *Purchased*.
10. In cell A8, type the words *Currently on Hand*.
11. In cell A9, type the words *Product Cost from Vendor*.
12. In cell A10, type the words *Product Price That We Charge*.
13. In cell A11, type the words *Retail Sales Potential*.
14. In cell A12, type the words *Value of Products on Hand*.
15. Leave cell A13 blank.
16. In cell A14, type the words *Profit Potential on This Product*.
17. Select cells B4 through B8, and make the format of them to be **Comma** (by clicking on the comma [,] in the **Home Ribbon** in the **Number Format Group**).
18. Select cells B9 through B12 and cell B14. Make the format **Accounting** (by clicking on the dollar sign [$] in the **Home Ribbon** in the **Number Format Group**).
19. Place a button that will store a **Visual Basic** program that extends from cells D2 to F8.
20. Place the **Visual Basic Code** in that button that will:
    a. Clear the contents of cells B2 through B14.
    b. Ask for input using an **input box** for cells B3 through B7 and cells B9 and B10 **ONLY**.

21. Make sure the **Visual Basic Code** will also enter formulas into cells as follows:
    a. B8 should have the formula that reads =B4+B6-B7.
    b. B11 should have the formula that reads =B8*B10.
    c. B12 should have the formula that reads =B9*B8.
    d. B14 should have the formula that reads =B11-B12.

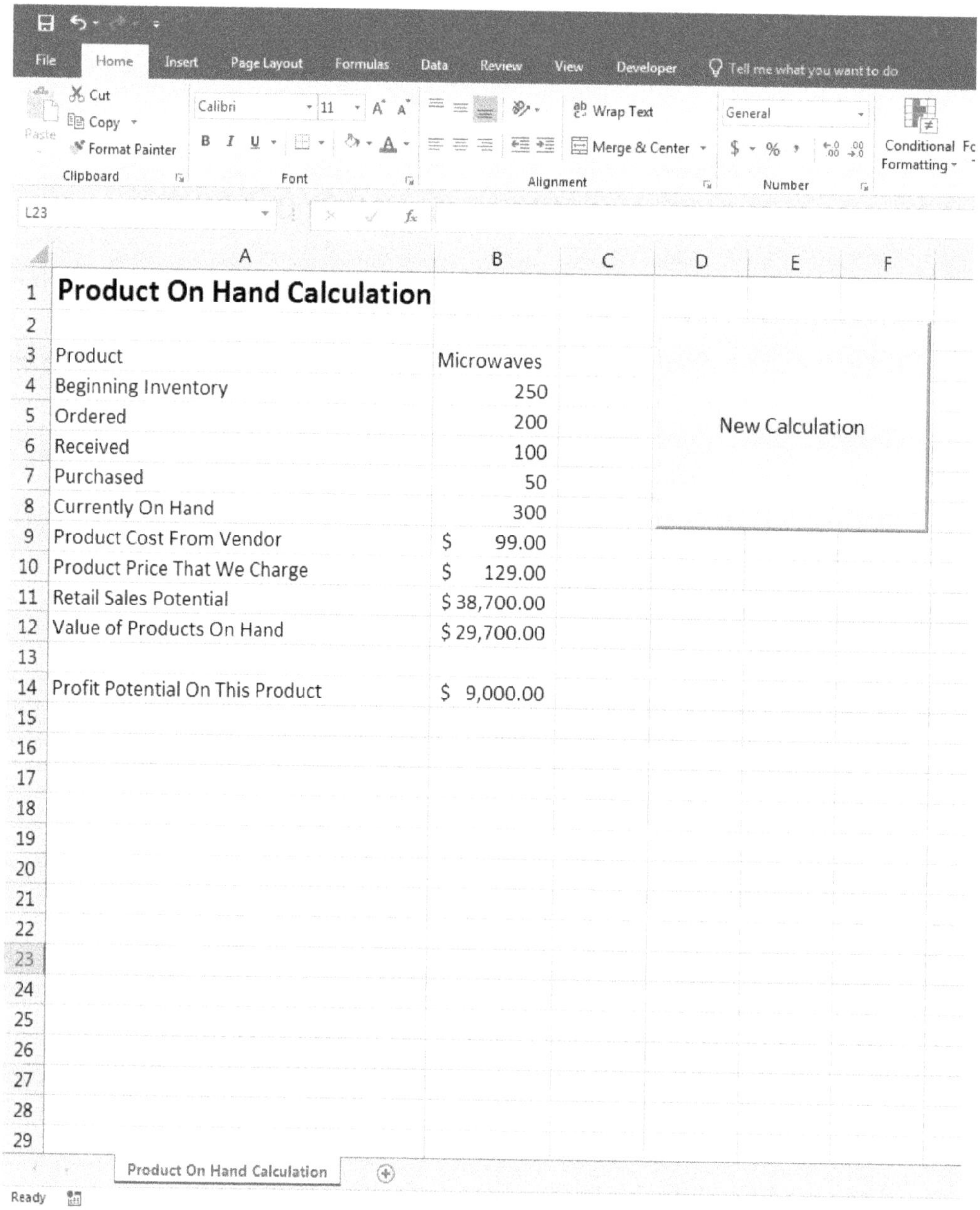

FIGURE 8-10

22. Enter the following data when prompted.
    a. When asked for *Product*, enter *Microwaves.*
    b. When asked for *Beginning Inventory*, enter *250.*
    c. When asked for *Ordered*, enter *200.*
    d. When asked for *Received*, enter *100.*
    e. When asked for *Purchased*, enter *50.*
    f. When asked for *Product Cost from Vendor*, enter $99.
    g. When asked for *Product Price That We Charge*, enter $129.

Your screen should look like Figure 8-10.

23. Save the workbook as **Visual Basic Homework II.xlsm** on your desktop and submit it.

Index

CPSIA information can be obtained
at www.ICGtesting.com
Printed in the USA
LVHW112000231121
704263LV00005B/15

9 781516 536887